Hands-on Morphological
Image Processing

Tutorial Texts Series

Hands-on Morphological Image Processing

Edward R. Dougherty • Roberto A. Lotufo

Tutorial Texts in Optical Engineering
Volume TT59

Arthur R. Weeks, Jr., Series Editor
Invivo Research Inc. and University of Central Florida

SPIE PRESS
A Publication of SPIE—The International Society for Optical Engineering
Bellingham, Washington USA

Library of Congress Cataloging-in-Publication Data

Dougherty, Edward R.
   Hands-on morphological image processing / Edward R. Dougherty, Roberto A. Lotufo.
     p. cm. — (Tutorial texts in optical engineering ; v. TT 59)
   Includes bibliographical references and index.
   ISBN 0-8194-4720-X (pbk.)
     1. Image processing–Mathematics. 2. Morphisms (Mathematics) I. Lotufo, Roberto A.
   II. Title. III. Series.

TA1637.D65 2003
621.36'7—dc21   2003054262
                                                                                 CIP

Published by

SPIE—The International Society for Optical Engineering
P.O. Box 10
Bellingham, Washington 98227-0010 USA
Phone: 360.676.3290
Fax: 360.647.1445
Email: spie@spie.org
www.spie.org

Printed in the United States of America.

Introduction to the Series

Since its conception in 1989, the Tutorial Texts series has grown to more than 60 titles covering many diverse fields of science and engineering. When the series was started, the goal of the series was to provide a way to make the material presented in SPIE short courses available to those who could not attend, and to provide a reference text for those who could. Many of the texts in this series are generated from notes that were presented during these short courses. But as stand-alone documents, short course notes do not generally serve the student or reader well. Short course notes typically are developed on the assumption that supporting material will be presented verbally to complement the notes, which are generally written in summary form to highlight key technical topics and therefore are not intended as stand-alone documents. Additionally, the figures, tables, and other graphically formatted information accompanying the notes require the further explanation given during the instructor's lecture. Thus, by adding the appropriate detail presented during the lecture, the course material can be read and used independently in a tutorial fashion.

What separates the books in this series from other technical monographs and textbooks is the way in which the material is presented. To keep in line with the tutorial nature of the series, many of the topics presented in these texts are followed by detailed examples that further explain the concepts presented. Many pictures and illustrations are included with each text and, where appropriate, tabular reference data are also included.

The topics within the series have grown from the initial areas of geometrical optics, optical detectors, and image processing to include the emerging fields of nanotechnology, biomedical optics, and micromachining. When a proposal for a text is received, each proposal is evaluated to determine the relevance of the proposed topic. This initial reviewing process has been very helpful to authors in identifying, early in the writing process, the need for additional material or other changes in approach that would serve to strengthen the text. Once a manuscript is completed, it is peer reviewed to ensure that chapters communicate accurately the essential ingredients of the processes and technologies under discussion.

It is my goal to maintain the style and quality of books in the series, and to further expand the topic areas to include new emerging fields as they become of interest to our reading audience.

Arthur R. Weeks, Jr.
University of Central Florida

Contents

Preface

Morphological image processing has become a standard part of the imaging scientist's toolbox and today is applied daily to a wide range of industrial applications, including (and certainly not limited to) inspection, biomedical imaging, document processing, pattern recognition, metallurgy, microscopy, and robot vision. Because the morphological operations can serve as a universal language for image processing, their application is only limited by the ability to design effective algorithms and efficient computational implementation.

In the last decade, since the publication of *An Introduction to Morphological Image Processing* by SPIE Press, there have been many developments in morphological imaging, both in theory and practice. This book concentrates on applications. In keeping with the outlook of the previous book, we aim to provide a handbook that instructs how to analyze a problem and then how to develop successful algorithms based on the analysis. We take a holistic approach by showing how generic methods can be used in combination to solve practical problems. We include demonstrations to show how various morphological techniques can be combined to produce complete and effective algorithms.

In concentrating on applications we have not sacrificed careful definitions and explicit statement of operation properties. Indeed, skillful application requires that one understand the general filtering effects of an operation, and efficient implementation requires that one have knowledge of the operational simplifications available. Filter properties are integrated into the applications and their relevance is emphasized.

We have heard recently the comment that "morphology is an industrial subject." To a great extent this is true, although one should extend this to include research and development laboratories of all kinds. Several months ago, when asked to develop a high-throughput algorithm for analyzing genetic cell arrays, we suggested to our biological colleagues that we would use an algorithm known as the watershed. "Of course," they replied.

The book is *hands-on* in a very real sense. Most of the techniques used in the book are available in the **Morphology Toolbox**, and a great majority of the images shown in the text have been processed by the toolbox. These images, along with a demonstration version of the toolbox, are downloadable free from the web (`http://sourceforge.net/projects/pymorph`), so that the reader can actually process the images according to the examples and demonstrations in the text. There is a brief discussion in the first chapter as to how the toolbox correlates to the text, but we do not emphasize this relation throughout the text since we do not want to digress from the imaging essentials. Detailed use of the toolbox can be learned from the downloadable version. To assist the user, each chapter concludes with a list of the toolbox operations used in the chapter, and the detailed demonstration sections include corresponding toolbox implementations. Each chapter also has its own bibliography.

We briefly describe the structure of the book. The first two chapters discuss the basic binary morphological operations, erosion and its dual dilation in the first chapter, and opening and its dual closing in the second. All of morphological image processing rests on these operations. Therefore, we treat them in great detail. The third and fourth chapters discuss the processing of binary images, the third focusing on applications of the primary operators, and the fourth on the hit-or-miss transform, which processes an image directly in terms of the foreground-background relation. We pay particular attention to morphological reconstruction in the third chapter because we believe that it is a very powerful tool for the development of algorithms based on inherent image structure.

Gray-scale morphological operators are discussed in the fifth chapter, and their application is treated in the following chapter. Once again we pay much attention to the role of reconstruction in applications. These tools have been more recently developed and we expect that they will lead to an ever-expanding range of application.

The seventh chapter is devoted to watershed-based segmentation. There are many variants of watershed segmentation. Our approach is to articulate the underlying principles while at the same time providing real-world applications. The key to successful segmentation is marker construction, and this issue is to some extent the focus of the chapter.

In his original (and highly mathematical) work, *Random Sets and Integral Geometry*, Georges Matheron, who along with Jean Serra founded the subject of mathematical morphology, comments, "Despite the purely mathematical nature of the present treatise, the formulation and the very choice of problems for solution are directly inspired by experimental techniques of texture analysis." It is not surprising, therefore, that morphological imaging is fundamental to a core understanding of texture. The sixth chapter discusses granulometric filters, which play a key role in describing and classifying texture and particle distributions. Whereas Matheron formalized their definition mathematically, granulometric-type methods are well known in sedimentology and the study of porous media.

The final chapter of the book concerns the automatic design of morphological operators. Morphological image processing is based on probing an image with structuring elements, and these determine the relationships within image structure that an algorithm can ascertain. Again quoting Matheron, "In general, the structure of an object is defined as the set of relationships existing between elements or parts of the object... Hence, this choice [of relationships]... determines the relative worth of the concept of structure at which we will arrive." In many cases it is possible to obtain satisfactory structuring elements by human ingenuity; however, when successful filtering requires hundreds or even thousands of structuring elements, automatic design from training data becomes essential. It is here that morphological image processing meets computational learning.

Before closing this preface, we would like to acknowledge some of the people who helped make this book possible: Junior Barrera, Rubens C. Machado, Roberto Hirata, Jr., Nina S. T. Hirata, Marcel Brun, Yidong Chen, Seungchan Kim, Artyom Grigoryan, Ulisses Braga Neto, and the graduate students who attended the second semester-2002 Morphological Image Analysis course at University of Campinas, Brazil, testing a preliminary version of this book. We especially thank our wives Terry and Valéria for their ongoing support.

Finally, we hope you find this book both enjoyable and useful to your imaging work, in whatever your field.

Edward R. Dougherty
College Station, Texas, U.S.A.

Roberto A. Lotufo
Campinas, São Paulo, Brazil

List of Symbols

$A \cap B$	Intersection between two sets
$A \cup B$	Union between two sets
$f \wedge g$	Pixelwise minimum between two images
$f \vee g$	Pixelwise maximum between two images
$f \leq g$	Image f is beneath g
$f \geq g$	Image f is above g
f^c	Negation of image f
$f^{\dot{c}}$	Bounded negation of image f
$f - g$	Subtraction between two images
$f -^* g$	Bounded subtraction between two images
$f + g$	Addition between two images
$f +^* g$	Bounded addition between two images
$f \triangle g$	Symmetrical difference between f and g
$f \ominus g$	Erosion of f by g
$f \oplus g$	Dilation of f by g
$f \dot{\ominus} g$	Bounded erosion of f by g
$f \dot{\oplus} g$	Bounded dilation of f by g
$f \circ g$	Open of f by g
$f \bullet g$	Close of f by g
$f \hat{\circ} g$	Open top-hat of f by g
$f \hat{\bullet} g$	Close top-hat of f by g
$\breve{f}$	Reflection of f
$ASF^n_{oc,g}(f)$	Alternating sequential filter of stage n, closing and opening
$f \oplus_g D$	Conditional dilation of f conditioned to g with connectivity given by the structuring element D
$f \ominus_g D$	Conditional erosion of f conditioned to g with connectivity given by the structuring element D
$(f \oplus_g D)^n$	N-conditional dilation of f conditioned to g with connectivity given by the structuring element D
$(f \ominus_g D)^n$	N-conditional erosion of f conditioned to g with connectivity given by the structuring element D
$f \triangle_D g$	Inf-reconstruction of f from the marker g using the connectivity given by the structuring element D
$f \nabla_D g$	Sup-reconstruction of f from the marker g using the connectivity given by the structuring element D
$f \circ_D g$	Reconstructive opening by the structuring element g using the connectivity given by the structuring element D
$f \bullet_D g$	Reconstructive closing by the structuring element g using the connectivity given by the structuring element D

$f \, \hat{\circ}_D \, g$	Reconstructive opening top-hat by the structuring element g using the connectivity given by the structuring element D
$f \, \hat{\bullet}_D \, g$	Reconstructive closing top-hat by the structuring element g using the connectivity given by the structuring element D
$\Lambda(A)_D$	Labeling of a binary image A with the connectivity given by the structuring element D
$f \circ (D)_a$	Area open of D-connected components of area less or equal than a
$f \bullet (D)_a$	Area close of D-connected holes of area less or equal than a
$ASF^n_{oc,g,D}(f)$	Reconstructive alternating sequential filter of stage n, closing and opening, connectivity given by D
$X_t(f)$	Threshold of image f at level t
$grad_{g1,g2}(f)$	Morphological gradient of f with the external structuring element $g1$ and internal structuring element $g2$
$HMAX_{h,D}(f)$	H-maxima of image f with contrast h and the connectivity given by D
$HMIN_{h,D}(f)$	H-minima of image f with contrast h and the connectivity given by D
$RMAX_D(f)$	Regional maximum of image f with the connectivity given by D
$RMIN_D(f)$	Regional minimum of image f with the connectivity given by D
$A \circledast T$	Hit-or-miss of binary image A by the template T
$A \odot T$	Thinning of binary image A by the template T
$A \odot_n T$	N-thinning of binary image A by the template T

Chapter 1

Binary Erosion and Dilation

The first chapter introduces the primary morphological operations on binary images, erosion and dilation. Erosion represents the probing of an image to see where some primitive shape fits inside the image, and all of mathematical morphology depends on this notion. Dilation is the dual operation to erosion, and is defined in terms of it relative to image complementation. Also discussed are the basic properties of dilation and erosion.

1.1   Introduction

In biology the term **morphology** refers to the study of form and structure in both plants and animals; in imaging, the term is not used so generically. **Mathematical morphology** refers to a branch of nonlinear image processing and analysis developed initially by Georges Matheron and Jean Serra that concentrates on the geometric structure within an image. That structure may be of a macro nature, where the goal is analysis of shapes such as tools or printed characters, or may be of a micro nature, where one might be interested in particle distributions or textures generated by small primitives. The original theory developed by Matheron and Serra was restricted to binary images, and we will concentrate there first, later proceeding to gray-scale morphology.

The scope of morphological methods is as wide as image processing itself. These include enhancement, segmentation, restoration, edge detection, texture analysis, particle analysis, feature generation, skeletonization, shape analysis, compression, component analysis, curve filling, and general thinning. There are many areas where morphological methods have been successfully applied, including robot vision, inspection, microscopy, medical imaging, remote sensing, biology, metallurgy, and digital documents. Both techniques and application areas continue to expand.

When we say that morphological processing is geometrically based, we mean this in a very specific sense. The basic idea, arising out of stereology, is to probe an image with a **structuring element** and to quantify the manner in which the structuring element fits (or does not fit) within the image. In Fig. 1.1 we see a binary image and a square structuring element (probe). The structuring element is placed in two different positions. In one location it fits; in the other it does not fit. By marking the locations at which the structuring element fits within the image, we derive structural information concerning the image. This information depends on both the size and shape of the structuring element, and, as emphasized by Matheron, the nature of that information is therefore dependent on the choice of the structuring element. To paraphrase Matheron, as is typically the case with

1

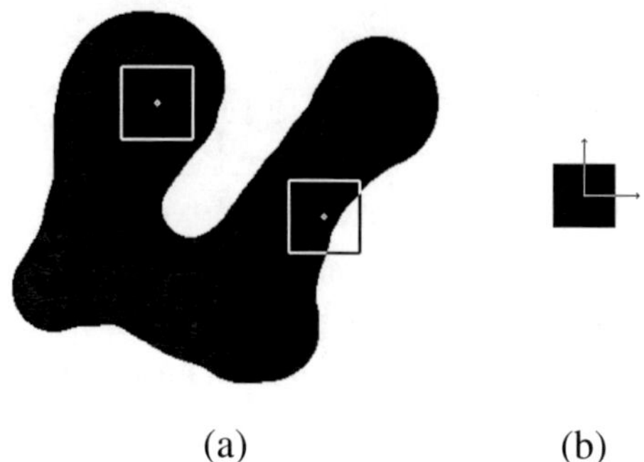

(a)   (b)

Figure 1.1 (a) Probing an image, (b) square probe.

scientific knowledge, the knowledge we have concerning an image is relative to the manner in which we probe (observe) it, and all consequent relationships are dependent on our choices regarding the manner of our observations. Even if we apply machine algorithms to select appropriate structuring elements, the criteria by which the algorithms make their selections are ultimately determined by the kind of information we desire.

All morphological processing depends on the concept of fitting structuring elements. There is, in fact, only one primary operation in mathematical morphology, and that operation is simply a formal characterization of the probing concept outlined and illustrated in Fig. 1.1.

1.2   Euclidean and Discrete Binary Images

A binary image is composed of two types of pixels: foreground and background. Figure 1.2 shows two typical binary images displayed in black and white. Morphological processing is based on set theory. The set of all black pixels constitutes a complete description of the binary image.

We will consider two types of binary images: Euclidean and discrete. A **Euclidean binary image** is a subset of n-dimensional Euclidean space. For signal processing, $n = 1$, and for image processing, $n = 2$. For the most part we will focus on image processing, so a Euclidean image will be a subset of the Euclidean plane. Sets in the Euclidean plane denote the foreground regions in binary images. For digital implementation we will consider a **discrete image** to be a subset of the 2D Cartesian grid. While introducing the basic operations, we will state the definitions, discuss fundamental properties, and illustrate the geometry by employing figures in the Euclidean setting; we will provide numerical examples in discrete space. For the most part, we will use the letters A, B, C, and D to represent Euclidean images and the letters S, T, U, and V to represent digital images. We will usually use B and E to denote structuring elements (which themselves are small images).

We will represent a digital image by a **bound matrix**, which is simply a matrix

(a)					(b)

Figure 1.2 Two binary images.

representing some portion of the Cartesian grid. Whereas it is typical to represent images by matrices, here we use a convention that the position of the origin $(0,0)$ is at the center of the matrix. To facilitate the visualization, the position of the origin relative to the matrix must be marked. This will be done by using bold print for the image value located at the origin. For the present, our attention is restricted to binary images, and we will denote activated (foreground) and unactivated (background) pixels by 1 and 0, respectively. By assumption, when we are dealing with binary images, any pixel not in the frame of the matrix will be assumed to have value 0.

As an illustration of a digital image and a structuring element in bound matrix format, let

$$
S = \begin{pmatrix} 0 & 0 & 0 & 0 & 0 \\ 0 & 0 & 1 & 0 & 0 \\ 0 & 0 & \mathbf{1} & 1 & 0 \\ 0 & 0 & 1 & 1 & 0 \\ 0 & 0 & 0 & 0 & 0 \end{pmatrix}, \quad E = \begin{pmatrix} 0 & 1 & 0 \\ 1 & \mathbf{1} & 1 \\ 0 & 1 & 0 \end{pmatrix} . \tag{1.1}
$$

The image S consists of five pixels, and it is represented within a 5×5 matrix with origin at the center of the grid. The structuring element E represents a 3×3 elementary diamond with the center pixel at coordinates $(0,0)$. The set representations of image S and structuring element E are

$$
\begin{aligned}
S &= \{(-1,0), \quad (0,0), \quad (0,1), \quad (1,0), \quad (1,1)\}, \\
E &= \{(0,-1), \quad (-1,0), \quad (0,0), \quad (1,0), \quad (0,1)\},
\end{aligned} \tag{1.2}
$$

where co-ordinates are relative to counting down (rows) and to the right (columns) relative to the center pixel.

The structuring element E can be created by the function `mmsecross` that creates a 3×3 diamond structuring element. The function `mmseshow` returns a matrix form of the structuring element useful for displaying in numeric format. `mmseshow` always returns an odd-size image, and by convention its center is the structuring element origin. To visualize small binary structuring elements, there is available the option 'EXPAND' in mmseshow. The following script generates the graphics of Fig. 1.3(b). Note that the origin is marked by a shaded square.

```
                            Example 1.1
>>>   E = mmsecross(1)      # create 3x3 diamond
>>>   print mmseshow( E)    # display E into matrix format
      [[0 1 0]
       [1 1 1]
       [0 1 0]]
>>>   mmshow( mmseshow( E, 'EXPAND') # visualization
```

In most of this chapter there will be no distinction between images and structuring elements, so S could be seen as a structuring element and E as an image. Later on, we will see that for practical use, there will be a distinction between images and structuring elements.

There are several ways to build the image S using the **Morphology Toolbox** (**MT**). The one illustrated below creates by initializing directly a binary matrix. Its visualization, created by mmseshow, is shown in Fig. 1.3(a). Note that the redundant zeros around the image were eliminated, resulting in a 3×3 image that is equivalent to image S of Eqs. (1.1) and (1.2).

```
                            Example 1.2
>>>   # Creating matrix S
>>>   S = mmbinary([          # binary image
        [0,  0,  0,  0,  0],
        [0,  0,  1,  0,  0],
        [0,  0,  1,  1,  0],
        [0,  0,  1,  1,  0],
        [0,  0,  0,  0,  0]])
>>>   print S                 # display S in numeric format
      [[0 0 0 0 0]
       [0 0 1 0 0]
       [0 0 1 1 0]
       [0 0 1 1 0]
       [0 0 0 0 0]]
>>>   # image visualization [Fig.  1.3(a)]
>>>   mmshow(mmseshow(S,'EXPAND'))
```

1.3 Erosion

Characterization of fitting depends on one basic Euclidean-space operation. The **translation** of a set A by a point x is denoted by A_x and is defined by

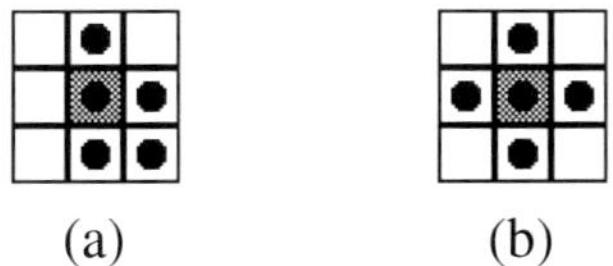

(a)   (b)

Figure 1.3 (a) Discrete image S, (b) discrete structuring element E.

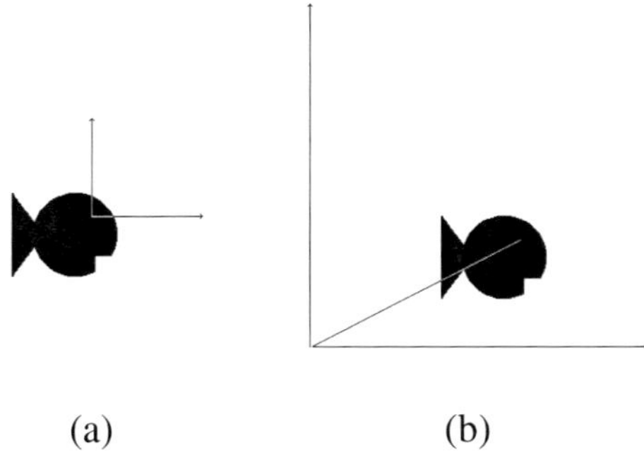

(a)   (b)

Figure 1.4 (a) Image A, (b) A translated by the vector x.

$$A_x = \{a + x : a \in A\}. \tag{1.3}$$

Geometrically, as illustrated in Fig. 1.4, A_x is A translated along the vector x. The nature of probing is to mark the positions (translations) of a structuring element where it fits into an image.

The fundamental operation of mathematical morphology is erosion. The **erosion** of set A by set B is denoted by $A \ominus B$ and is defined by

$$A \ominus B = \{x : B_x \subset A\}, \tag{1.4}$$

where $\subset$ denotes the subset relation. Erosion is also denoted by $\varepsilon_B(A)$ using the functional notation. In the **MT**, the erosion is obtained by `mmero(A,B)`. Relative to erosion, we call A the input image and B the structuring element. $A \ominus B$ consists of all points x for which the translation of B by x fits inside of A. If we treat B as a template, then $A \ominus B$ consists of all template origin positions for which the translated template fits inside A. Erosion can also be used in robot path planning. The structuring element is the robot shape. The input image is the region where the robot can move (the robot cannot rotate). The erosion consists of the loci of all robot coordinates where the robot can fit. These coordinates are related to the robot (structuring element) origin. This is illustrated in Fig. 1.5, where the structuring element is a half-rounded square [Fig. 1.5(b)]. Geometrically, the robot has been moved around inside the input image (shown in light gray in Fig. 1.5(a); the positions of the origin have been marked in black so as to produce the eroded image, i.e., places where the robot can be translated to.

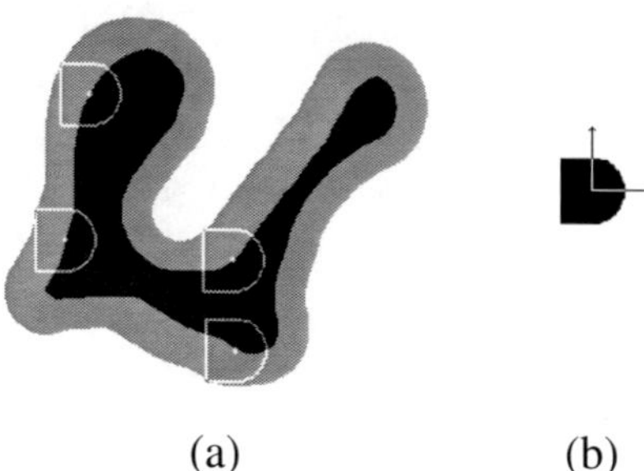

(a)   (b)

Figure 1.5 Erosion as robot path planning: (a) input image in gray and erosion in black (region where the center of the robot can move), (b) structuring element (the robot).

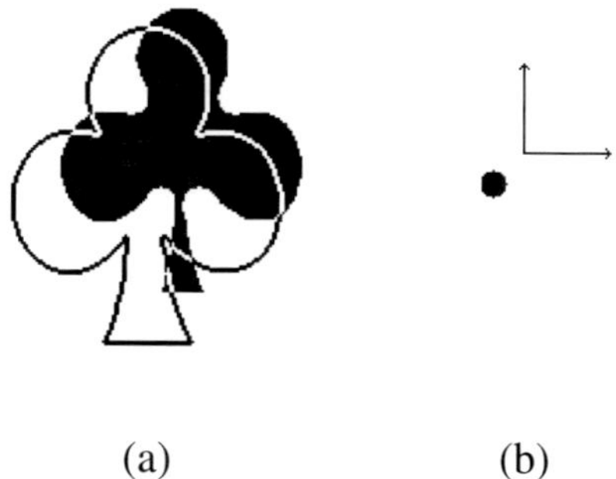

(a)   (b)

Figure 1.6 Erosion is not a subimage: (a) original in outline, eroded in black, (b) structuring element that does not contain the origin.

If the origin lies inside of the structuring element, then erosion has the effect of shrinking the input image, as illustrated by the robot example. Formally, we can state the following property: if the origin is contained within the structuring element, then the eroded image is a subset of the input image. Should the origin not lie within the structuring element, then it may not be that the eroded image lies within the input image. This situation is illustrated in Fig. 1.6.

Erosion can be formulated in other ways besides the fitting characterization of Eq. (1.4). Of particular importance is its representation by an intersection of image translations:

$$A \ominus B = \bigcap_{b \in B} A_{-b}. \tag{1.5}$$

Here, the erosion is found by intersecting all translations of the input image by negatives of points in the structuring element. The method is illustrated in Fig. 1.7. While the fitting definition of erosion is paramount for image-processing insight, the formulation of Eq. (1.5) can be useful for both computation and theory.

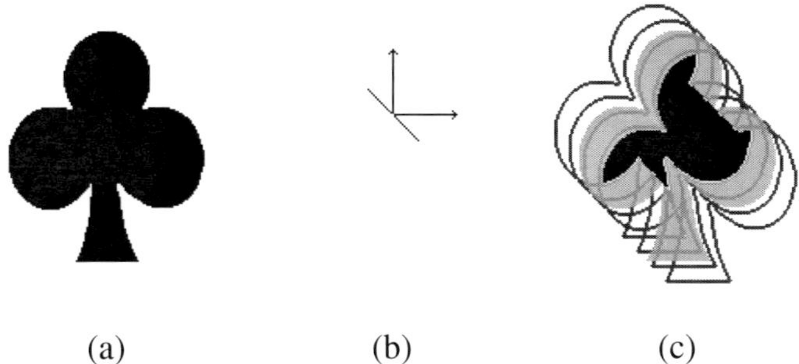

(a) (b) (c)

Figure 1.7 Erosion as intersection of translations: (a) input image, (b) structuring element, (c) erosion in black.

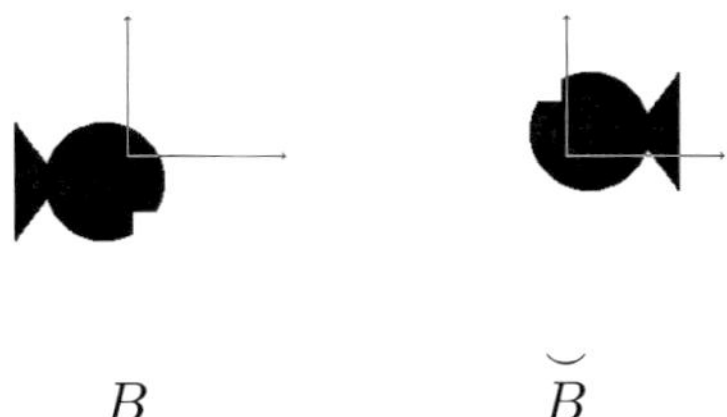

B $\breve{B}$

Figure 1.8 (a) Structuring element reflection.

The erosion formulation of Eq. (1.5) is closely related to a classical set operation first studied by Minkowski: relative to erosion and its formulation via Eq. (1.5), **Minkowski subtraction** of A by B is defined by

$$A \ominus \breve{B} = \bigcap_{b \in B} A_b, \tag{1.6}$$

where

$$\breve{B} = \{-b : b \in B\} \tag{1.7}$$

is the **reflection** of B or a 180-deg rotation of B about the origin (see Fig. 1.8). In words, Minkowski subtraction is erosion by the reflected structuring element. When reading the literature, one must show care with regard to the notation $\ominus$, since in many cases it refers to Minkowski subtraction, and erosion is defined relative to Minkowski subtraction. Here we stay with current practice and use $\ominus$ to denote erosion.

The fitting characterization of Eq. (1.4) applies directly to digital space [as does the intersection formulation of Eq. (1.5)]. Consider the following example using the digital image S and structuring element E:

$$\begin{aligned} S &= \{(-1,-1),(0,-1),(1,-1),(0,0),(1,0),(-1,1),(1,1),(0,2)\}, \\ E &= \{(0,0),(1,0),(1,1)\}. \end{aligned} \tag{1.8}$$

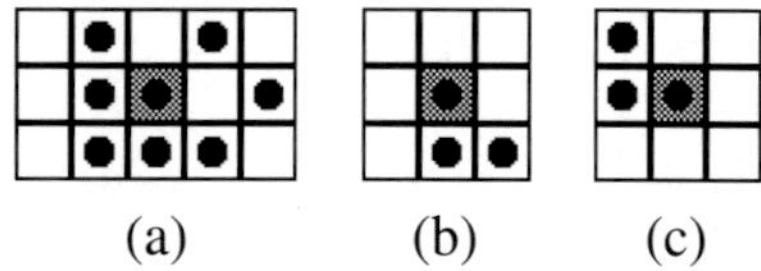

(a)   (b)   (c)

Figure 1.9 Erosion for digital images: (a) input image S, (b) structuring element E, (c) erosion $S \ominus E$.

The erosion is stored in image T. The three digital images are graphically represented in Fig. 1.9. The eroded image contains three points: $(-1, -1)$, $(0, -1)$, and $(0, 0)$, which are the points that a translation of the structuring element E by them yields a fit in the input image S.

```
                          Example 1.3
>>>   S = mmbinary([                    # create S
        [0, 0, 0, 0, 0, 0, 0],
        [0, 0, 1, 0, 1, 0, 0],
        [0, 0, 1, 1, 0, 1, 0],
        [0, 0, 1, 1, 1, 0, 0],
        [0, 0, 0, 0, 0, 0, 0]])
>>>   E = [ # create E for struct.  element
        [0, 0, 0],
        [0, 1, 0],
        [0, 1, 1]]
>>>   T = mmero( S, E)                  # erosion of S by E
>>>   print T
      [[0 0 0 0 0 0 0]
       [0 0 1 0 0 0 0]
       [0 0 1 1 0 0 0]
       [0 0 0 0 0 0 0]
       [0 0 0 0 0 0 0]]
```

1.4   Dilation

The second most basic operation of binary mathematical morphology is dilation. It is a dual operation to erosion, meaning that it is defined via erosion by set complementation. The **dilation** of set A by B is denoted by $A \oplus B$ and is defined by

$$A \oplus B = (A^c \ominus \breve{B})^c, \tag{1.9}$$

where A^c denotes the set-theoretic complement of A. Dilation is also denoted by $\delta_B(A)$ using the functional notation. In the **MT**, the dilation is obtained by

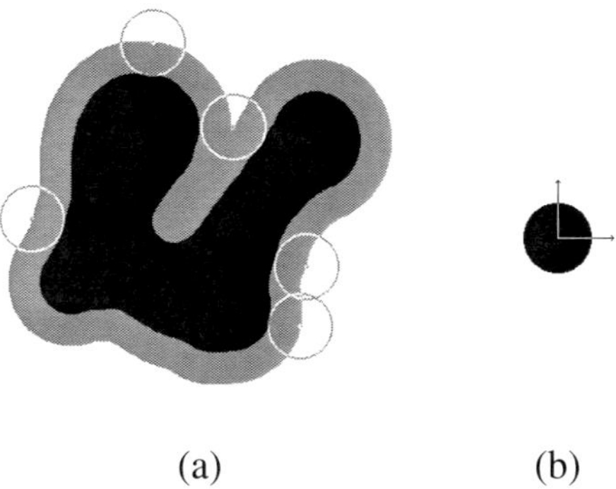

(a) (b)

Figure 1.10 Dilation as the dual of erosion (dilation as expansion): (a) input image in black, dilation in gray, (b) structuring element.

`mmdil(A,B)`. To dilate A by B, B is rotated around the origin to obtain $\breve{B}$, A^c is eroded by $\breve{B}$, and then the complement of the erosion is taken. As indicated by Fig. 1.10 where B is a disk, if B contains the origin, then dilation of A by B results in an expansion of A. Since dilation involves a fitting into the complement of an image, it represents a filtering on the outside, whereas erosion represents a filtering on the inside. Recalling the robot path interpretation of erosion of Fig. 1.5, dilation is the place where the reflected robot cannot move from outside. Another key difference is that dilation is commutative:

$$A \oplus B = B \oplus A. \tag{1.10}$$

If we think of a disk structuring element, then dilation fills in small (relative to the disk) holes and protrusions into the image, whereas erosion eliminates small components and extrusions of the image into its complement.

Two equivalent formulations of dilation deserve mention. First,

$$A \oplus B = \bigcup_{b \in B} A_b, \tag{1.11}$$

so that the dilation can be found by translating the input image by all points in the structuring element and then taking the union. Written in the form of Eq. (1.11), dilation has historically been called **Minkowski addition**. Because dilation is commutative, Eq. (1.11) can be rewritten as

$$A \oplus B = \bigcup_{a \in A} B_a. \tag{1.12}$$

Equation (1.12) is useful for characterizing the effects of dilation and will be extensively employed; however, it is computationally more burdensome because it involves an image translation for every point in the input image, whereas Eq. (1.11) only requires a translation for every point in the structuring element. Geometrically

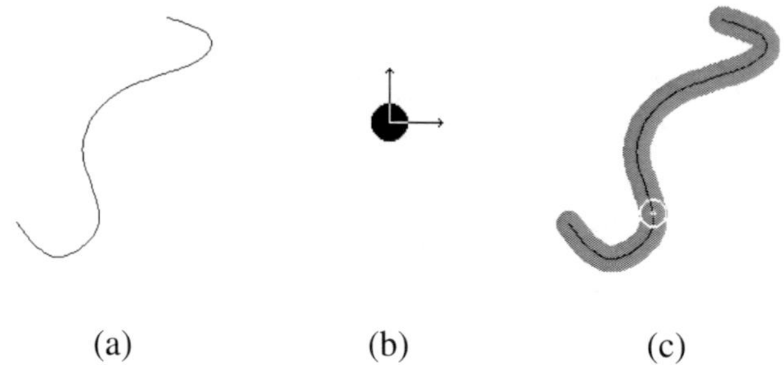

(a)   (b)   (c)

Figure 1.11 Dilation as structuring element stamping: (a) input image, (b) structuring element, (c) dilation as union of translations [Eq. (1.12)].

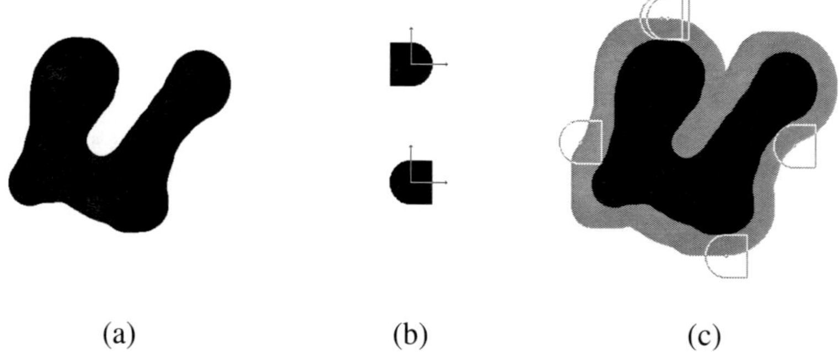

(a)   (b)   (c)

Figure 1.12 Dilation as reflected robot path planing from outside: (a) input image; (b) structuring element, normal and reflected; (c) illustration of Eq. (1.13), the reflected structuring element slides along the shape from outside.

this formulation resembles stamping the structuring element on each activated pixel of the input image. Figure 1.11 illustrates this. The input image is a line and the stamp is a circle. This is also similar to a painting process where the structuring element is the brush shape and the pixels of the input image indicate the places to apply the brush.

Another formulation of dilation involves translations of the rotated structuring element that "hit" (intersect) the input image:

$$A \oplus B = \{x : (\breve{B})_x \cap A \neq \emptyset\}. \tag{1.13}$$

This approach is illustrated in Fig. 1.12.

Using the image and structuring element of Fig. 1.9 [Eq. (1.8)], we illustrate digital dilation using the Minkowski addition formulation of Eq. (1.12). The structuring element is translated to all image pixels, and the resulting structuring element translations are unioned. Referring to Fig. 1.13, $(-1, -1)$ is an activated pixel of S,

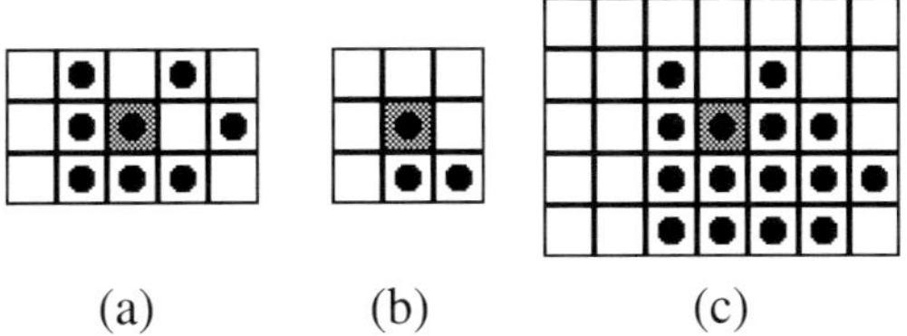

(a)　　　　(b)　　　　(c)

Figure 1.13 Dilation for digital images: (a) input image S, (b) structuring element E, (c) dilation $S \oplus E$.

so the structuring element is translated by $(-1, -1)$, the net effect being to place the structuring element origin at $(-1, -1)$. Owing to this translation, pixels $(-1, -1)$, $(0, -1)$, and $(0, 0)$ are activated in the dilated image. Upon forming the union of all such translations, the dilated image is given by

```
                        Example 1.4
>>   T = mmdil( S, E)     # dilation of S by E
>>   print T
     [[0 0 0 0 0 0 0]
      [0 0 1 0 1 0 0]
      [0 0 1 1 1 1 0]
      [0 0 1 1 1 1 1]
      [0 0 1 1 1 1 0]]
```

The effects on pixel geometry of both dilation and erosion are depicted in Fig. 1.14, where the image S is both dilated and eroded by the structuring element E consisting of four pixels, the center pixel being situated at the origin. Dilation has the expected expanding effect, filling in small intrusions into the image. Note that a one-pixel hole and the small intrusions in the middle of the character "g" have disappeared. Erosion has a shrinking effect, eliminating small extrusions. Two side effects of erosion are to fully eliminate the horizontal line at the bottom of the im-

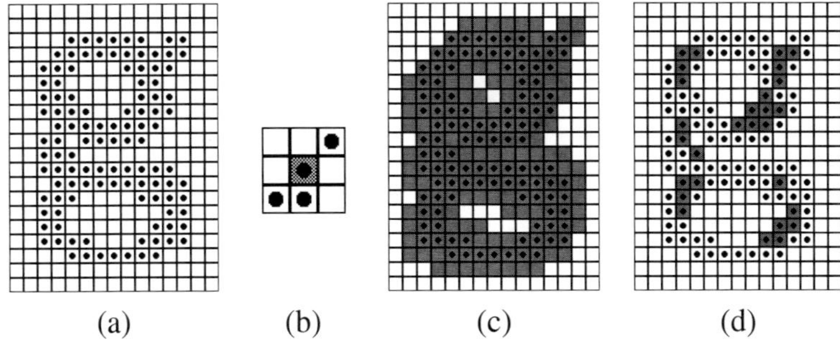

(a)　　　　(b)　　　　(c)　　　　(d)

Figure 1.14 Pixel geometry for dilation and erosion of digital images: (a) input image, (b) structuring element, (c) dilation, (d) erosion.

age (because it is too thin for the structuring element to fit) and to disconnect the image.

1.5 Algebraic Properties

One appealing feature of mathematical morphology is the existence of a system of algebraic relations involving erosion, dilation, and the basic set-theoretic operations. Taken together, these relations constitute the **Minkowski algebra**. In this and the next two sections we discuss some basic propositions within the Minkowski algebra. The sections divide the properties into three categories: those concerning general algebraic properties of dilation and erosion, those concerning the operations as image filters, and those concerning the relations between the morphological and set-theoretic operations.

Two basic relationships are commutativity and associativity of dilation, the former being given already in Eq. (1.10), and the latter being given by

$$A \oplus (B \oplus C) = (A \oplus B) \oplus C. \tag{1.14}$$

Associativity allows us to perform iterated dilations without having to concern ourselves with which dilation is performed first. It is legitimate to write expressions such as $A \oplus B \oplus C$ and $A \oplus B \oplus C \oplus D$ without having to worry about parentheses.

An operation that plays a central role in sizing and particle analysis is the scalar multiplication of a set by a real number. For any real number t, we define $tA = \{ta : a \in A\}$. A special case of tA is 180-deg rotation about the origin, $-A$, where $t = -1$. Scalar multiplication satisfies a type of distributivity relative to both dilation and erosion:

$$t(A \oplus B) = tA \oplus tB, \tag{1.15}$$

$$t(A \ominus B) = tA \ominus tB. \tag{1.16}$$

When employing special-purpose hardware to perform erosions, it is typical to be confronted by a limitation on the size of the structuring element. There is a property whose digital counterparts have played a central role in circumventing this dilemma, namely,

$$A \ominus (B \oplus C) = (A \ominus B) \ominus C. \tag{1.17}$$

Rather than erode by $B \oplus C$, which might be too large, we can iteratively erode by B and then by C.

Referring to Fig. 1.15, consider eroding by the disk $2B$ of radius 2, B being the disk of radius 1. Since $2B = B \oplus B$, we can perform the operation in the following manner:

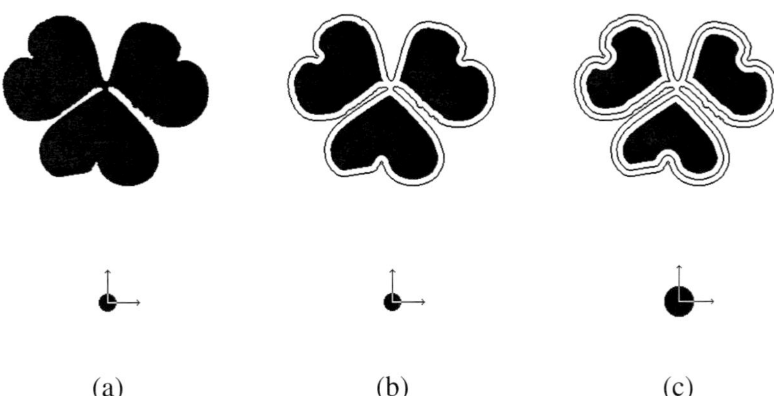

Figure 1.15 Structuring element decomposition: (a) input image A and structuring element B, (b) $A \ominus B$, (c) $A \ominus 2B = (A \ominus B) \ominus B$.

$$A \ominus 2B = A \ominus (B \oplus B) = (A \ominus B) \ominus B. \tag{1.18}$$

The key to Eq. (1.18) is not that B is a disk, but that $2B = B \oplus B$, the latter representing a decomposition of $2B$. Moreover, the iterative evaluation applies to nB, where n is any positive integer, with $0B$ being simply the origin:

$$nB = B \oplus B \oplus \ldots \oplus B. \tag{1.19}$$

For instance, for a disk of radius 3,

$$A \ominus 3B = ((A \ominus B) \ominus B) \ominus B. \tag{1.20}$$

For a practical digital illustration, suppose we wish to erode by the 5 by 5 square. This square admits a decomposition into a dilation of two 3 by 3 squares, so that the desired erosion can be accomplished by iteration. Decomposition of digital structuring elements is not generally possible and one must often settle for an approximate result. Owing to practical hardware limitations, the problem has been well studied.

1.6 Filter Properties

Certain fundamental notions concerning image operators (filters) are useful for the description of basic properties pertaining to elementary morphological operations, and they will be key to our discussions regarding morphological filtering. At present our main concern is with operations on binary images that yield binary images, as opposed to those that yield gray-level images (such as labeling) or those that yield numerical values (such as perimeter measurement and component counting). If A is the input image, then we let $\Psi(A)$ denote the output image. For instance, if Ψ is erosion by B, then $\Psi(A) = A \ominus B$; if Ψ is dilation by B, then $\Psi(A) = A \oplus B$. Throughout the book we will be considering many operators and Ψ will therefore take many forms. To fully appreciate the algebraic structure of mathematical morphology, especially as it relates to filtering, one must recognize that there are certain properties that an operator may or may not possess that make it useful or not useful for a certain task. For instance, in linear processing operators are required to be linear, and the degree to which linearity is or is not an appropriate constraint determines the efficacy of it being required. Here we consider properties relevant to morphological processing.

An operator Ψ is said to be **translation invariant** if the same output results from translating the input image and then operating by Ψ as would result from operating by Ψ and then translating:

$$\Psi(A_x) = (\Psi(A))_x. \tag{1.21}$$

Both dilation and erosion are translation invariant. For dilation, this means that the same output results from first translating the image and then dilating by a given structuring element as would result by first dilating the image by the structuring element and then translating:

$$A_x \oplus B = (A \oplus B)_x. \tag{1.22}$$

For erosion, translation invariance takes the form

$$A_x \ominus B = (A \ominus B)_x. \tag{1.23}$$

When considering translation invariance, one must be careful to recognize that it applies to translating the image, not the structuring element. Nevertheless, because dilation is commutative it is straightforward to deduce from Eq. (1.22) that

$$A \oplus B_x = (A \oplus B)_x, \tag{1.24}$$

so that dilation is translation invariant relative to the structuring element. Such is not the case with erosion. In fact, it can be shown that

$$A \ominus B_x = (A \ominus B)_{-x}, \tag{1.25}$$

so that translating the structuring element prior to eroding is equivalent to eroding and then translating the eroded image by the same amount in the opposite direction.

An operator Ψ is said to be **monotonically increasing** if, whenever A_1 is a subset of A_2, then $\Psi(A_1)$ is a subset of $\Psi(A_2)$, so that Ψ preserves order. A standard operator that is not increasing is the boundary operator: the fact that one set is a

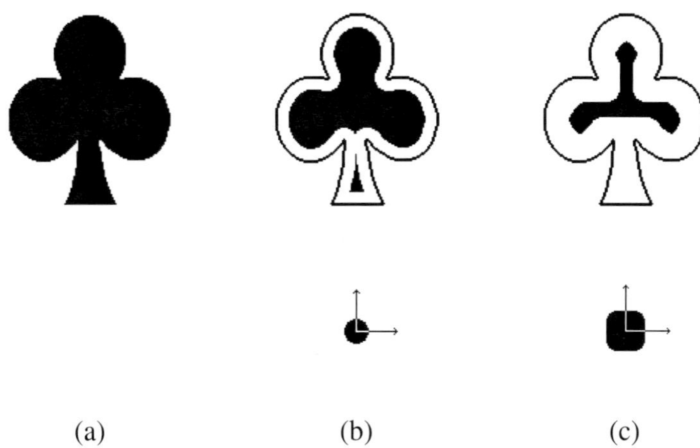

(a)   (b)   (c)

Figure 1.16 Effect of structuring element order: (a) original, (b) erosion by $B1$, (c) erosion by $B2$.

subset of another does not imply a similar order relationship between their boundaries. For a fixed structuring element, both dilation and erosion are increasing: if $A_1 \subset A_2$, then $A_1 \oplus B \subset A_2 \oplus B$ and $A_1 \ominus B \subset A_2 \ominus B$.

Increasing monotonicity for erosion is relative to a fixed structuring element and input images ordered by set inclusion; quite a different phenomenon occurs if the input image is kept constant and two ordered structuring elements are employed. If A is a fixed image and B_1 is a subset of B_2, then it is easier to fit B_1 inside of A than it is to fit B_2. Consequently, $A \ominus B_1$ contains $A \ominus B_2$. This property is illustrated in Fig. 1.16 and it is central to the design of morphological filters involving multiple erosions.

We have spoken of the fact that dilation is dual to erosion because it is found by eroding the complement of an image. The notion of duality is central to morphological processing. In general, given an operator Ψ, the **dual operator** is denoted by Ψ^* and is defined by

$$\Psi^*(A) = (\Psi(A^c))^c. \tag{1.26}$$

According to Eq. (1.9), dilation by B is the dual of erosion by $\breve{B}$. If we apply Eq. (1.9) to the dilation of A^c by $\breve{B}$ (instead of A by B) and then take the complement of each side (recognizing that $(A^c)^c = A$ and $\breve{\breve{B}} = B$), we obtain

$$A \ominus B = (A^c \oplus \breve{B})^c, \tag{1.27}$$

which says that erosion by B is the dual of dilation by $\breve{B}$.

It is no accident that Eqs. (1.9) and (1.27) appear as a pair, that is, that dilation is the dual of erosion and vice versa. In general, the dual of the dual is the original operator: $\Psi^{**} = \Psi$. We will see this relationship between the dual and the original operator again in the next chapter, where we treat opening and (its dual) closing.

1.7 Relationship to Set Operations

Dilation and erosion satisfy certain distributivity properties with respect to the set-theoretic operations, but we must be careful. Dilation distributes over union,

$$A \oplus (B_1 \cup B_2) = (A \oplus B_1) \cup (A \oplus B_2), \tag{1.28}$$

but it does not distribute over intersection. Regarding the latter, all we can say is that

$$A \oplus (B_1 \cap B_2) \subset (A \oplus B_1) \cap (A \oplus B_2). \tag{1.29}$$

In terms of structuring elements, dilating by a union of structuring elements is equivalent to dilating by each structuring element individually and then forming the union of the individual dilations; dilating by an intersection of structuring elements yields a subset of the intersection of the individual dilations.

As the dilation is commutative, it also distributes from the right over union:

$$(A_1 \cup A_2) \oplus B = (A_1 \oplus B) \cup (A_2 \oplus B). \tag{1.30}$$

Unioning two images and then dilating is equivalent to dilating each and then unioning. Any operator that is right distributive over union is called an **algebraic dilation**.

Turning to the relationship between erosion and the set operations, we might expect some subtleties arising from the fact that erosion is not commutative. In fact, here we must distinguish between right and left distributivity. Erosion distributes from the right over intersection:

$$(A_1 \cap A_2) \ominus B = (A_1 \ominus B) \cap (A_2 \ominus B). \tag{1.31}$$

Thus, intersecting two images and then eroding is equivalent to eroding each and then intersecting. Similarly to algebraic dilation, any operator that is right distributive over intersection is called an **algebraic erosion**.

Relative to union, erosion satisfies left antidistributivity:

$$A \ominus (B_1 \cup B_2) = (A \ominus B_1) \cap (A \ominus B_2). \tag{1.32}$$

In terms of structuring elements, eroding by a union of structuring elements is equivalent to eroding by each structuring element individually and then forming the intersection of the individual erosions.

1.8 Bounded Operators

When implementing images in bound matrix format, there are some necessary conventions relative to the operations described so far. One difficulty is implementation of the set-theoretical complement. To do so would require a special data structure to indicate that the matrix elements are complemented and the pixels outside the bound matrix are one instead of zero. In this book, as implemented in the **MT**, images and the structuring elements are represented in the bound matrix format, which is convenient because all computer languages support this simple data structure.

It is customary to have a large input image and a small structuring element. Although these are mathematically equivalent, they are implemented differently. Hence, operations require variants for the two different data structures. From the operations described thus far, the following operate on structuring elements exactly as described in the previous sections and have the algebraic properties seen in Sec. 1.5: translation (mmsetrans), union (mmseunion), intersection (mmseintersec), dilation (mmsedil), and erosion (mmseero). For instance, for structuring elements E_1 and E_2, the implementation of dilation $E_1 \oplus E_2$ behaves exactly as defined, with the accompanying algebraic properties. If E_1 and E_2 are both 3×3, then $E_1 \oplus E_2$ is 5×5. Below we show a simple numerical example with the dilation. Recall that the structuring element origin is always in the matrix center.

```
                           Example 1.5
>>>   E1 = mmsecross(1)      # create 3x3 diamond
>>>   print mmseshow( E1)    # convert E into matrix format
      [[0 1 0]
       [1 1 1]
       [0 1 0]]
>>>   E2 = mmsedil(E1,E1)    # dilation
>>>   print mmseshow( E2)    # display E2 into matrix format
      [[ 0 0 1 0 0]
       [ 0 1 1 1 0]
       [ 1 1 1 1 1]
       [ 0 1 1 1 0]
       [ 0 0 1 0 0]]
```

For processing images, the implementation differs from the theory because the input and output images are represented by bound matrices of the same size. Specifically, the union (mmunion), intersection (mmintersec) and complement (mmneg) require the input and output matrices be of the same domain (size). As a consequence, pixels at the image border must be processed differently, with a corresponding effect on operator properties. In particular, it is not possible to have the translation invariance property for dilation and erosion.

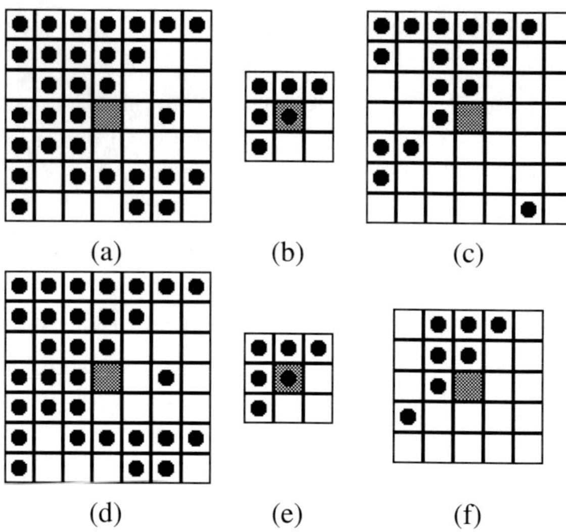

Figure 1.17 Bounded erosion: (a) input image, (b) structuring element, (c) bounded erosion. Ordinary erosion: (d) input image, (e) structuring element, (f) ordinary erosion.

Corresponding to the image region represented by a bound matrix, we define a **view** of a binary image S, denoted $V[S]$, to be a bounded region such that S and all operations concerning S are confined to the region. Practically, this means that, owing to a finite image frame, we are only able to observe and operate within the view. It could be that the real-world image from which the set-theoretic (or digital) image has been formed extends beyond the view, but that is unknown to us and irrelevant to image processing. It is necessary to define translation, complement, erosion and dilation operations relative to the view. All translations must be truncated to the view and the outputs of operations are confined to the view. Under this convention, the operations are implemented via bound matrices.

The **bounded complement** of the image S is defined as

$$S^{\dot{c}} = S^c \cap V[S]. \qquad (1.33)$$

For image S and structuring element E, the **bounded erosion** is defined as

$$S \dot{\ominus} E = \{x : E_x \cap V[S] \subset S\}. \qquad (1.34)$$

Comparing this to Eq. (1.4), after translation the structuring element is restricted to the view before verifying that it fits within the input image. If all translations of the structuring element are subsets of the view, then the bounded erosion equals the ordinary erosion. Note that the view of the output image is the same as the input image. Figure 1.17 illustrates the bounded and the ordinary erosions.

Analogously, for image S and structuring element E, the **bounded dilation** is defined as

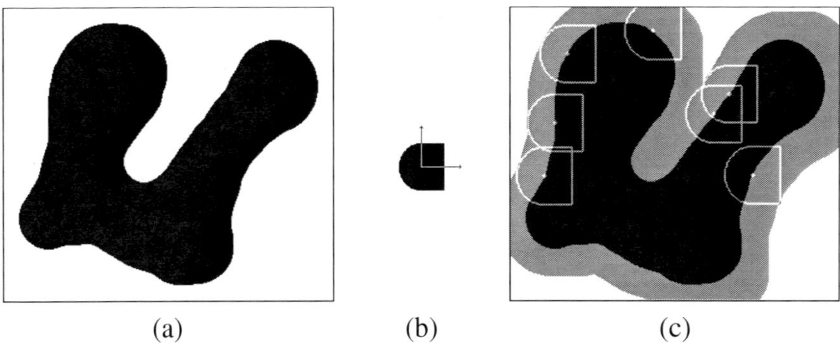

(a)   (b)   (c)

Figure 1.18 Bounded dilation: (a) input image, (b) structuring element, (c) bounded dilation.

$$S \dot{\oplus} E = \bigcup_{x \in E} S_x \cap V[S].\tag{1.35}$$

Like bounded erosion, the translated structuring element is restricted to the view of the input image. In this case, there is a difference in the treatment of the image and the structuring element making bounded dilation not commutative. However, duality remains between bounded dilation and bounded erosion with respect to the bounded complement:

$$S \dot{\oplus} E = (S^{\dot{c}} \dot{\ominus} \breve{E})^{\dot{c}}.\tag{1.36}$$

In analogy to bounded erosion, if all translations of the image are subsets of the view, then the bounded dilation equals the ordinary dilation. In fact, there is a simple relation between the bounded and ordinary dilations. The bounded dilation is obtained by applying the ordinary dilation of the intersection of the input image and the view and then intersecting the result with the view:

$$S \dot{\oplus} E = ((S \cap V[S]) \oplus E) \cap V[S].\tag{1.37}$$

From the duality equation it is possible to write the relation between the bounded and ordinary erosions, which is not obvious. To obtain the bounded erosion from the ordinary erosion, the image is unioned with the complement of the view, then the ordinary erosion is applied, and finally the result is intersected with the view:

$$S \dot{\ominus} E = ((S \cup (V[S])^c) \ominus E) \cap V[S].\tag{1.38}$$

In the **MT**, the functions `mmneg`, `mmdil` and `mmero` implement bounded complement, bounded dilation and bounded erosion, respectively. Figure 1.18 illustrates the border processing and the bound matrix fixed size in bounded dilation. Figure 1.19 shows the operators of the **MT** discussed in this chapter.

name	description
mmseunion:	union
mmsetrans:	structuring element translation [Eq. (1.3)]
mmsereflect:	structuring element reflection [(Eq. (1.7)]
mmsedil:	dilation [Eq. (1.11)]
mmsesum:	n-composition of struct. el. [Eq. (1.19)]
mmneg:	bounded complement [Eq. (1.33)]
mmunion:	bounded union
mmintersec:	bounded intersection
mmero:	bounded erosion [Eq. (1.34)]
mmdil:	bounded dilation [Eq. (1.35)]

Figure 1.19 MT operators presented in this chapter.

1.9 Exercises

1. Let

$$
\begin{aligned}
S &= \{(0,0), \quad (0,2), \quad (2,0), \quad (2,2)\} \\
E &= \{(0,0), \quad (1,1), \quad (0,2), \quad (0,3)\} \;. \\
T &= \{(0,0), \quad (-1,0), \quad (-2,0), \quad (1,1)\}
\end{aligned}
\tag{1.39}
$$

Find each of the following sets (in set notation and pictorially):

   (a) $S + (0, -1)$

   (b) $S \oplus E$

   (c) $(S \oplus E) \oplus T$

   (d) $S \ominus E$

   (e) $E \ominus S$

   (f) $2S$

   (g) $(S \oplus 2E) \cup T$

2. Let

$$
\begin{aligned}
A &= \{(x,y) : 0 \le x \le 2, 0 \le y \le 3\} \\
B &= \{(x,y) : x^2 + y^2 \le 1\} \\
C &= \{(x,y) : y = x \text{ and } 0 \le x \le 1\}
\end{aligned}
$$

Give graphical representation of A, B, and C, and the following:

   (a) $A \oplus B$

   (b) $A \oplus C$

   (c) $B \oplus C$

   (d) $A \ominus B$

   (e) $A \ominus C$

   (f) $2A \ominus B$

(g) $2A \ominus C$

(h) $B \oplus 2B$

(i) $A \oplus 2B$

(j) $2A \oplus C$

3. Consider the images

$$S = \begin{pmatrix} 1 & 1 & 1 & 1 & 1 & 1 & 0 \\ 1 & 1 & 1 & 1 & 0 & 0 & 0 \\ 1 & 1 & 1 & 1 & 1 & 0 & 0 \\ 0 & 0 & 1 & \mathbf{1} & 1 & 1 & 1 \\ 0 & 0 & 0 & 1 & 1 & 1 & 1 \\ 0 & 0 & 0 & 1 & 1 & 1 & 0 \\ 0 & 0 & 0 & 0 & 1 & 1 & 1 \end{pmatrix},$$

$$T = \begin{pmatrix} 1 & 1 & 1 & 1 & 1 \\ 0 & 0 & \mathbf{1} & 1 & 1 \\ 1 & 1 & 0 & 0 & 1 \end{pmatrix}, \text{ and } E = \begin{pmatrix} 0 & 1 & 1 \\ 0 & \mathbf{1} & 1 \\ 0 & 0 & 0 \end{pmatrix}.$$

Find

(a) S^c

(b) $S \cap T$

(c) $S \cup T$

(d) $E_{1,2}$

(e) $S \oplus E$

(f) $T \oplus E$

(g) $S \ominus E$

(h) $T \ominus E$

4. Prove the equality of Eqs. (1.4) and (1.5).

5. Prove the equality of Eqs. (1.9) and (1.11).

6. Prove the equality of Eqs. (1.9) and (1.13).

7. Prove Eq. (1.17).

8. Prove Eq. (1.28).

9. Give an example to show that equality does not generally hold in Eq. (1.29).

10. Prove Eq. (1.31).

11. Let E be the 3×3 diamond structuring element composed of the origin together with its vertical and horizontal neighbors. Show that the dilation $S \oplus E$ can be expressed as a union of five erosions, each having a structuring element consisting of a single pixel.

12. Simplify the following expressions:

    (a) $(\breve{A} \oplus B)^{\smile}$

    (b) $(A \cup (B)_x)^{\smile}$

    (c) $(A \cup \breve{B})_x$

13. Show how to obtain the bounded erosion from the ordinary erosion equation [Eq. (1.38)]. Use the bounded dilation [Eq. (1.37)], its duality [Eq. (1.36)] and the bounded complement [Eq. (1.33)] equations.

14. Let
$$
S = \begin{pmatrix} 0 & 0 & 0 & 0 & 0 \\ 0 & 0 & 1 & 0 & 1 \\ 0 & 0 & 1 & 1 & 1 \end{pmatrix}, \quad E = \begin{pmatrix} 1 & \mathbf{0} & 0 \end{pmatrix}.
$$

    Compute the bounded erosion $S \dot\ominus E$. Compare the result with the ordinary erosion $S \ominus E$ and give a geometric interpretation of the results.

1.10 Laboratory Experiments

1. Study the implementation of the ordinary dilation and erosion (`mmsedil`, `mmseero`). Suggest and implement a more efficient code.

2. Study the implementation of the bounded dilation and erosion (`mmdil`, `mmero`). Which dilation and erosion equations were used for these implementations?

3. Compare the speed of dilating a large image by an 11×11 structuring element and two successive dilations of the image by first a vertical and then a horizontal 11-point structuring element.

4. Write a program to illustrate the duality between the bounded erosion and dilation. Use a 9×9 image and a 3×3 structuring element generated randomly several times to confirm if the implementation of the bounded erosion and dilation is correct.

5. Write a function to overlay a text on a binary image. Use the `mmtext` function to create a structuring element from a text and dilate it on a single image pixel of the size of the input image. Union both images to obtain the overlaid image.

6. Write a program to create a graphical representation of a binary image similar to the one shown in Fig. 1.3(a). Create the grid by placing isolated pixels at the graphical pixel center and dilating it with a square frame with the desired pixel size. Create the "on" pixels by dilating the center of these pixels by a circle. Finally, mark the origin by dilating it with a small chessboard-like structuring element.

7. Write a program to create another type of graphical representation of binary images of your own style.

References

1. J. Barrera, G. J. F. Banon, R. A. Lotufo, and R. Hirata, Jr. MMach: a mathematical morphology toolbox for the Khoros system. *Journal of Electronic Imaging*, 7(1):233–260, 1998.

2. E. R. Dougherty, editor. *Mathematical Morphology in Image Processing*. Marcel Dekker, New York, 1993.

3. E. R. Dougherty and J. T. Astola, editors. *Nonlinear Filters for Image Processing*. SPIE and IEEE Presses, Bellingham, WA, 1999.

4. J. Goutsias and S. Batman. Morphological methods for biomedical image analysis. In M. Sonka and J. M. Fitzpatrick, editors, *Handbook of Medical Imaging, Volume 2. Medical Image Processing and Analysis*, pages 175–272. SPIE, Bellingham, WA, 2000.

5. E. R. Dougherty and J. Barrera. Logical image operators. In E. R. Dougherty and J. T. Astola, editors, *Nonlinear Filters for Image Processing*, pages 1–60. SPIE and IEEE Presses, Bellingham, WA, 1999.

6. C. R. Giardina and E. R. Dougherty. *Morphological Methods in Image and Signal Processing*. Prentice-Hall, Englewood Cliffs, NJ, 1988.

7. H. J. A. M. Heijmans. *Morphological Image Operators*. Academic Press, Boston, 1994.

8. G. Matheron. *Random Sets and Integral Geometry*. John Wiley & Sons, New York, 1975.

9. J. Serra. *Image Analysis and Mathematical Morphology*. Academic Press, London, 1982.

10. J. Serra, editor. *Image Analysis and Mathematical Morphology. II: Theoretical Advances*. Academic Press, London, 1988.

11. P. Soille. *Morphological Image Analysis*. Springer-Verlag, Berlin Heidelberg New York, 2nd edition, 2003.

Binary Opening and Closing

Besides the two primary operations of erosion and dilation, there are two secondary operations that play key roles in morphological image processing, these being opening and its dual, closing. We focus mostly on opening, the properties of closing usually being analogous via complementation. Although opening is defined in terms of erosion and dilation, it possesses a more geometric formulation in terms of structuring element fits that is the basis for its application.

2.1 Opening

The **opening** of image A by image B is denoted by $A \circ B$ and is defined as a composition of erosion and dilation by

$$A \circ B = (A \ominus B) \oplus B. \tag{2.1}$$

A functional notation for opening is $\gamma_B(A)$, and the command for opening in the **MT** is mmopen (A, B). For a better appreciation of the role of opening in processing, we state an equivalent formulation:

$$A \circ B = \bigcup \{B_x : B_x \subset A\}. \tag{2.2}$$

Here, the opening results from the union of all translations of the structuring element that fit inside the input image. Each fit is marked and the opening results from taking the union of the structuring element translations to each marked location. Indeed, this is precisely what is meant by eroding and then dilating.

The expression of opening as erosion followed by dilation is illustrated in Fig. 2.1, where a rectangle is eroded and then dilated by a disk. It is also possible to discern the effect of fitting, as expressed in Eq. (2.2): opening the rectangle has resulted in it being rounded from the inside, this rounding resulting from the manner in which the disk has been "rolled around" inside the rectangle to achieve a union of the fits. Had the structuring element been a small square with horizontal base, then there would have been no rounding and the opened image would have been the same as the original.

We see at once two applications of opening in Fig. 2.1. Opening by a disk results in a filter that smooths from the inside; that is, it rounds corners extending into the background. The effect is quite different with a square structuring element.

Rather than view the opened image itself as the final output of the processing, we can take a different view. We can consider the set-theoretic subtraction of the opening from the input image. This operator is called **opening top-hat**:

$$A \hat{\circ} B = A - (A \circ B). \tag{2.3}$$

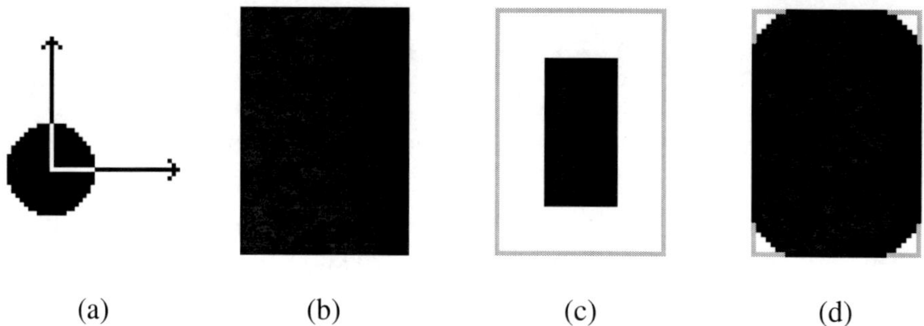

(a)   (b)   (c)   (d)

Figure 2.1 (a) Structuring element, (b) input image, (c) erosion, (d) opening.

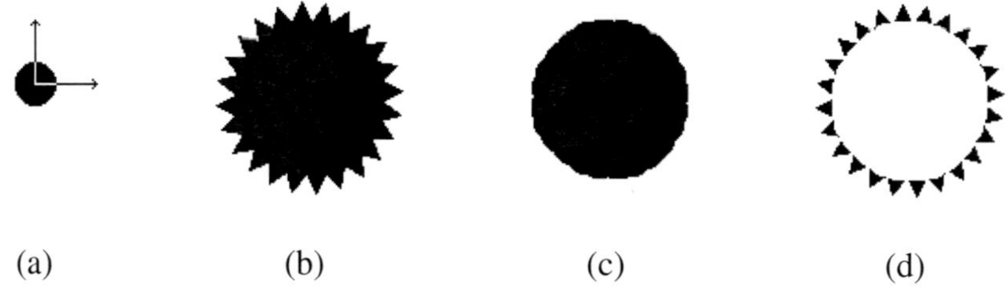

(a)   (b)   (c)   (d)

Figure 2.2 (a) Structuring element, (b) input image, (c) opening, (d) opening top-hat.

In Fig. 2.1, the opening top-hat consists of input-image corners that protrude into the background, and it can be employed for recognition purposes. Figure 2.2 shows another example of the opening top-hat to detect the teeth of a gear. Use of a disk is common because its shape-effect is rotationally invariant; however, there are many instances when it is beneficial to employ other kinds of structuring elements.

As for a digital example, if S and E are the image and structuring element of Fig. 1.9 [Eq. (1.8)], then

```
                         Example 2.1
>>>    T=mmopen(S,mmimg2se(E))    # opening of S by E
>>>    print T
       [[0 0 0 0 0 0]
        [0 1 0 0 0 0]
        [0 1 1 0 0 0]
        [0 1 1 1 0 0]
        [0 0 0 0 0 0]]
```

If we view S as a triangle that has some background noise, then opening by E has restored the triangle. We will shortly have much more to say on this type of restoration.

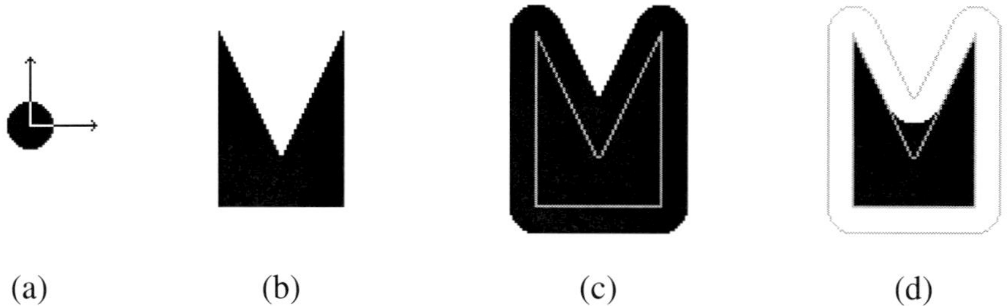

(a)   (b)   (c)   (d)

Figure 2.3 (a) Structuring element, (b) input image, (c) dilation, (d) closing.

2.2 Closing

The dual operation to opening is closing, which is defined as a dilation followed by an erosion. The **closing** of A by B is denoted by $A \bullet B$ and is defined by

$$A \bullet B = (A \oplus B) \ominus B. \tag{2.4}$$

Closing is also denoted by $\phi_B(A)$ in functional notation. In the **MT**, the close function is obtained by `mmclose(A,B)`. Figure 2.3 illustrates closing. The effect can be seen in the manner in which the closing has filtered from the outside, smoothing only corners that protrude into the image.

Closing is the dual operator of the opening because

$$A \bullet B = (A^c \circ \breve{B})^c. \tag{2.5}$$

Because closing is dual to opening, opening is dual to closing: replacing A by A^c in Eq. (2.5) and complementing yields

$$A \circ B = (A^c \bullet \breve{B})^c. \tag{2.6}$$

Note that the structuring element used in the closing is reflected. If B is a disk or any symmetrical shape, reflection plays no role. Rather than employ the iteration of Eq. (2.4) we could employ duality in conjunction with the union formulation of opening given in Eq. (2.2), thereby fitting, or "rolling the ball," around the outside of the image. Figure 2.4 illustrates the duality between open and close with a nonsymmetrical structuring element.

As the closing contains the input image, the set-theoretical subtraction of the input image from the closing gives the **closing top-hat** operator:

$$A \hat{\bullet} B = (A \bullet B) - A. \tag{2.7}$$

Figure 2.5 shows an application of the closing and the closing top-hat. The closing of a shape by a large disk approximates its convex hull and the closing top-

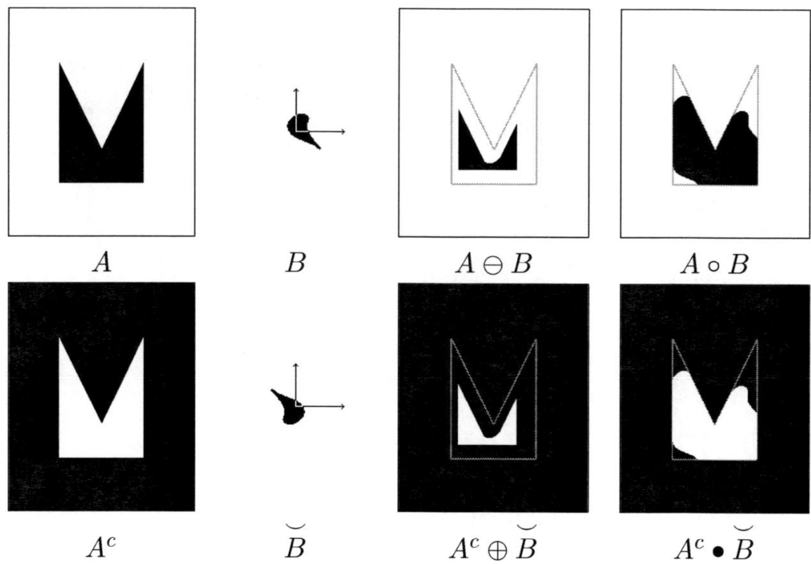

Figure 2.4 Duality between open and close with a nonsymmetrical structuring element.

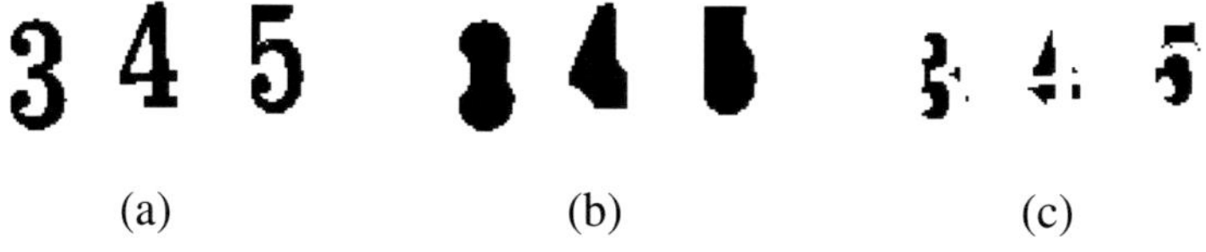

(a)   (b)   (c)

Figure 2.5 (a) Input image, (b) closing by a disk, (d) closing top-hat.

hat approximates its convex hull deficiencies. These deficiencies are commonly used in character recognition because of their powerful discriminant property.

For a digital example we close image S by structuring element E of Fig. 1.9 [Eq. (1.8)] to obtain

```
                          Example 2.2
>>>   T = mmclose(S, mmimg2se(E))    # closing of S by E
>>>   print T
      [[ 0 0 0 0 0 0]
       [ 0 1 1 1 0 0]
       [ 0 1 1 1 1 0]
       [ 0 1 1 1 0 0]
       [ 0 0 0 0 0 0]]
```

To see the pixel effect of opening and closing, consider the image and structuring element of Fig. 1.14, the opening and closing being shown in Fig. 2.6. As

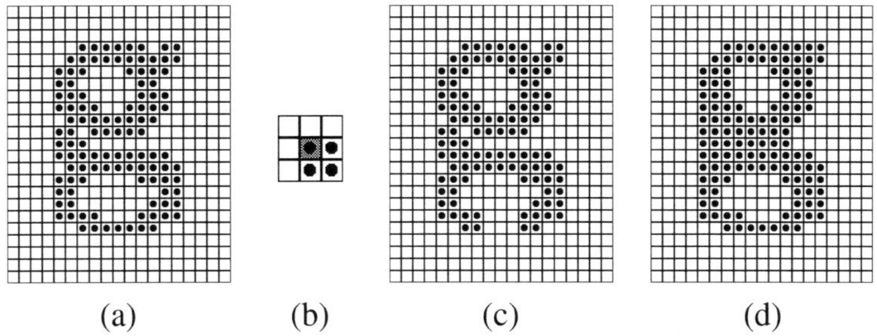

(a)   (b)   (c)   (d)

Figure 2.6 (a) Input image, (b) structuring element, (c) opening, (d) closing.

a filter, opening has cleaned the boundary by eliminating small extrusions; however, it has done this in a much finer manner than erosion, the net effect being that the opened image is a much better replica of the original than the eroded image. Analogous remarks apply to the closing, the difference being the filling of small intrusions. Note that whereas the position of the origin relative to the structuring element has a role in both erosion and dilation, it plays no role in opening and closing.

2.3 Filter Properties

Like erosion and dilation, opening and closing are both translation invariant:

$$A_x \circ B = (A \circ B)_x, \tag{2.8}$$

$$A_x \bullet B = (A \bullet B)_x. \tag{2.9}$$

They are also increasing: if A_1 is a subset of A_2, then $A_1 \circ B$ is a subset of $A_2 \circ B$. They also satisfy some very special properties.

An operator Ψ is said to be **antiextensive** if $\Psi(A)$ is always a subset of A, and it is said to be **extensive** if $\Psi(A)$ always contains A. Opening is antiextensive: $A \circ B$ must be a subset of A. This follows from the fact that the opening is a union of translations lying within the input image. Closing is extensive: $A \bullet B$ must contain A. Therefore, the order relation

$$A \circ B \subset A \subset A \bullet B \tag{2.10}$$

is always satisfied. This property is the basis of the top-hat operators defined by Eqs. (2.3) and (2.7).

An operator Ψ is said to be **idempotent** if, for any set A, $\Psi(\Psi(A)) = \Psi(A)$. In words, operating twice by Ψ is equivalent to operating once by Ψ. Both opening and closing are idempotent:

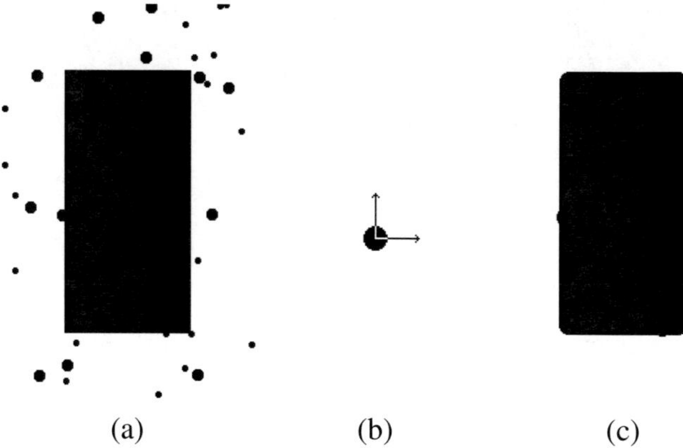

Figure 2.7 Rectangle degraded by pepper noise: (a) input image, (b) disk structuring element, (c) opening.

$$(A \circ B) \circ B = A \circ B, \tag{2.11}$$

$$(A \bullet B) \bullet B = A \bullet B. \tag{2.12}$$

The importance of idempotence is that once an image has been opened (closed), successive openings (closings) produce no further effects. This is quite different than erosion or, if we think of linear processing, moving-average filters.

In sum, whereas erosion and dilation satisfy two basic operator properties, translation invariance and increasing monotonicity, opening (closing) satisfies another two, namely antiextensivity (extensivity) and idempotence, and these play central roles in applying both opening and closing.

2.4   Application of Opening and Closing Filters

We examine the kind of restoration that can be effected by opening and closing. Consider the rectangle degraded by pepper noise in Fig. 2.7. Opening by the disk B has a restorative effect because the disk does not fit into the small components strewn about the background. Except for the rounded corners and the slight bubbles along the left and bottom edges, the rectangle has been restored. Perfect restoration could have been achieved by opening with a square; however, this would most likely be impractical, since it would require that we horizontally orient the noisy rectangle prior to filtering. Using a disk makes the filtering insensitive to rotation.

Careful consideration of the fitting formulation of opening [Eq. (2.2)] shows the manner in which the opening acts as a filter: treating the structuring element as a shape primitive, it passes only those portions of the image that are part of some translation of the shape primitive that fits inside the image. Put rigorously, a point

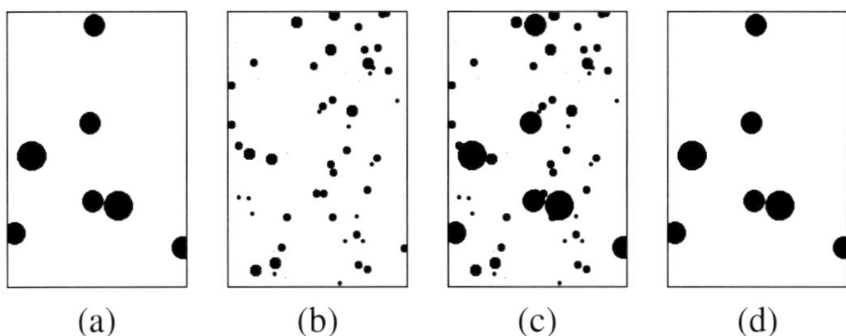

Figure 2.8 Restoration of corrupted grain-type image: (a) uncorrupted image S, (b) noise image N, (c) corrupted image $S \cup N$, (d) restoration by opening.

lies in $A \circ B$ if and only if there exists some translation of B containing the point and itself being contained in A. If an image is made up entirely of translations of the shape primitive (that is, if it is a union of such translations), then it is fully passed by the opening.

We can, in a straightforward manner, characterize the filtering of Fig. 2.7 with an image-noise model. There is an underlying uncorrupted image S, a noise image N, and a corrupted image $S \cup N$ formed by the union of S with N. Owing to translation invariance, the filter is spatially invariant, and because of increasing monotonicity and antiextensivity,

$$S \circ B \subset (S \cup N) \circ B \subset S \cup N, \tag{2.13}$$

so that the filtered image lies between the opened uncorrupted image and the noisy image. In Fig. 2.7 the filter has performed well because the structuring element has passed most of the image, while passing very little noise.

The filter effect becomes more transparent if we consider the uncorrupted circular grain-type image S of Fig. 2.8(a), the noise image N of Fig. 2.8(b), and the corrupted image $S \cup N$ of Fig. 2.8(c). Here, the largest noise grain is smaller than the smallest uncorrupted-image grain. Opening with a disk B whose radius is between that of the largest noise grain and the smallest uncorrupted-image grain will yield close to perfect restoration so long as there is little grain overlap [See Fig. 2.8(d)]. In the absence of overlap, restoration is perfect.

The situation is complicated if the amount of overlap is large or if some noise-grain radii exceed some uncorrupted-image-grain radii. Perfect restoration is not possible and the choice of the radius can be found from treating the image and noise as random processes and proceeding with a statistical optimization analysis.

Opening can be used to filter pepper noise; closing can be used to filter salt noise. Figure 2.9 shows a text image, the image degraded by pepper noise, and the result of filtering the pepper degraded image by a 2×2 structuring element.

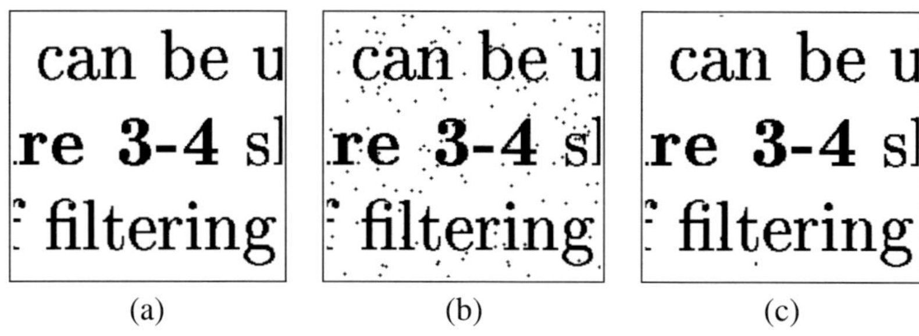

(a)   (b)   (c)

Figure 2.9 Filtering pepper noise: (a) uncorrupted image, (b) image corrupted with pepper noise, (c) restoration by opening.

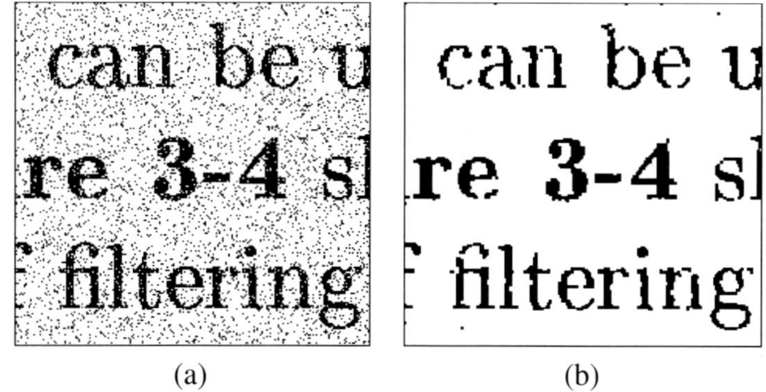

(a)   (b)

Figure 2.10 Filtering salt-and-pepper noise: (a) corrupted image, (b) restoration by open-close.

2.5 Alternating Sequential Filters

When there is both union and subtractive noise, one strategy is to open to eliminate union noise in the background and then close to fill subtractive noise in the foreground. The resulting filter is called an **open-close**. Open-close is illustrated in Fig. 2.10, which shows a salt-and-pepper degraded realization of the original text image of Fig. 2.9(a) and the result of open-close with a 2×2 structuring element applied to the noisy image. One can also close and then open, the filter then being called a **close-open**. These two filters are not dual, and they tend to give similar but different results.

A potential pitfall of the open-close strategy occurs when large noise components need to be eliminated but a direct attempt to do so will destroy too much of the original image. This is illustrated in Figs. 2.11. Both filters fail to restore the uncorrupted image satisfactorily. If the radius of the structuring element is small compared to the noise grains, the salt-and-pepper noise is not removed; and if the radius is large, the filter will destroy too much of the original image.

One way around this problem is to employ an **alternating sequential filter (ASF)**. Open-close (or close-open) filters are performed iteratively, beginning with

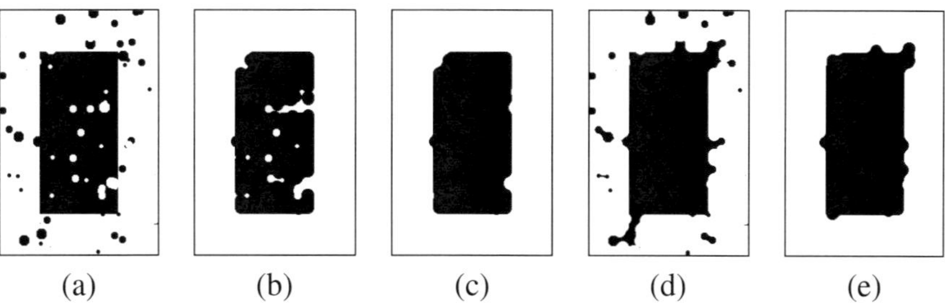

(a)   (b)   (c)   (d)   (e)

Figure 2.11 Open-close and close-open filters: (a) S, image corrupted with salt-and-pepper noise, (b) $S_o = S \circ B$, (c) $S_{co} = (S \circ B) \bullet B$, (d) $S_c = S \bullet B$, (e) $S_{oc} = (S \bullet B) \circ B$.

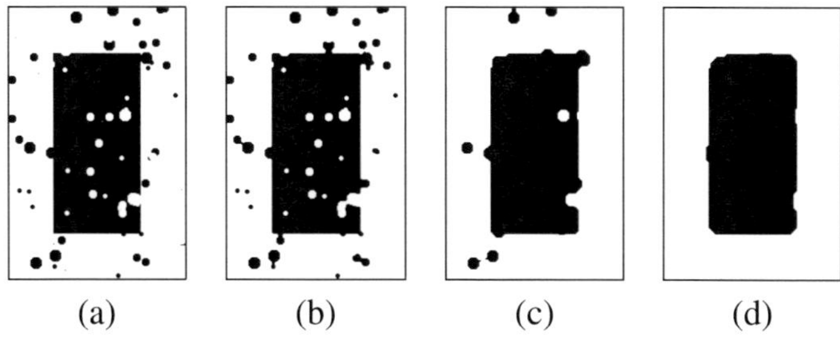

(a)   (b)   (c)   (d)

Figure 2.12 Alternating sequential open-close filters: (a) image corrupted with salt-and-pepper noise, (b) single-stage ASF open-close, (c) three-stage ASF open-close, (d) seven-stage ASF open-close.

a very small structuring element and then proceeding with ever-increasing structuring elements. The close-open filter is given by

$$ASF_{oc,B}^{n}(S) = (((((S \bullet B) \circ B) \bullet 2B) \circ 2B) \ldots \bullet nB) \circ nB, \qquad (2.14)$$

and the open-close filter by

$$ASF_{co,B}^{n}(S) = (((((S \circ B) \bullet B) \circ 2B) \bullet 2B) \ldots \circ nB) \bullet nB. \qquad (2.15)$$

The strategy is to eliminate small salt-and-pepper components, thereby allowing the larger structuring elements to more likely fit when they are eventually applied in the process. Figure 2.12 shows the image corrupted by both union and subtractive noise; and the results of single-stage, three-stage, and seven-stage alternating-sequential filters.

2.6 Invariance

As noted previously, if an image A is formed from a union of translations of some shape primitive B, then the opening of A by B yields A; that is, $A \circ B = A$. In

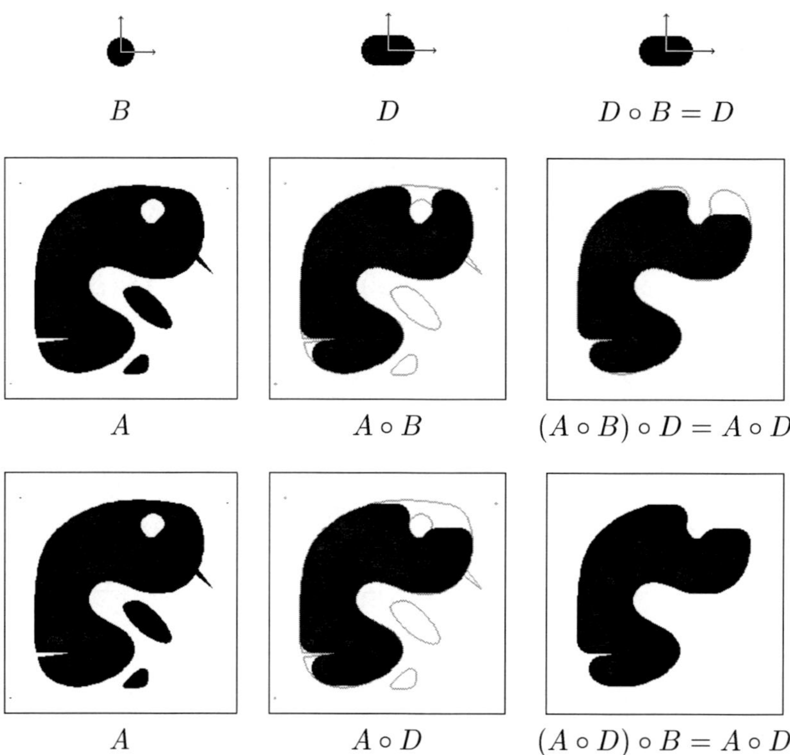

Figure 2.13 Illustration of opening property of Eq. (2.16).

such a case we say that A is **open with respect to** B, or that A is B-**open**. From the perspective of filtering, A is **invariant** when filtered by the opening with structuring element B. In fact, not only is an image formed as a union of translations of B invariant when opened by B, but unions of translations of B are the only images invariant when opened by B. Examples abound—perhaps the most important one being that a disk is always open relative to a disk possessing a smaller radius.

A union of translations of an image is, in fact, a dilation of the image. Thus, the preceding formulation of B-openness can be reformulated: A is B-open if and only if there exists some image D such that $A = B \oplus D$. This formulation proves to be useful when we wish to construct granulometries (see Chapter 8).

Iterated openings by two structuring elements, one that is open with respect to the other, produce the same result as if only a single opening were applied, the order of iteration being irrelevant: if D is B-open, then

$$(A \circ D) \circ B = (A \circ B) \circ D = A \circ D. \tag{2.16}$$

This property is illustrated in Fig. 2.13.

A key property relative to the analysis of particles and texture, and one that will play a central role in the chapter on granulometries, regards the effect of opening by different structuring elements when one is open with respect to the other: if D is B-open, then $A \circ D$ is a subset of $A \circ B$. This subset relation can be seen in Fig. 2.13, where D is B-open.

If we consider the image-noise model, $S \cup N$, of the preceding section, a very special relationship occurs if the uncorrupted image is open relative to the opening structuring element B used for the filter. In such a case, $S \circ B = S$, so that Eq. (2.13) reduces to

$$S \subset (S \cup N) \circ B \subset S \cup N, \tag{2.17}$$

so that the filtered image lies between the uncorrupted image and the noisy image.

2.7 τ-Openings

An opening passes only those portions of an image that conform to the shape of the structuring element. Suppose one wishes to pass portions of an image conforming to any one of a number of primitive shapes, not simply a single primitive. This effect can be accomplished by using a filter comprised of a number of openings, one for each desirable shape primitive. The final filter output is the union of the individual openings. A filter Ψ is called a τ-**opening** if there exists some class $\mathbf{B} = \{B_1, B_2, \ldots, B_n\}$ of structuring elements such that

$$\Psi(A) = \bigcup_{k=1}^{n} A \circ B_k. \tag{2.18}$$

$\mathbf{B}$ is called a **base** for Ψ. A base is not unique: different bases can produce the same filter; however, our desire is to use a base with a small number of primitives. Design of a τ-opening requires finding an appropriate base.

The **invariant class** of a filter Ψ consists of all images that are invariant under Ψ, namely, those images A for which $\Psi(A) = A$. If possible, we would like a filter to pass unchanged those images that are considered to be uncorrupted, while at the same time removing noise from corrupted images. τ-openings are particularly straightforward when it comes to invariance. The invariant class of a τ-opening consists precisely of those images that are formed as unions of translations of the base primitives. This is in complete accord with the situation for a single opening, for we have already noted that the invariants of a single opening are those images that are unions of translations of the opening structuring element.

For design purposes in the presence of union noise, if we can express an image as a union of desirable primitives, then it would behoove us to construct a τ-opening that passes only those primitives. Of course, should the noise also be in part made up of some of the same primitives, then some of it will also be passed. Just as in the case of linear filtering, τ-opening filtering requires a trade-off: it is usually necessary to filter out some of the image and to pass some of the noise in order

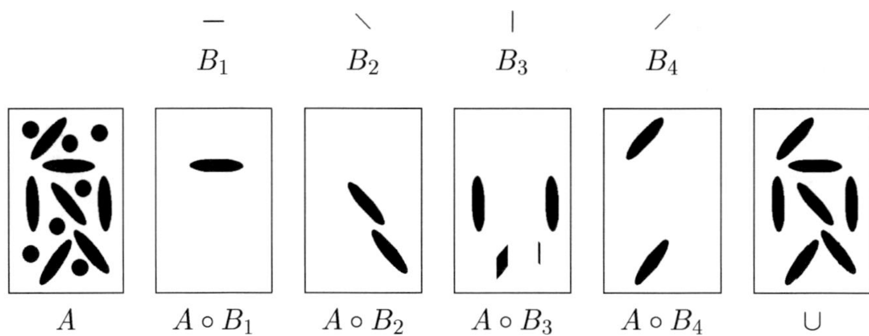

Figure 2.14 Radial opening.

to obtain an optimally filtered image. Such an optimization analysis requires the image and noise to be treated as random processes, and we will not examine the issue here; instead, we will confine ourselves to consideration of image geometry.

A typical τ-opening filter is the radial opening. The **radial opening** is a τ-opening filter with a set of linear structuring elements of varying angles. Figure 2.14 shows an example of the application of a radial opening with four different directions, 0, 45, 90, and 135 deg: $B_1 = L_0, B_2 = L_{45}, B_3 = L_{90}, B_4 = L_{135}$. The figure is composed of elliptical signal and circular noise grains. Since the radii of the noise grains are smaller than the major axes of the signal grains, filtering can be achieved by applying the radial opening to mainly pass the signal grains. Parts (b) through (e) of the figure show the result of applying the opening for each structuring elements and part (f) shows the radial opening result. Each opening passes one part of the signal. Note also that the individual openings diminish the signal grains they pass on account of the structuring elements and grains possessing different geometries. This problem will be corrected when we consider reconstructive operators.

We have seen that openings satisfy four fundamental filter properties: translation invariance, antiextensivity, increasing monotonicity, and idempotence. Because τ-openings are formed as unions of openings, it might be conjectured that they, too, satisfy the four properties. In fact, they do. Moreover, the converse holds: if a filter satisfies the four properties, then it must be a τ-opening. From an algebraic perspective, the four properties characterize a class of filters, i.e., those filters satisfying the four properties. Based on what we have just said, this class is comprised solely of unions of openings. A developmental note: Matheron actually defined a τ-opening as any filter satisfying the four properties and then proceeded to show that all such filters are unions of openings over some base.

Dual comments apply to closings. An intersection of closings is called a τ-**closing** and all such filters are translation invariant, increasing, extensive, and idempotent. Conversely, any filter satisfying these four properties must be an intersection of closings. Whereas τ-openings are good for filtering union (pepper) noise corrupting the background, τ-closings perform in a dual manner and are good for filtering subtractive (salt) noise corrupting the foreground. Design strategies are analogous.

2.8 Demonstration

Here we present a complete demonstration of a real-world example using the **MT**. The input image is a binary image of a printed circuit board (PCB). The goal is to identify the various components of the PCB: holes, different types of islands and different types of tracks. The shape decomposition is created mainly using openings by structuring elements that depend on the geometry of the components.

The **MT** functions used in this demonstration are shown below.

function name	description
mmopen:	open [Eq. (2.2), but using bounded operators)
mmdil:	dilation [Eq. (1.35)]
mmsubm:	subtraction
mmsebox:	3×3 box structuring element
mmsedisk:	disk structuring element
mmsedil:	dilation of structuring elements [Eq. (1.11)]
mmseline:	line structuring element
mmclohole:	close of holes [Eq. (3.11)]
mmreadgray:	read image file

The program is shown in Fig. 2.15 and all the images are shown in Figs. 2.16, 2.17, and 2.18. The steps used to detect all the PCB components are below. The program line numbers and figure parts are shown in brackets.

Image reading (line 1) The binary image of a printed circuit board is read (a).

Detecting holes (lines 2–3) A new image is created by filling the holes (b), which will be described in the next chapter. The input image is subtracted from this new image without holes. The resulting residues are the holes (c).

Detecting square islands (lines 4–5) The square islands are detected using an opening by a square of size 17×17 (d). The result is dilated and intersected with the input image to show the square PCB pads (e).

Detecting circle islands (lines 6–8) A residues image (f) is created by subtracting the already detected square pads from the input image. The circle islands are detected using an opening by a Euclidean disk on the residues image (g). The result is dilated and intersected with the input image (h).

```
                         Program 2.1
01   a = mmreadgray( 'pcb1bin.tif')
02   b = mmclohole( a)
03   c = mmsubm( b, a) # holes
04   d = mmopen( b, mmsebox( 8))
05   e = mmintersec( mmdil( d, a)) # square pads
06   f = mmsubm( b, d)
07   g = mmopen( f, mmsedisk( 8))
08   h = mmintersec( mmdil( g, a)) # circle pads
09   i = mmsubm( f, g)
10   j = mmopen( i, mmsedil(mmseline(8,90),mmseline(25)))
11   k = mmintersec( mmdil( j, a))
12   l = mmsubm( i, j)
13   m = mmopen(l, mmsebox( 2))
14   n = mmintersec( mmdil( m, a)) # thick tracks
15   o = mmsubm( l, m)
16   p = mmopen( o, mmsebox( 1))
17   q = mmintersec( mmdil( p, a)) # thin tracks
18   r = mmunion(mmgray(c,'uint8',1),mmgray(e,'uint8',2),
                 mmgray(h,'uint8',3),mmgray(k,'uint8',4),
                 mmgray(n,'uint8',5))
19   r = mmunion( r, mmgray( q, 'uint8', 6))
```

Figure 2.15 Python code for the PCB component identification. The images from (a) to (r) are shown in Figs. 2.16, 2.17, and 2.18.

Detecting rectangular islands (lines 9–11) The rectangular islands are detected using an opening by a rectangle of size 25×8 on a residues image (k). The rectangle structuring element is built from the composition of vertical and horizontal lines.

Detecting thick connections (lines 12–14) The thick connections are detected using an opening by a 5×5 square on a residues image (n).

Detecting thin connections (lines 15–17) The thin connections are detected using an opening by a 3×3 square on a residues image (q).

Combining all together (lines 18–19) The main components of the circuit are combined in a single image. Each image is converted to gray scale using a unique label and unioned afterwards.

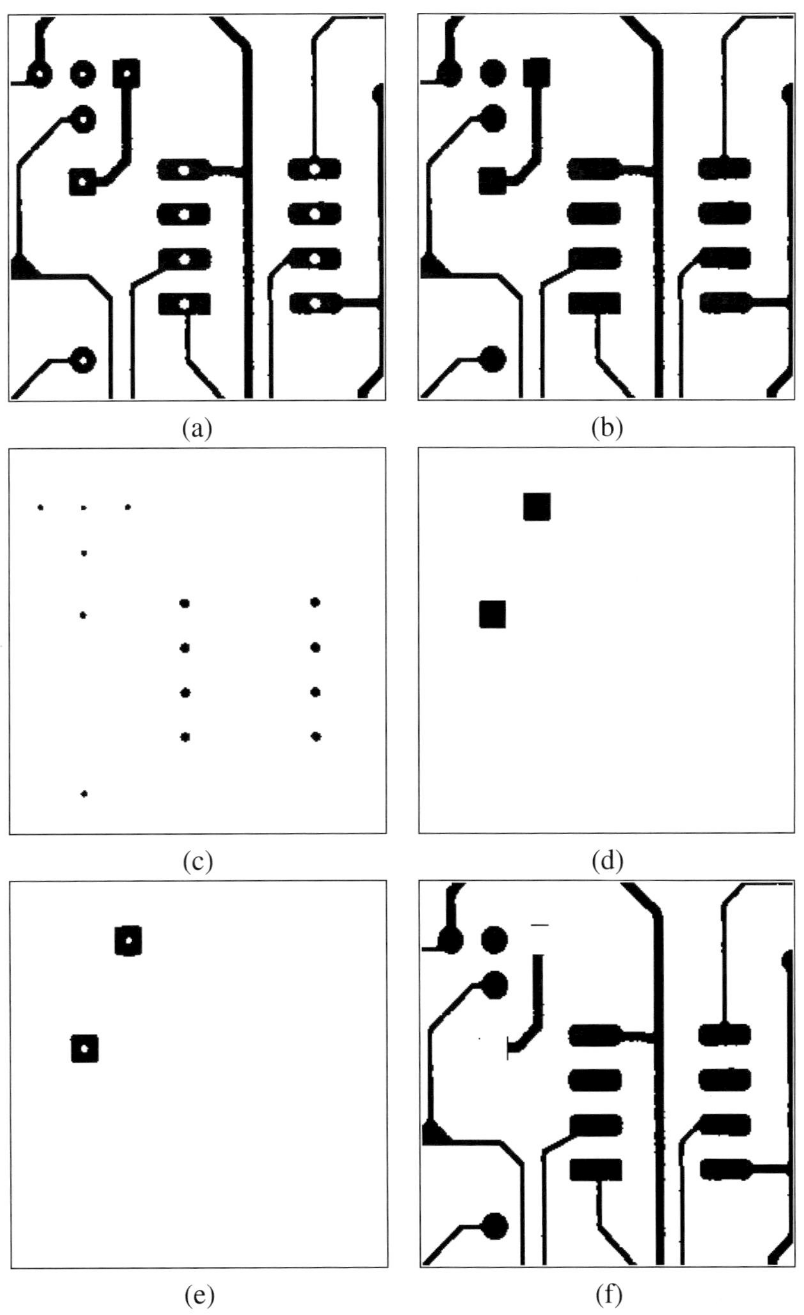

Figure 2.16 PCB component detection. These images refer to the program of Fig. 2.15.

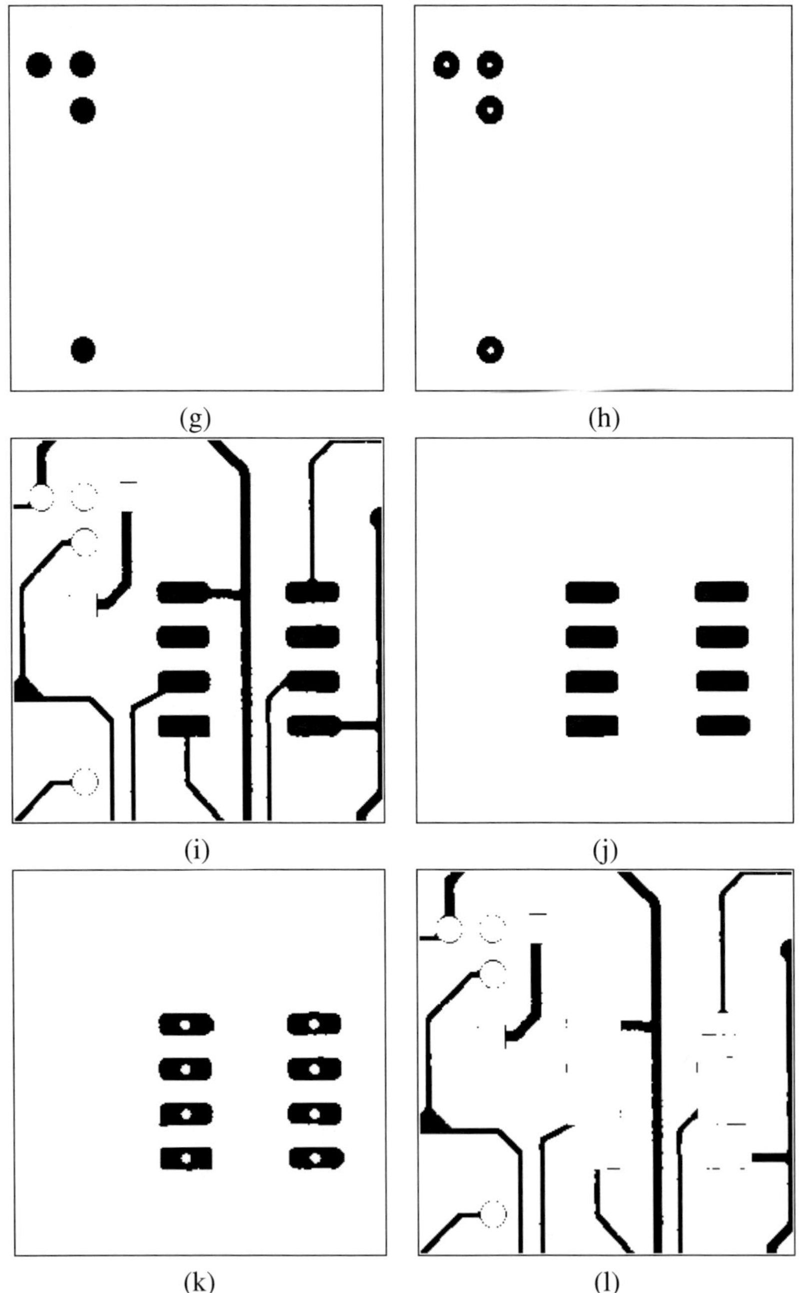

Figure 2.17 PCB component detection. These images refer to the program of Fig. 2.15.

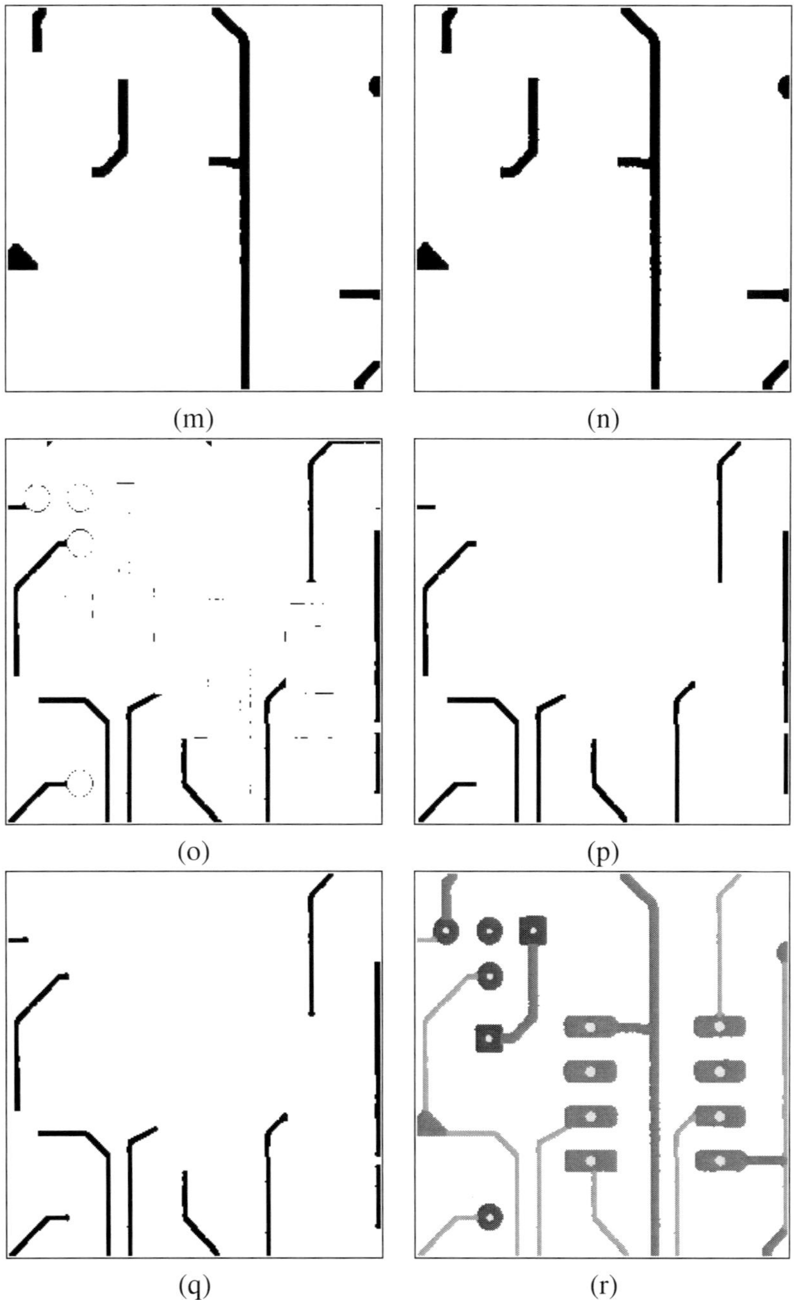

Figure 2.18 PCB component detection. These images refer to the program of Fig. 2.15.

2.9 Exercises

1. Let
$$\begin{aligned}
A &= \{(x,y) : 0 \le x \le 2, 0 \le y \le 3\} \\
B &= \{(x,y) : x^2 + y^2 \le 1\} \\
C &= \{(x,y) : y = x \text{ and } 0 \le x \le 1\} \\
D &= \{(x,y) : 0 \le y \le |2x| \text{ and } -1 \le x \le 1\}
\end{aligned}$$

Give graphical representation of the following:

   (a) $A \circ B$

   (b) $A \circ C$

   (c) $2B \circ C$

   (d) $2A \circ B$

   (e) $A \circ \frac{1}{4}B$

   (f) $D \circ B$

   (g) $D \bullet B$

   (h) $D \bullet 2B$

2. Let
$$\begin{aligned}
S &= \{(-1,-1), (0,0), (0,1), (1,0), (1,1)\}, \\
E &= \{(0,0), (1,1)\}.
\end{aligned}$$
(2.19)

Find each of the following sets (in set notation and pictorially):

   (a) $S \circ E$

   (b) $S \hat{\circ} E$

   (c) $2S \circ E$

   (d) $S \bullet E$

3. For the images S, T, and E of Exercise 3, find

   (a) $S \circ E$

   (b) $T \circ E$

   (c) $S \bullet E$

   (d) $T \bullet E$

4. Is the opening commutative?

5. Although the erosion typically reduces the size of the shape, it is not an antiextensive operator. Why?

6. Prove that a point is in the opening if and only if there exists a translation of the structuring element that contains the point while at the same time being a subset of the image.

7. Prove opening is idempotent.

8. Knowing that opening is idempotent, usc duality to show that closing is idempotent.

9. Knowing that opening is antiextensive, use duality to show that closing is extensive.

10. Construct Euclidean and digital examples where neither closing nor opening give perfect reconstruction but open-close does give perfect reconstruction.

11. Construct an example in which neither open-close nor close-open give perfect reconstruction but a two-stage ASF will give perfect reconstruction.

12. Prove that if an image is constructed as a union of translations of B, then opening by B is invariant.

13. An operator Ψ_1 is **complementary** to Ψ_2 if $\Psi_1(A) = \Psi_2(A^c)$. Is the opening top-hat complementary to closing top-hat? Give the mathematical equation of this relationship.

14. Is the opening top-hat an increasing operator? Give a simple example.

15. What is the effect of translating the structuring element in the opening? Discuss the case of using bounded and ordinary operators.

16. Show that opening by the structuring element (0 **0** 0) can be written as the union of three erosions.

17. Define the digital mapping Ψ in the following way: pixel p lies in $\Psi(S)$ if and only if there exists a pixel in S that is adjacent to p, either vertically, horizontally, or diagonally. Find $\Psi(S)$ for the image

$$
S = \begin{pmatrix}
0 & 0 & 0 & 0 & 1 & 1 \\
0 & 0 & 0 & 0 & 1 & 1 \\
0 & 0 & 1 & 1 & 0 & 0 \\
1 & 1 & 1 & 0 & 0 & 0 \\
0 & 1 & 0 & 0 & 0 & 0 \\
1 & 0 & 0 & 0 & 0 & 1
\end{pmatrix}.
$$

Find a base for Ψ. Describe the invariant class for Ψ.

```
name          description
mmopen:       opening [Eq. (2.1)]
mmopenth:     opening top-hat [Eq. (2.3)]
mmclose:      closing [Eq. (2.4)]
mmcloseth:    closing [Eq. (2.7)]
mmasf:        alternating sequential filters
              [Eqs. (2.14) and (2.15)]
mmropen:      radial opening (Fig. 2.14)
```

Figure 2.19 MT operators presented in this chapter.

2.10   Laboratory Experiments

The **MT** functions introduced in this chapter are presented in Fig. 2.19.

1. Write a function $\texttt{isopen(A,B)}$ to verify if A is B-open.

2. Illustrate the duality between the opening and closing using bounded operators.

3. Discuss the advantages and disadvantages of implementing the opening from the erosion and dilation [Eq. (2.1)], or directly, following Eq. (2.2).

4. Modify the holes detecting part of the demonstration of PCB component detection to use closings and openings.

References

1. C. Ronse and H. J. A. M. Heijmans. The algebraic basis of mathematical morphology — part II: Openings and closings. *Computer Vision, Graphics and Image Processing: Image Understanding*, 54:74–97, 1991.

Morphological Processing of Binary Images

The present chapter covers some morphological algorithms for binary images. These include boundary detection, conditional dilation, curve filling, thinning, segmentation, and restoration. The next chapter explores algorithms employing the hit-or-miss transform. The intent is not to produce a complete compilation of morphological algorithms, but rather to provide a geometrically grounded introduction to some of the most useful techniques. In all cases, it should be recognized that actual use of a particular methodology will often require preprocessing to put the image into a form suitable for application of the algorithm and postprocessing to provide an acceptable output image.

3.1 Pixel Regions

Appreciation of operators concerned with pixel regions requires familiarity with basic concepts having to do with pixel topology. We briefly review the ones necessary for the present text.

Two pixels are said to be **4-neighbors** if they are vertically or horizontally adjacent. They are said to be **8-neighbors** if they are 4-neighbors or diagonally adjacent. In this vein, Figs. 3.1 (a) and (b) depict the 4-neighbors and 8-neighbors masks, respectively, for the origin.

When concerned with geometric understanding, we often refer to a collection of pixels as a region. A region is said to be **4-connected** if for any two pixels p and q in the region there exists a sequence of pixels also in the region such that the first pixel is p, the last is q, and each pixel in the sequence is a 4-neighbor of the next. A region is said to be **8-connected** if the same definition applies with "8-neighbor" in place of "4-neighbor." Figures 3.2 (a), (b), and (c) depict 4-connected, 8-connected, and disconnected (not connected) regions, respectively. Note that a 4-connected region is *ipso facto* 8-connected.

Every binary image can be expressed as the union of connected regions. If each of these regions is maximally connected, which means that it is not a proper subset

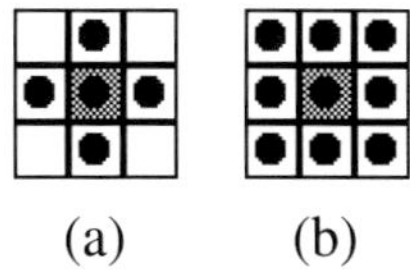

(a) (b)

Figure 3.1 (a) 4-neighbors mask, (b) 8-neighbors mask.

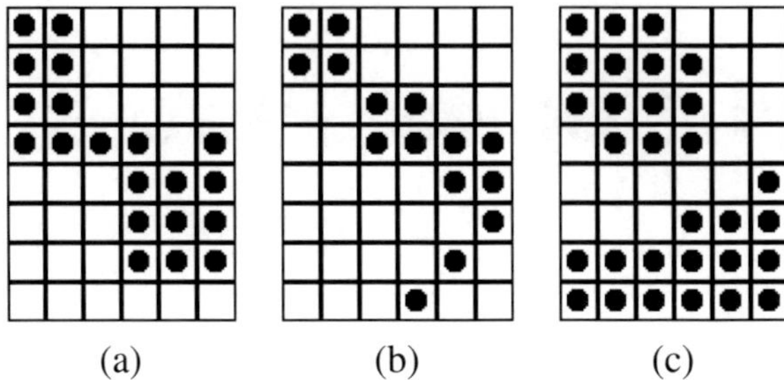

(a)   (b)   (c)

Figure 3.2 Region connectedness: (a) 4-connected, (b) 8-connected, (c) disconnected region.

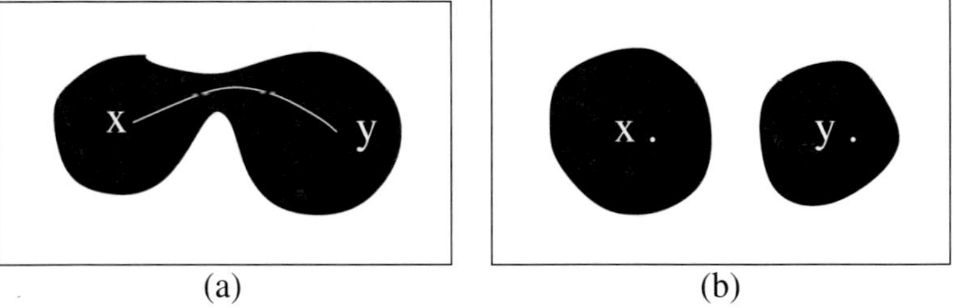

(a)   (b)

Figure 3.3 (a) Connected, and (b) disconnected sets.

of a larger connected region within the image, then the regions are called **connected components** of the image. For instance, the image of Fig. 3.2 (c) consists of two connected components; Fig. 3.2 (b) has one 8-connected component and four 4-connected components. A connected image has one component.

The notion of digital connectivity provides a digital analogue for the usual Euclidean connectivity. There, a set is path connected if for any two points in the set there exists a continuous path between the points lying fully in the set. The definition depends on the definition of a continuous path, which is a continuous function on the closed interval $[0, 1]$. Figure 3.3 shows (a) a connected set, and (b) a disconnected set, where there is no path with endpoints x and y lying in the set.

If we look at the definitions of 4-neighbor and 8-neighbor, we see that pixels p and q are 4-neighbors if q lies in the 5-pixel diamond structuring element (E_4) centered at p, and vice versa. An analogous statement holds for 8-connectivity using the 9-pixel square structuring element (E_8). In other words, p and q are 4-neighbors if and only if p lies in the dilation $q \oplus E_4$. The alternative relation $q \in p \oplus E_4$ holds automatically since the structuring element is symmetric. Analogous statements hold for 8-connectivity. Focusing on 8-connectivity, this means that if C_1 and C_2 are two distinct connected components, then $(C_1 \oplus E_8) \cap C_2 = \emptyset$ and $(C_2 \oplus E_8) \cap C_1 = \emptyset$.

For morphological image processing, it is useful to extend the definition of connectivity in the discrete setting (and this could also be done in the Euclidean setting but that would require changing the usual Euclidean topology, and we will

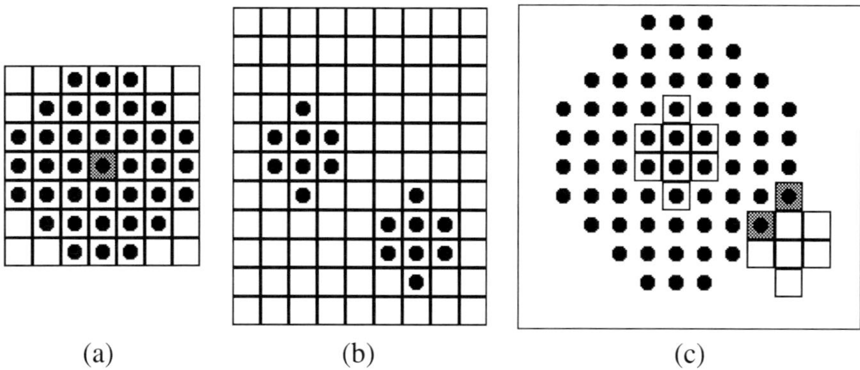

(a) (b) (c)

Figure 3.4 (a) E_{36} structuring element, (b) two 8-connected components and a single 36-connected component, (c) dilation of one component by E_{36} intersects the other component.

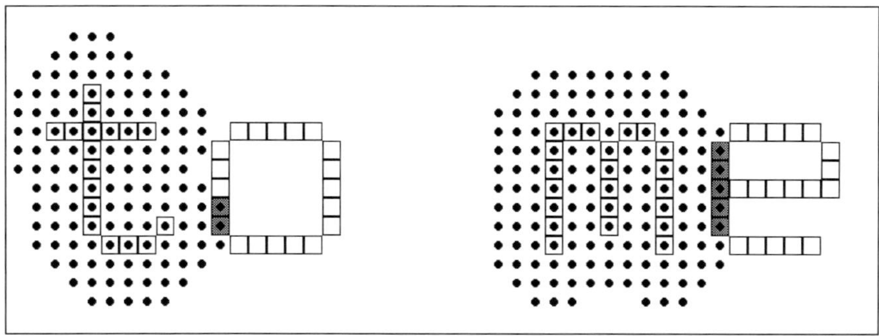

Figure 3.5 Square pixels: four 8-connected letters but two 36-connected words; circle pixels: dilation of letters "t" and "m" by E_{36}; shaded pixels: intersection of the dilation with letters "o" and "e."

not pursue that here). If E is any structuring element that contains the origin and is symmetric about the origin, pixels p and q are E-neighbors if $p \in q \oplus E$. Using this concept of neighbor, we can define connectivity in the same manner as before. Again, the connected components are the maximally connected regions. Also again, if C_1 and C_2 are two distinct connected components, then $(C_1 \oplus E) \cap C_2 = \emptyset$ and $(C_2 \oplus E) \cap C_1 = \emptyset$. Figure 3.4(b) shows an image in which there are two 8-connected components but a single 36-connected component, based on the structuring element made of the Euclidean disk of radius 3 (E_{36}), shown in part (a). Note in part (c) how dilation of one of the 8-connected components by E_{36} results in a nonnull intersection. Motivation for this kind of extended connectivity is shown in Fig. 3.5, where the distinct letters of the words "to" and "me," depicted by square pixels, constitute the 8-connected components, whereas the dilations, depicted by circle pixels, show that words themselves constitute the 36-connected components.

Unless otherwise specified, the default connectivity will be 8-connectivity,

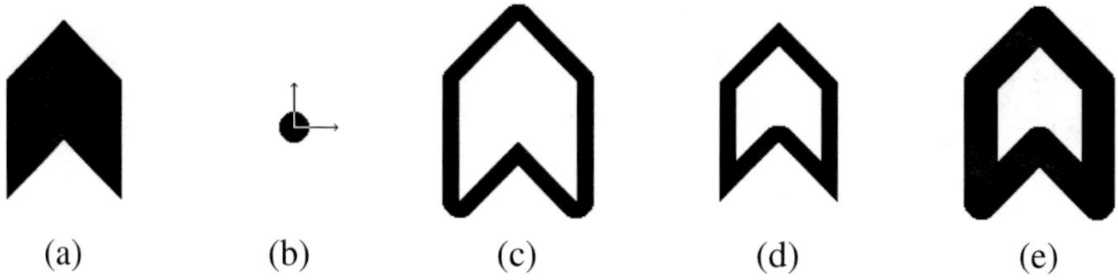

Figure 3.6 Morphological boundaries: (a) input image, (b) structuring element, (c) external, (d) internal, (e) morphological gradient.

which keeps us geometrically close to the usual Euclidean connectivity.

3.2 Boundary Detection

Dilation by a disk expands an image and erosion by a disk shrinks an image. Both can be used for finding boundaries for binary images. Three possibilities are shown in Fig. 3.6: (a) image A; (b) disk B; (c) $(A \oplus B) - A$ gives an external boundary; (d) $A - (A \ominus B)$ an internal boundary; and (e) $(A \oplus B) - (A \ominus B)$ a boundary that straddles the actual Euclidean boundary and is known as the **morphological gradient**.

For digital implementation the type of boundary depends on the selection of a digital structuring element to take the place of the disk in the Euclidean procedure. Figure 3.7 illustrates boundaries resulting from the 8-neighbors structuring element E_8 [see Fig. 3.1(b)]: (a) image S, (b) the dilation $S \oplus E_8$, (c) the erosion $S \ominus E_8$, (d) the external boundary, (e) the internal boundary, and (f) the morphological gradient. Figure 3.8 does the same for the 4-neighbors mask [see Fig. 3.1 (a)]. Notice how the square mask has yielded 4-connected boundaries, whereas the 4-neighbors mask has yielded only 8-connected boundaries. While there may sometimes be an advantage in having a 4-connected boundary, for small image components it can look "cluttered," especially the internal boundary. Thicker boundaries can be obtained by using larger structuring elements.

3.3 Reconstruction

One of the most important operations in morphological image processing is reconstruction from markers. The operation involves an input image A decomposed into the union of its connected components,

$$A = \bigcup_{k=1}^{n} C_k, \tag{3.1}$$

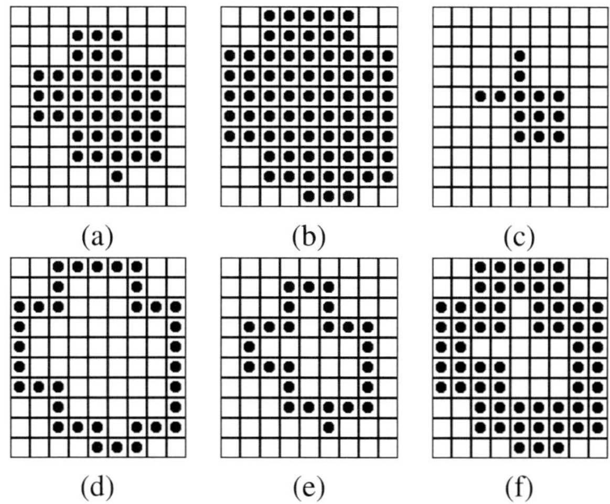

Figure 3.7 4-connected boundaries, using 8-connected structuring element: (a) input image, (b) dilation, (c) erosion, (d) 4-connected external boundary, (e) 4-connected internal boundary, (f) morphological gradient.

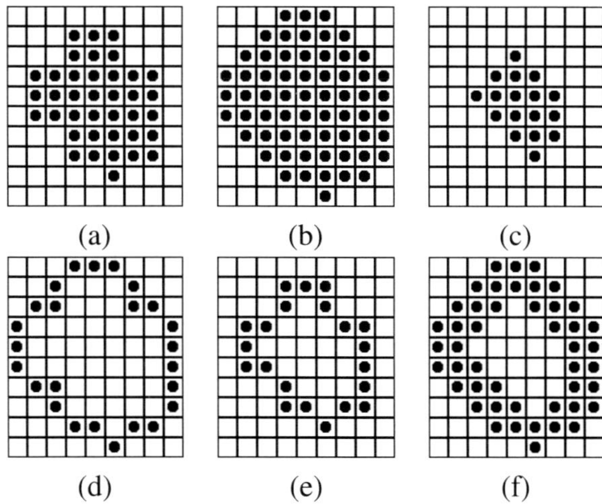

Figure 3.8 8-connected boundaries, using 4-connected structuring element: (a) input image, (b) dilation, (c) erosion, (d) 8-connected external boundary, (e) 8-connected internal boundary, (f) morphological gradient.

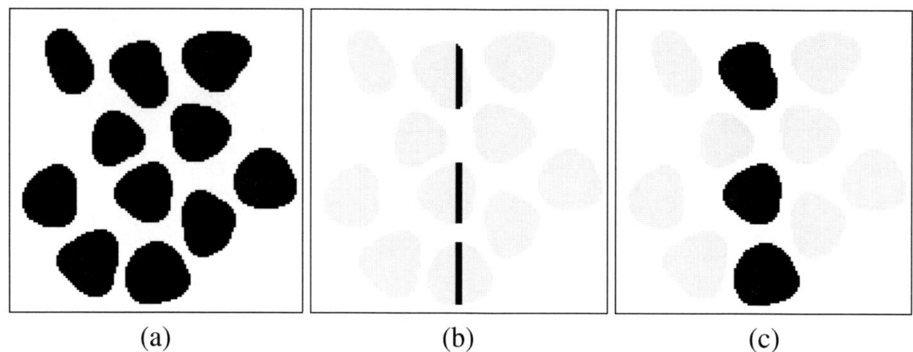

(a)   (b)   (c)

Figure 3.9 Reconstruction from markers: (a) input image A, (b) marker image M, (c) reconstructed image $A \vartriangle_{E_4} M$.

and a marker image M that is a subset of A. The output image, called the **morphological reconstruction** of A from the **marker** M, denoted by $A \vartriangle M$, is the union of all components of A that intersect M:

$$A \vartriangle M = \bigcup \{C_k : C_k \cap M \neq \emptyset\}. \tag{3.2}$$

In the discrete case, one must keep in mind the kind of connectivity being used. If T is being reconstructed from S, then we write $T \vartriangle_B S$, where the structuring element B defines the connectivity. The default connectivity is 8-connectivity.

Notice that morphological reconstruction is defined abstractly, not operationally. In practice, some algorithm must perform the reconstruction from markers. We will discuss one of these in the next section; in this, we remain with the abstract formulation.

An example of reconstruction from markers, based on 8-connectivity, is shown in Fig. 3.9: (a) the input image, which is a collection of grains; (b) the marker image made of a central vertical line intersecting the grains; and (c) the reconstruction from the markers, which extracts the three central components from the original image. An example using 36-connectivity is shown in Fig. 3.10. In part (a) the letter "t" of the first line and the letter "m" of the second line are marked. In part (b), reconstruction yields the word "to," in the first line and the word "me" in the second line.

To find all the connected components of an image, one can iteratively find any pixel of the image, use it to reconstruct its connected component, remove the component from the image, and iteratively repeat the same extraction until no more pixels are found in the image. This operation is called **labeling**. The labeling decomposes an image into its n connected components:

$$\Lambda(A) = \{C_1, C_2, \ldots C_n\}. \tag{3.3}$$

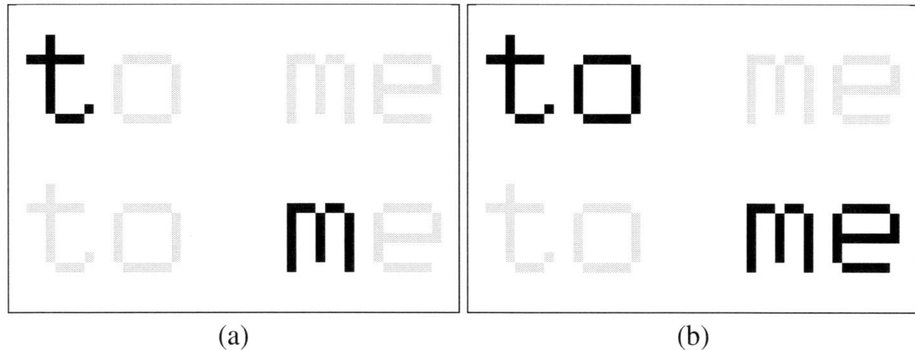

(a)   (b)

Figure 3.10 Reconstruction from markers using 36-connectivity: (a) letter "t" of the first line and letter "m" of the second line are marked, (b) reconstruction using 36-connectivity.

The union of all connected components recovers the input image and the intersection of any two connected components is empty. The result of the labeling is usually stored in a numeric image with each pixel value associated to its connected component number.

Figure 3.11 illustrates the labeling algorithm based on reconstruction: (a) the first pixel as marker, (b) extraction of the first component, (c) the first pixel in the image without the first component, (d) second component extraction, and (e) labeled image in which each pixel has the value associated to the connected component identification. The labeling can also compute the number of connected components in the image. Particularly, it is common to implement the labeling using sequential values so that the largest label in the output image gives the number of components.

Reconstruction has some properties of the opening: increasing monotonicity, idempotence, and antiextensivity, but it is not translation invariant. One must be careful when interpreting these properties relative to reconstruction: the implicit assumption is that the markers are given *a priori*, so that the properties apply to reconstruction from a given set of markers. To emphasize this point, suppose the markers consist of all connected components of the input image whose areas are less than some specified threshold. If we view marking and reconstruction as a single operation, then for this marking procedure we lose increasing monotonicity. Idempotence can also be lost by combining marking and reconstruction into a single operation—for instance, when a component is marked if and only if there exists another component within some specified distance of it whose area is below some specified threshold. One should not pay too much attention to these anomalies. We will soon discuss perhaps the most important class of reconstructive filters, and for these the operation consisting of both marking and reconstruction is monotonically increasing, idempotent, antiextensive, and translation invariant. More generally, one should consider the properties of individual classes of reconstructive filters.

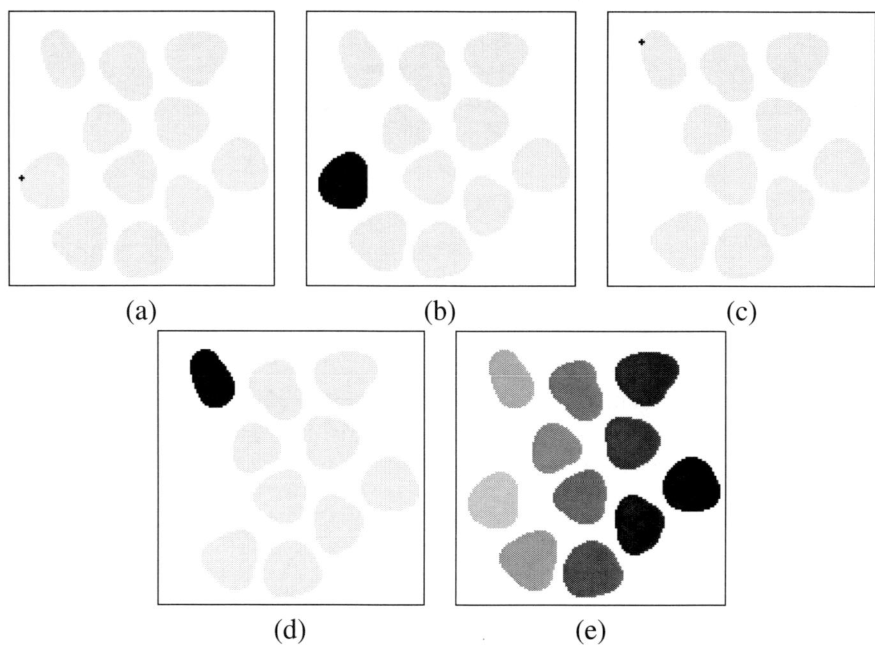

Figure 3.11 Labeling: (a) first pixel, (b) reconstruction from the first pixel, (c) second pixel without the first component, (d) extraction of the second component, (e) unique label for each connected component.

3.4 Conditional Dilation

If an image is dilated by a structuring element containing the origin, it is expanded, and the manner of the expansion depends only on the shape of the structuring element. If the dilation is successively repeated, the original image grows without bound. Sometimes it is important to restrict the growth. This can be accomplished by **conditioning** the dilation. Referring to the Minkowski addition form of dilation [Eq. (1.12)], conditioning is accomplished by restricting in some manner the translations comprising the union.

A common form of conditioning restricts the union-forming translations to a superset of the input image: if image A is a subimage of C, and B is a structuring element, then the **conditional dilation** of A by B relative to C is defined by restricting the translations to C, the result being

$$A \oplus_C B = \bigcup_{a \in A} B_a \cap C, \qquad (3.4)$$

where the notation $\oplus_C$ instead of $\oplus$ indicates there is conditioning. Keep in mind that to appreciate the meaning of $\oplus_C$ in any particular context, one must recognize the type of conditioning being employed. Figure 3.12 illustrates conditional dilation by a disk: (a) input image, (b) disk, (c) conditioning image, (d) overall con-

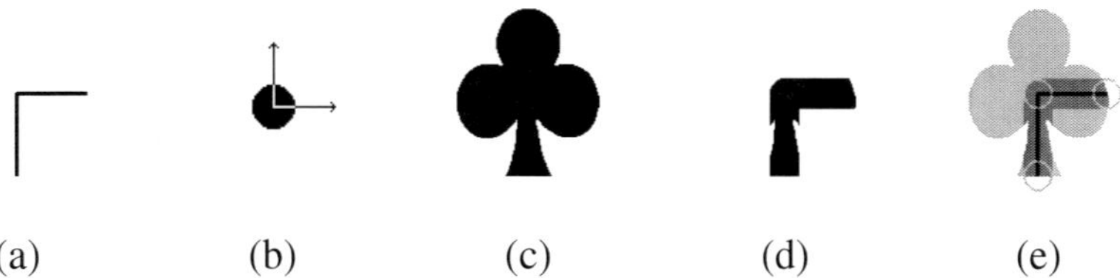

(a)   (b)   (c)   (d)   (e)

Figure 3.12 Conditional dilation: (a) input image, (b) structuring element, (c) conditioning image, (d) output of conditional dilation, (e) illustration.

ditional dilation, and (e) illustration of the conditioning effect on three translations of the disk on A. Note that, in the present context, Eq. (3.4) is equivalent to

$$A \oplus_C B = (A \oplus B) \cap C. \tag{3.5}$$

Referring to Eqs. (3.5) and (1.36), we see that bounded dilation is conditional dilation relative to the view of the image.

A sequence of n conditional dilations of S relative to T using the structuring element B is called a **size-n geodesic dilation**:

$$(S \oplus_T B)^n = (((S \oplus_T B) \oplus_T B) \oplus_T \ldots \oplus_T B). \tag{3.6}$$

Viewing S as a marker, reconstruction of T from S is accomplished by repeating the geodesic dilation until stability is reached. In this case, the geodesic dilation is denoted by $(S \oplus_T B)^\infty$ and we have

$$T \,\triangle_B\, S = (S \oplus_T B)^\infty. \tag{3.7}$$

Note that the preceding relation provides a specific implementation of reconstruction, but not the only one.

An illustration of reconstruction using geodesic dilation is shown in Fig. 3.13: (a) three-component image with one component marked by a pixel; (b) one conditional dilation; (c) two conditional dilations; (d) three conditional dilations; (e) four conditional dilations; (f) five conditional dilations, with stability being reached.

A drawback of the conditional-dilation approach to morphological reconstruction is that it can be time consuming; indeed, it is possible for the number of conditional dilations to equal the number of pixels in the input image. There are, however, very efficient reconstruction algorithms that work recursively and require only a few image scannings.

It is important to recognize the role played by the structuring element in geodesic dilation. This has already been illustrated in Fig. 3.10, which shows how a larger structuring element, in this the 37-pixel disk, can be used to "jump" 8-connected components, thereby joining nearby 8-connected components. Figure 3.14 shows this effect on labeling: (a) the input image; (b) labeling using 8-connectivity; and

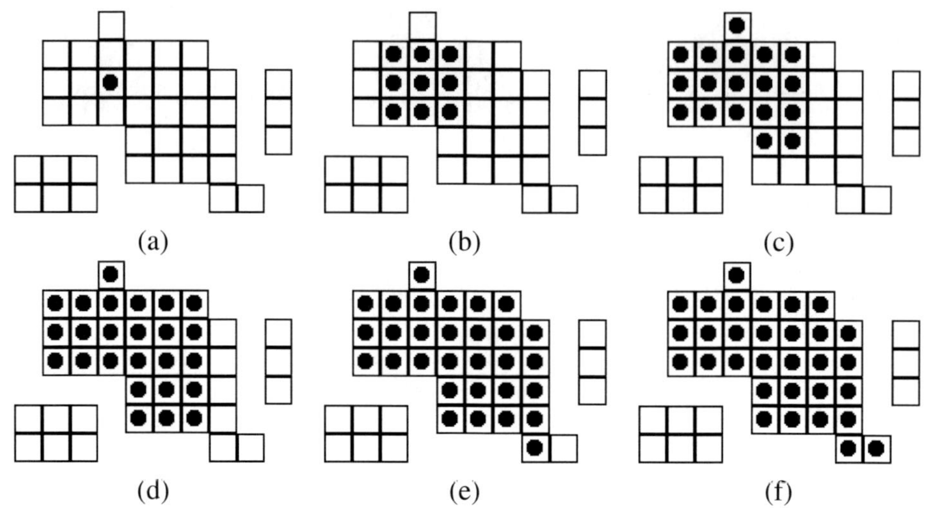

Figure 3.13 Reconstruction using geodesic dilation.

(c) labeling using 68-connectivity. Note in part (c) how there are three labeled regions, these corresponding to 68-connectivity components.

3.5   Marker Selection in Reconstruction

The reconstruction operation requires the input image, the marker, and the structuring element. The marker informs which component of the input image will be extracted, and the structuring element specifies the connectivity. There are typically three ways to design the marker placement: (i) *a-priori* selection; (ii) selection from the opening; (iii) or by means of some rather complex operation.

An example of the *a-priori* selection is the labeling algorithm where the marker is given by the first pixel in the image. Two further examples in this category are removing blobs from the image frame and hole filling. In both of these examples, the marker is placed at the image frame. Figure 3.15 illustrates the process to remove blobs connected to the image frame: (a) the input image, (b) the marker at the image frame, (c) the result of reconstruction, all blobs touching the image frame have been selected, and (d) the subtraction of the reconstructed image from the input image, which gives all blobs not connected to the marker.

Another example of *a-priori* marker design is hole filling. To better describe this, we will introduce the dual of conditional dilation and geodesic reconstruction. The **conditional erosion** of A by B relative to C is given by

$$A \ominus_C B = (A \ominus B) \cup C, \tag{3.8}$$

and a sequence of n conditional erosions by

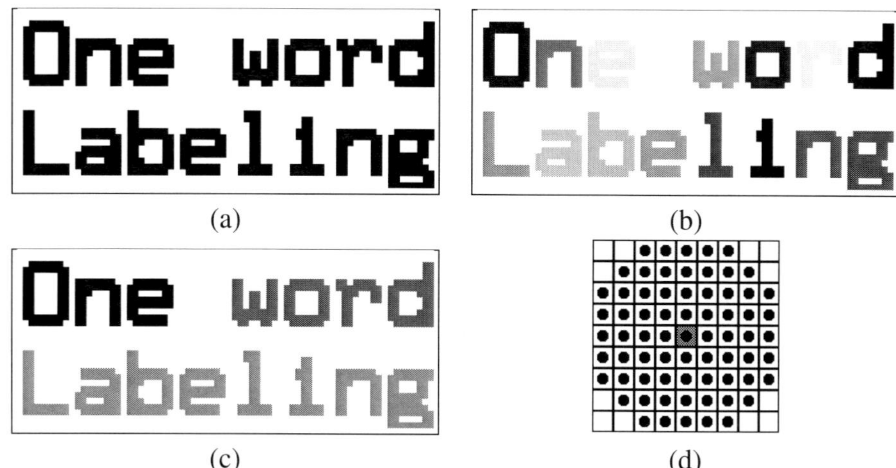

(a)   (b)   (c)   (d)

Figure 3.14 Labeling: (a) input image, (b) letter labeling, with 3×3 square structuring element, (c) word labeling, with E_{68} disk, (d) E_{68} structuring element.

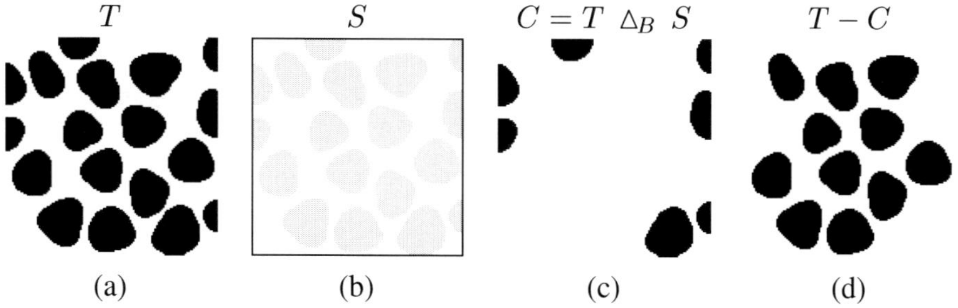

(a)   (b)   (c)   (d)

Figure 3.15 Remove blobs touching the image frame: (a) input image, (b) marker at the image frame, (c) reconstruction of the input image from the marker, (d) subtraction from the input image.

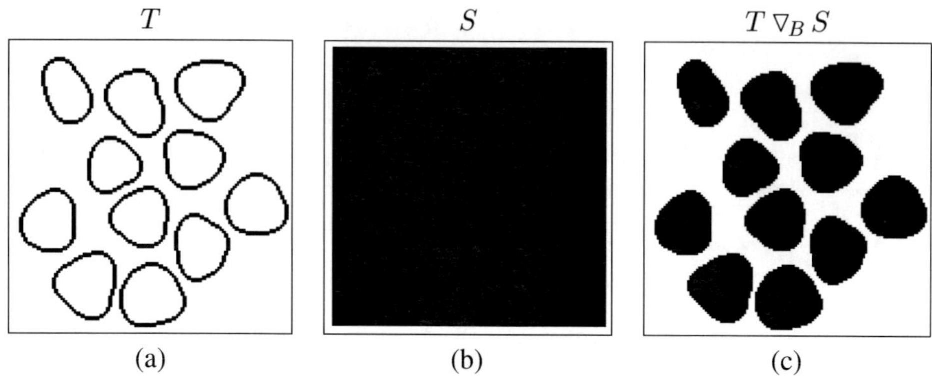

Figure 3.16 Filling holes: (a) input image, (b) dual marker at the image frame, (c) sup-reconstruction of the input image from the dual marker.

$$(A \ominus_C B)^n = (((A \ominus_C B) \ominus_C B) \ominus_C \ldots \ominus_C B). \qquad (3.9)$$

To differentiate the reconstructions resulting from conditional dilation and conditional erosion, we refer to Eq. (3.7) as **inf-geodesic reconstruction** and define **sup-geodesic reconstruction** of T from the marker S using the connectivity given by the structuring element B by

$$T \triangledown_B S = (S \ominus_T B)^\infty. \qquad (3.10)$$

Once the sup-reconstruction is defined, we can illustrate hole filling with the help of Fig. 3.16. The input is a contour image, shown in part (a). Now we will work on the dual, so the background pixels constitute the image of interest. The background pixels inside the contour are the only pixels not connected to the image frame. The sup-reconstruction from the dual marker placed at the image frame [part (b)] will detect as background only the background regions touching the frame. The result of the sup-reconstruction is shown in part (c), which is the input image with its holes detected. The mathematical equation for the **closing-of-holes** operator is given by

$$\Psi(A) = A \triangledown_E (\partial V[A])^c, \qquad (3.11)$$

where $\partial V[A]$ is the one-pixel-thick frame of image A.

The second common way to design the marker for reconstruction is from the opening: the marker is found by opening the input image by a connected structuring element. The result of the reconstruction detects all components where at least one pixel belongs to the opening. Owing to the assumed connectivity of the structuring element, this marking procedure detects connected components that contain at least one translation of the structuring element. The reconstruction from opening is also called **reconstructive opening** and is defined by

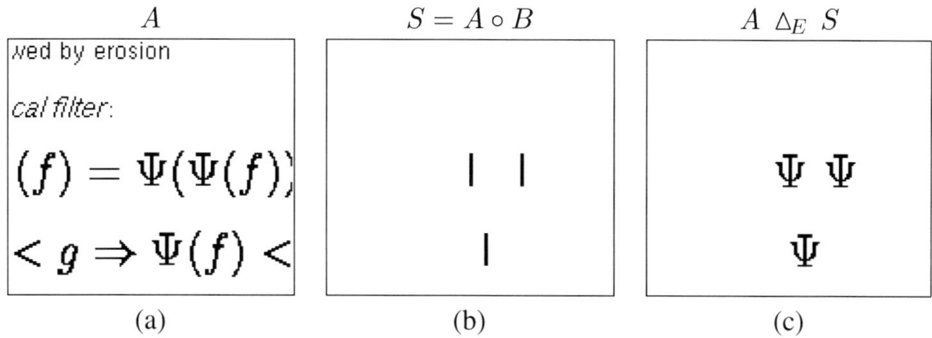

Figure 3.17 Component detection: (a) input image, (b) opening by vertical line of 15 pixels length, (c) reconstruction.

$$A \circ_E B = A \, \Delta_E \, (A \circ B). \tag{3.12}$$

The reconstructive opening requires two structuring elements: one to specify the shape of the fitting criterion (B), and the other to specify the connectivity (E). Unless otherwise specified, it is assumed that E is the 8-connected mask about the origin. This results in reconstructive openings that pass all 8-connected components that are not eliminated by the original opening by B. Observe that the reconstructive opening can be obtained by letting the marker be the erosion of the input image instead of the opening:

$$A \circ_E B = A \, \Delta_E \, (A \ominus B). \tag{3.13}$$

Figure 3.17 shows an example of the reconstructive opening being used to detect symbols in which a straight vertical line of length 15 can be fit: (a) input text image, (b) the opening by the vertical line, and (c) the reconstructive opening. The three symbols possessing the required geometry are detected.

Using the same mechanism of the reconstruction from the opening to detect objects with particular geometric features, more complex techniques can be designed to find the markers from combined operators. At the last step, the reconstruction reveals the objects that exhibit those features. We will see examples of this technique later in the text.

The top-hat concept can be applied to reconstructive opening producing the **reconstructive opening top-hat**:

$$A \, \hat{\circ}_E \, B = A - (A \circ_E B). \tag{3.14}$$

In this case, the operator reveals the objects that do not exhibit a fitting criterion. For instance, to detect thin objects, one can use a disk of diameter larger than the thickest of the thin objects. Figure 3.18 illustrates this situation. The reconstructive opening has detected the suite-objects of the cards as a disk can be fit into them,

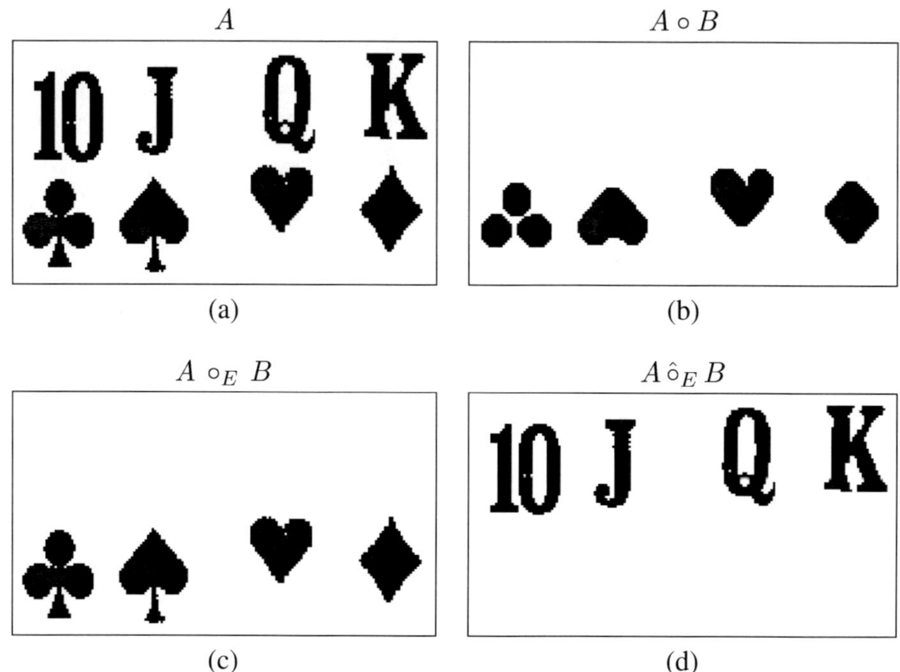

Figure 3.18 Reconstructive opening top-hat: detects objects that a disk does not fit: (a) input image, (b) opening by disk, (c) reconstructive opening by disk, (d) reconstructive opening top-hat by disk.

and the reconstructive opening top-hat has detected the number-objects as a disk cannot be fit.

3.6   Reconstructive τ-opening

As we have seen in Chapter 2, τ-openings are useful for filtering grain images, where the goal is to pass only desired grains. However, unless a passing grain is open with respect to one of the base primitives, its shape will be altered by the operation. To avoid this, one can apply reconstructive openings to form the operator.

A **disjunctive opening** (or **reconstructive τ-opening**) is defined by a union of reconstructive openings:

$$\Psi(A) = \bigcup_{k=1}^{n} A \circ_E B_k, \tag{3.15}$$

where E is the structuring element responsible for the selected connectivity.

A **conjunctive opening** is defined by an intersection of reconstructive openings:

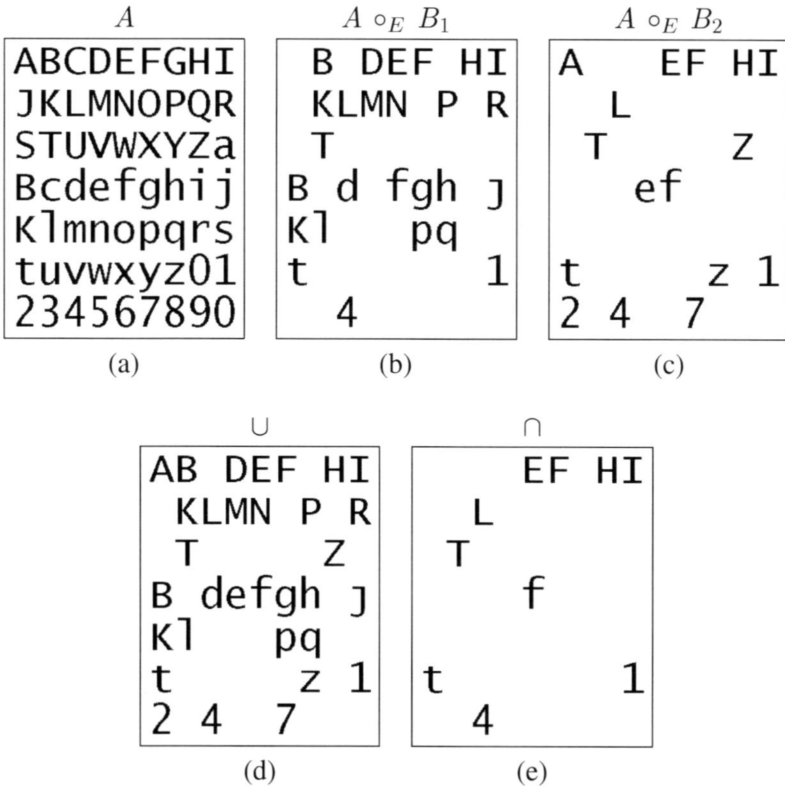

Figure 3.19 Disjunctive and conjunctive opening: (a) input image, (b) letters with long vertical features, (c) letters with short horizontal features, (d) letters with either vertical or horizontal features, (e) letters with both vertical and horizontal features.

$$\Psi(A) = \bigcap_{k=1}^{n} A \circ_E B_k. \tag{3.16}$$

Disjunctive and conjunctive opening are illustrated in Fig. 3.19: (a) the input image, (b) the reconstructive opening by a long vertical element B_1, (c) the reconstructive opening by a short horizontal element B_2, (d) disjunctive opening, and (e) conjunctive opening. The output of the disjunctive operator consists of all characters containing either horizontal or vertical elements and the output of the conjunctive opening contains all characters containing both horizontal and vertical elements.

If the base of the disjunctive opening is a set of linear structuring elements of varying angles, then the filter is called **reconstructive radial opening**. Figure 3.20 shows an illustration of the application of a radial opening and a reconstructive radial opening with angles varying from 0 to 180 deg with 2-deg steps. The filter

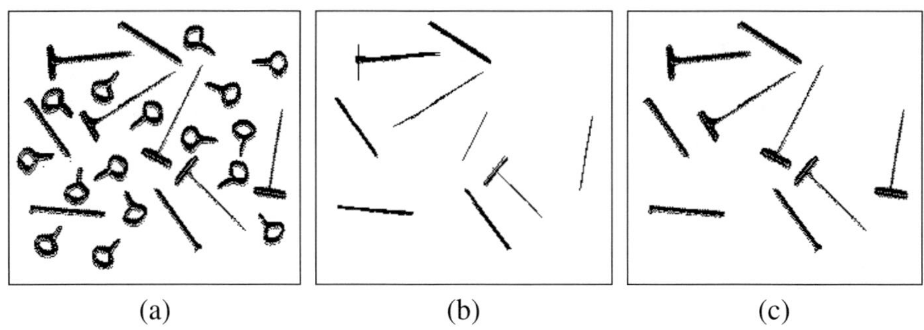

(a)   (b)   (c)

Figure 3.20 Reconstructive radial opening to detect objects with long linear features. (a) input image, (b) radial opening, (c) reconstructive radial opening.

detects all the pieces containing a long straight line.

3.7 Logical Openings

Conjunction and disjunction can be combined to form a more general form of reconstructive opening, called a **logical opening**:

$$\Psi(A) = \bigcup_{k=1}^{n} \bigcap_{j=1}^{m_k} A \circ_E B_{k,j}. \tag{3.17}$$

The kth union is formed from intersecting the reconstructive openings by $B_{k,1}$, $B_{k,2}, \ldots, B_{k,m_k}$. This means that component C is passed if and only if there is at least one k for which there is a translation of each of the structuring elements $B_{k,1}, B_{k,2}, ..., B_{k,m_k}$ that fits inside of C. The logical opening

$$\begin{aligned} \Psi(A) &= ((A \circ_E B_{1,1}) \cap (A \circ_E B_{1,2})) \cup \\ &\quad ((A \circ_E B_{2,1}) \cap (A \circ_E B_{2,2})) \end{aligned} \tag{3.18}$$

is illustrated in Fig. 3.21, where $B_{1,1}$ and $B_{1,2}$ are vertical and horizontal linear structuring elements, respectively, and $B_{2,1}$ and $B_{2,2}$ are diagonal structuring elements.

For a biomedical application of logical openings, we consider comparative genomic hybridization analysis (CGH), in which a test genome and a reference genome are simultaneously hybridized to normal metaphase target chromosomes [Fig. 3.22(a)]. We employ a logical opening using the four linear structuring elements (vertical, horizontal, 45 deg and −45 deg) to identify overlapping chromosomes. We use the logical opening of Eq. (3.18) and obtain the output image shown in Fig. 3.22(c). Because chromosome length is highly variable, the structuring-element length is a critical parameter for good filter performance. The length of

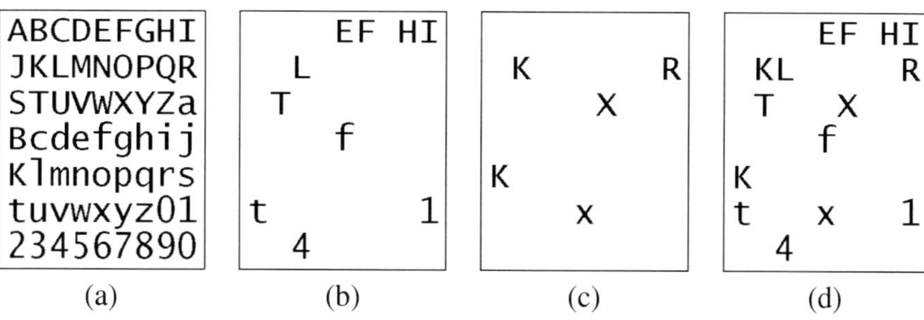

Figure 3.21 Logical opening: (a) input image, (b) first conjunctive opening: letters with both vertical and horizontal features, (c) second conjunctive opening: letters with both diagonal features, (d) union of both conjunctive openings.

30 pixels was obtained by an adaptation procedure trained with 5000 grains from a simulation program. It is interesting to note that, based on the simulation protocol, the logical opening of Eq. (3.18) is not the best choice. Among all logical openings composed of the four linear structuring elements, the best performing one is a conjunctive opening,

$$\Psi(A) = (A \circ_E B_{1,1}) \cap (A \circ_E B_{1,2}) \cap (A \circ_E B_{1,3}) \cap (A \circ_E B_{1,4}), \quad (3.19)$$

where $B_{1,1}, B_{1,2}, B_{1,3}$ and $B_{1,4}$ are the horizontal, vertical, and the two diagonal structuring elements, respectively. Using the best lengths for each, the error of the logical opening in Eq. (3.18) is 50% greater than the conjunctive opening in this case.

3.8 Logical Structural Filters

Logical openings involve unions of intersections of reconstructive openings. The logic can be pushed further by involving reconstructive opening top-hats.

To ease notation and facilitate the definition of the related operator, we introduce the notation $\langle C \circ_E B \rangle^1$ to denote the reconstructive opening of a connected set by B. The corresponding top-hat is defined by $\langle C \circ_E B \rangle^0$.

Logical openings pass or do not pass components based only on whether combinations of structuring elements fit. Using reconstructive opening top-hats, this can be extended to requiring nonfitting elements. For connected structuring elements $B_1, B_2, ..., B_n$ and a set $\mathbf{U} = \{\mathbf{u}_1, \mathbf{u}_2, ..., \mathbf{u}_m\}$ of binary logical vectors, $\mathbf{u}_j = (u_{j1}, u_{j2}, ..., u_{jn})$, a **logical structural filter (LSF)** is defined by

$$\Psi(A) = \bigcup_{j=1}^{m} \bigcap_{k=1}^{n} \langle A \circ_E B_k \rangle^{u_{jk}}. \quad (3.20)$$

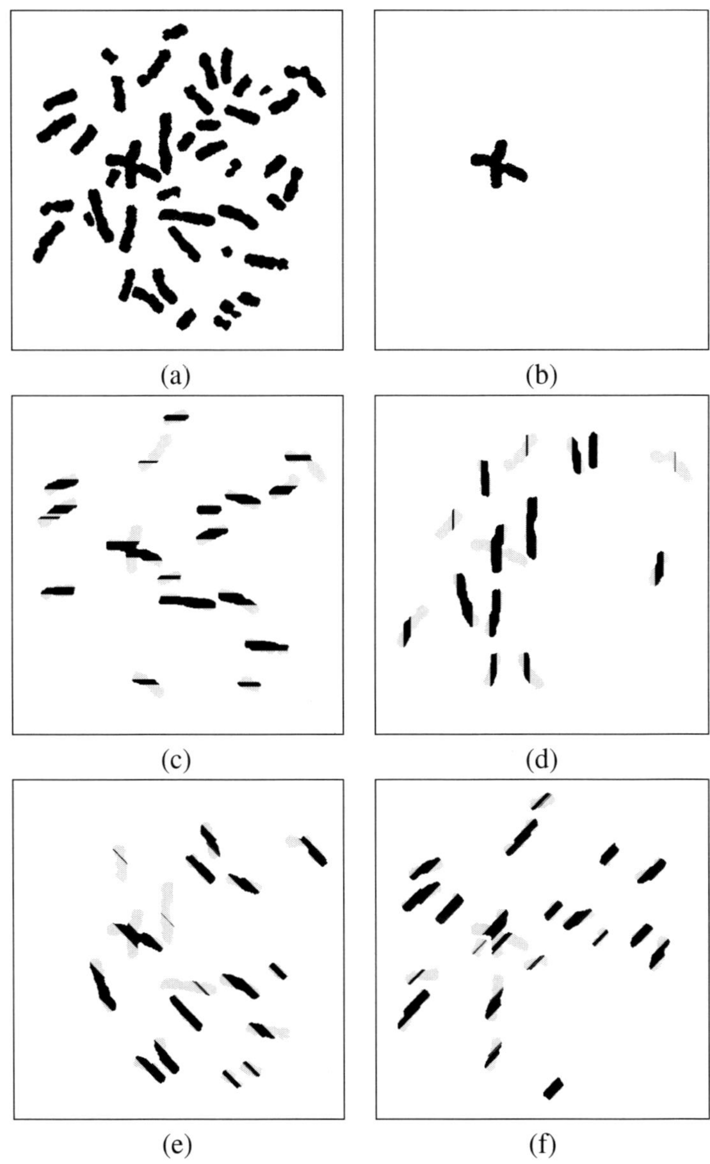

Figure 3.22 Logical opening: (a) input image, (b) conjunctive opening of four linear elements: horizontal, vertical, 45 deg and −45 deg, (c) opening by horizontal and its reconstruction (light gray), (d) vertical element, (e) 45-deg element, (f) −45-deg element.

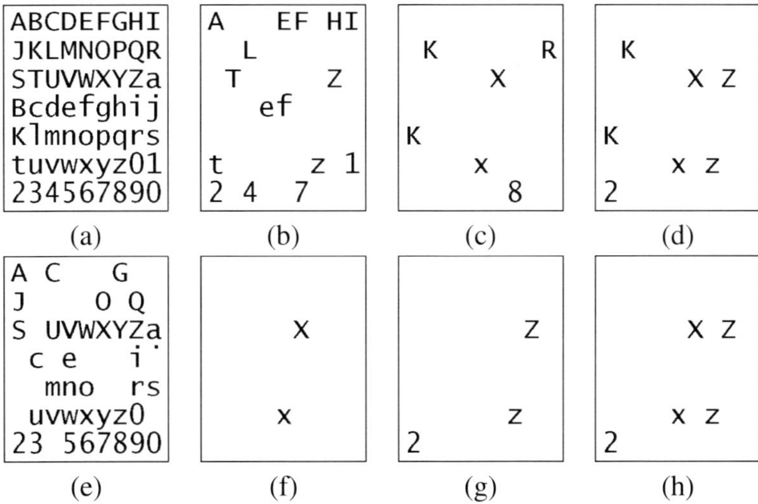

Figure 3.23 Logical structural filter of Eq. (3.21): (a) text image; (b) reconstructive opening by a short horizontal line; (c) reconstructive opening by a 45-deg diagonal line; (d) reconstructive opening by a −45-deg diagonal line; (e) reconstructive opening top-hat by a vertical line; (f) intersection of parts (c), (d), and (e); (g) intersection of parts (b) and (d); (h) union of parts (f) and (g).

Logical structural filters need not be increasing. If A is decomposed into its connected components, then the LSF can be evaluated on each component independently with the final filter being the union of the passed components. Component C is passed if there exists a vector $\mathbf{u}_j$ such that for each $u_{jk} = 1$ there is a translation of B_k that is a subset of C, and for each $u_{jk} = 0$ there is no translation of B_k that is a subset of C; otherwise, the component is eliminated.

For an LSF involving reconstructive opening top-hats, consider the LSF defined by

$$\Psi(A) = (\langle A \circ_E B_{1,1}\rangle^0 \cap \langle A \circ_E B_{2,1}\rangle^1 \cap \langle A \circ_E B_{2,2}\rangle^1) \cup (\langle A \circ_E B_{1,2}\rangle^1 \cap \langle A \circ_E B_{2,2}\rangle^1),$$

$$(3.21)$$

where $B_{1,1}, B_{1,2}, B_{2,1}$, and $B_{2,2}$ are a vertical line, a short horizontal line, a 45-deg diagonal, and a −45-deg diagonal, respectively. Application of this LSF to a text image is shown in Fig. 3.23: (a) text image; (b) reconstructive opening by a short horizontal line; (c) reconstructive opening by a 45-deg diagonal line; (d) reconstructive opening by a −45-deg diagonal line; (e) reconstructive opening top-hat by a vertical line; (f) intersection of parts (c), (d), and (e), which extracts any character with both diagonals but no vertical line; (g) intersection of parts (b) and (d), which extracts any character with horizontal and 45-deg diagonal; and (h) the final output of the LSF, which is the union of parts (f) and (g).

3.9 Connected Operators

One property of an LSF is that it does not introduce new boundary points within an image. The boundaries exist between connected components of the image and connected components of its complement. The effect of the filter is to act on the grains by removing them or leaving them. Hence, boundary points can only be lost; they cannot be gained. The matter can be explained more generally in terms of partitions.

A **partition**, $\mathcal{P}$, of the entire space is a collection of disjoint sets whose union equals the full space. For each point (pixel) z there exists a unique set in the partition containing z, which we will denote by $\mathcal{P}(z)$. Partition $\mathcal{P}'$ is said to be **finer** than partition $\mathcal{P}$ if $\mathcal{P}'(z) \subset \mathcal{P}(z)$ for all points z, and $\mathcal{P}$ is said to be **coarser** than $\mathcal{P}'$. In effect, this means that some sets within the partition $\mathcal{P}'$ have been joined to form the partition $\mathcal{P}$. A partition is said to be **connected** if all sets forming it are connected. Every binary image A induces a unique connected partition, $\mathcal{P}_A$, by taking the sets forming the partition to be the connected components of the image and its complement. A **connected operator** Ψ is one for which the partition $\mathcal{P}_{\Psi(A)}$ induced by $\Psi(A)$ is coarser than the partition $\mathcal{P}_A$ induced by A for every image A. This is indeed the case for LSFs and all reconstructive filters seen so far.

Figure 3.24 shows a simple illustration of the difference between a connected and a nonconnected filter. Part (a) shows the original image and part (b) shows the result of the labeling operator, seen in Fig. 3.11, that places an individual number to each connected component on the image. Figure 3.24(c) is the connected partition induced by the input image. Note there are 12 regions, 7 for the foreground and 5 for the background. Part (d) shows the result of the opening which is not a connected operator, and parts (e) and (f) show the result of the reconstructive opening and its associated partition, respectively. Note that 3 regions have been merged resulting in a coarser partition with 9 connected regions.

Connected operators can also be formed by adjoining connected components of the complement of the image. In effect, these "pores" are filled in. To this end, we can define the **reconstructive closing** as the sup-reconstruction from the closing,

$$A \bullet_E B = (A \bullet B) \, \nabla_E \, A. \tag{3.22}$$

The reconstructive closing passes no part of a component within the image complement not fully filled in by the closing itself. By duality, the reconstructive closing can be expressed in terms of the reconstructive opening,

$$A \bullet_E B = (A^c \circ_E \breve{B})^c. \tag{3.23}$$

The definition of reconstructive closing leads at once to a corresponding definition for **disjunctive closing**,

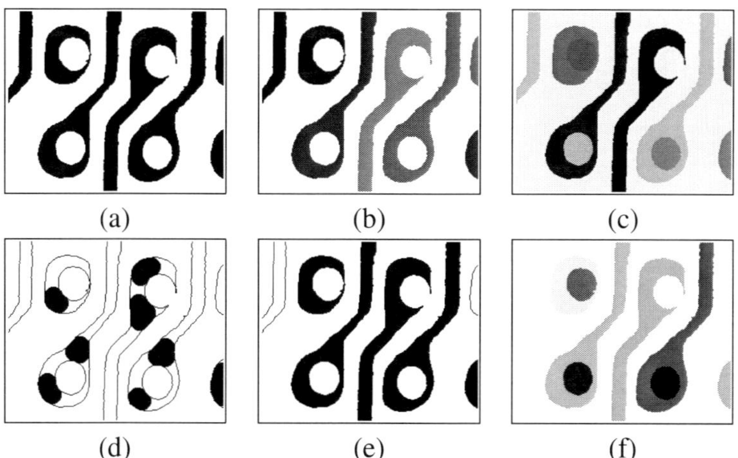

Figure 3.24 Connected operators: (a) input image, (b) 7 connected components of the foreground pixels (labeling), (c) 12 regions in the partition is made of foreground and background connected components, (d) opening is not a connected operator, (e) reconstructive opening is connected, (f) 9 regions of the reconstructive opening, 3 regions were merged.

$$\Psi(A) = \bigcup_{k=1}^{n} A \bullet_E B_k. \tag{3.24}$$

The disjunctive closing is the complement of a conjunctive opening on the complement image. A component of the complement is filled in by the disjunctive closing if and only if it is filled in by at least one of the reconstructive closings composing the disjunction. Analogously, the **conjunctive closing** is defined by

$$\Psi(A) = \bigcap_{k=1}^{n} A \bullet_E B_k. \tag{3.25}$$

A component of the complement is filled in by the conjunctive closing if and only if it is filled in by all of the reconstructive closings.

The definition of logical opening extends analogously to the definition of a logical closing. The definition of an LSF similarly extends. The net effect of these extensions is to produce connected operators that act at the level of the pores of an image.

The relationship between connected operators and both the grains and pores of an image can be seen in the following proposition: an operator Ψ is **connected** if and only if $A - \Psi(A)$ consists of grains of A and $\Psi(A) - A$ consists of grains of A^c. A connected operator passes in full or eliminates in full grains in both the image and its complement. In this way, the image boundary is not altered except to the extent that parts of it might be eliminated along with the components contributing

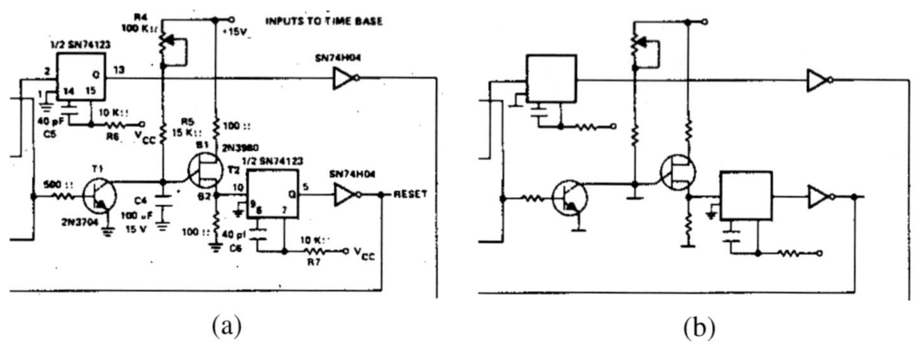

(a)   (b)

Figure 3.25 Area open: (a) input image, (b) only connected components with area greater or equal to 200 pixels.

to it that have themselves been eliminated. By combining logical openings and logical closings, one can produce operators that work on both the image and its complement.

Another τ-opening that is a connected operator is the **area open** filter. In this case, the base is composed of all the E-connected structuring elements with area equal to α. This definition is useful to understand the classification of this filter. In practice its efficient implementation is similar to the labeling algorithm earlier described. The area of each connected component C_i is measured and it is removed if it is less than α:

$$A \circ (\alpha)_E = \bigcup \{C_i, \; area(C_i) \geq \alpha\}. \tag{3.26}$$

Figure 3.25 shows a typical application of the area open to remove small blobs in the image. In this case, all disconnected symbols of a circuit diagram are removed.

3.10   Skeletonization

A standard problem in image processing is finding a thinned replica of a binary image to use in either a recognition algorithm or for data compression. For instance, one might wish to thin characters prior to applying an automatic character reading algorithm. A commonly employed thinning procedure is skeletonization, which is based on the concept of maximal disks.

Given a point interior to a Euclidean binary image, there exists a largest disk having the point at its center and also lying within the image. Regarding the largest disk at a point, there are two possibilities: either there exists another disk lying within the image and properly containing the given disk, or there does not exist another disk within the image properly containing the given disk. Any disk satisfying the second condition is called a **maximal disk**. The centers of all maximal disks comprise the **skeleton** (or **medial axis**) of the image.

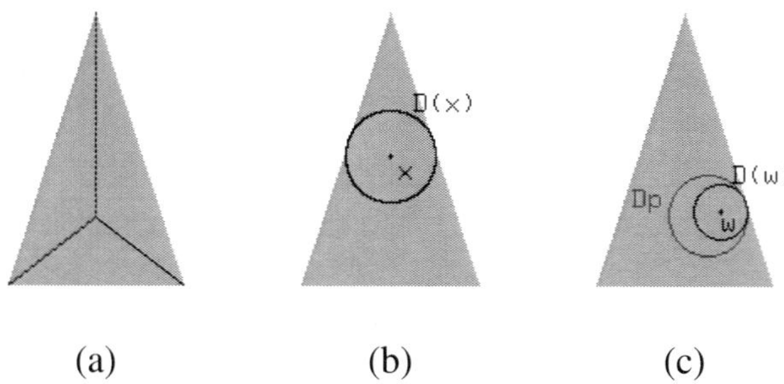

(a)   (b)   (c)

Figure 3.26 Triangle skeleton: (a) skeleton, (b) maximal disk $D(x)$ for skeleton point x, (c) w is not a skeleton point as $D(w)$ is contained in disk D_p.

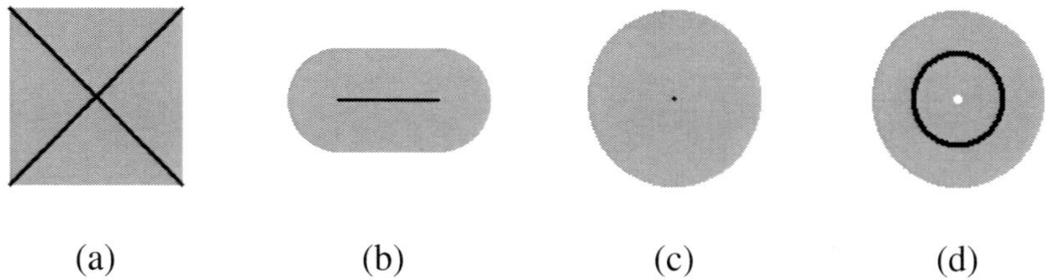

(a)   (b)   (c)   (d)

Figure 3.27 Various skeletons: (a) square, (b) rounded shape, (c) disk, (d) disk without a point in the center.

As an illustration, consider the isosceles triangle in Fig. 3.26, whose skeleton is depicted in part (a) of the figure. Part (b) shows a maximal disk $D(x)$ situated at point x, so that x lies in the skeleton. In part (c), $D(w)$ is the largest disk centered at w; however, it is not maximal since it is properly contained in D_p, which itself lies within the triangle. Thus, w does not lie in the skeleton. For an image A, we denote the skeleton by $Skel(A)$. Figure 3.27 illustrates some skeletons. Notice that different images can have the same skeleton, and that the skeleton is very sensitive to noise.

Relative to the skeleton of a set A, we define the **quench function** q_A on the skeleton in the following manner: for any x in the skeleton, $q_A(x)$ is the radius of the maximal disk centered at x.

As might be expected, adaptation of the skeleton to the digital setting requires some care because there is no analog to the Euclidean disk. To proceed, we begin with some "disk-like" digital primitive, and the actual skeleton depends on the choice of primitive. For the moment, let E denote the 3×3 square structuring element, and let nE be defined by the iterated dilation of Eq. (1.19). (We could just as easily have chosen E to be the 4-connected mask or the 2×2 square with the origin at one of the corners.) The notion of maximal disk is put into the digital setting by considering "disks" chosen from among $0E, E, 2E, 3E, \ldots$, where $0E$ is simply the origin. Figure 3.28(a) illustrates some maximal and nonmaximal disks

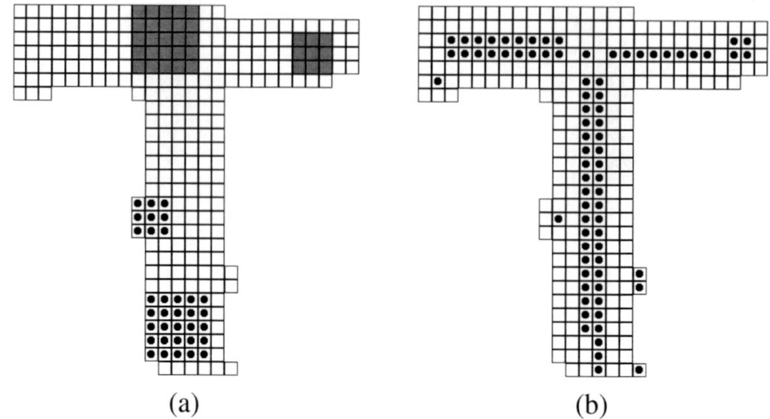

(a)   (b)

Figure 3.28 (a) Maximal (dot) and nonmaximal "disks" (in gray), (b) digital morphological skeleton.

relative to E and Fig. 3.28(b) shows the resulting skeleton.

As defined, the skeleton is prone to having numerous spurious branches owing to boundary irregularities and noise. In addition, in the discrete case it is also prone to disconnecting connected sets. Consequently, its direct use is limited.

The digital skeleton can be characterized morphologically. For $n = 0, 1, \ldots,$ we define the **skeletal subset** $Skel(S; n)$ to be the set of all pixels x in S such that x is the center of a maximal disk nE. Then it is evident from the definition of the skeleton that the skeleton is the union of all skeletal subsets:

$$Skel(S) = \bigcup_{n=0}^{\infty} Skel(S; n). \tag{3.27}$$

It can be shown that the skeletal subsets are given by

$$Skel(S; n) = (S \ominus nE) - [(S \ominus nE) \circ E]. \tag{3.28}$$

Together Eqs. (3.27) and (3.28) yield **Lantuejoul's formula** for the skeleton:

$$Skel(S) = \bigcup_{n=0}^{\infty} (S \ominus nE) - [(S \ominus nE) \circ E]. \tag{3.29}$$

As a transformation, the skeleton is not invertible; however, it can be shown that given all skeletal subsets, the original set must equal the union of the skeletal subsets dilated by the respective structuring elements nE:

$$S = \bigcup_{n=0}^{\infty} Skel(S; n) \oplus nE. \tag{3.30}$$

While it might seem at first from Eqs. (3.27) and (3.30) that derivation of the skeleton and reconstruction by means of the skeletal subsets involve infinite unions, such is not actually the case, since, assuming the initial set is bounded (which it will be for digital imaging applications), for sufficiently large n the skeletal subset is empty. Let N be the size of the largest nonempty skeletal subset. This leads at once to the definition of the **digital quench function** q_S: for any x in the skeleton of S, $q_S(x)$ is the value $n + 1$ for which $(nE)_x$ is maximal in S, or

$$q_S(x) = \begin{cases} n+1, \ x \in Skel(S;n) \\ 0, \text{otherwise} \end{cases}. \tag{3.31}$$

The exact reconstruction of the input image from the quench function is given by the union of dilations of the quench function cross sections:

$$S = \bigcup_{n=0}^{N} \Xi_{n+1}(q_S) \oplus nE, \tag{3.32}$$

where $\Xi_k(f) = \{z : f(z) = k\}$ is the cross section of f at level k. When performing skeletonization or reconstruction by skeletal subsets, we need only use unions up to the maximum value of the quench function.

The skeletonization procedure is illustrated in Fig. 3.29, and the reconstruction from the quench function is illustrated in Fig. 3.30.

As noted at the outset, the skeleton can be used for image compression, which is accomplished by only transmitting the quench function. There is an important property relating the quench function and the opening. Reconstruction from the quench function using only values above n is equivalent to opening the original image by nE:

$$S \circ nE = \bigcup_{k=n}^{N} \Xi_{k+1}(q_S) \oplus kE. \tag{3.33}$$

Transmitting all skeletal subsets results in lossless compression; greater savings are accomplished by only passing higher-order skeletal subsets, the result being the loss of small detail removed by the implicit opening.

Another limitation of the discrete skeleton characterized by Eq. (3.29) is the fact that its construction is based on disks of the form nE. The Euclidean disk cannot be generated in this way. The two most-used disks are based on the 4-connected and 8-connected structuring elements, which are not rotation invariant.

3.11 Distance Transform

A **distance** or **metric** d is a function that associates a nonnegative value to any two points and satisfies three conditions: i) the value is positive, or zero if the two

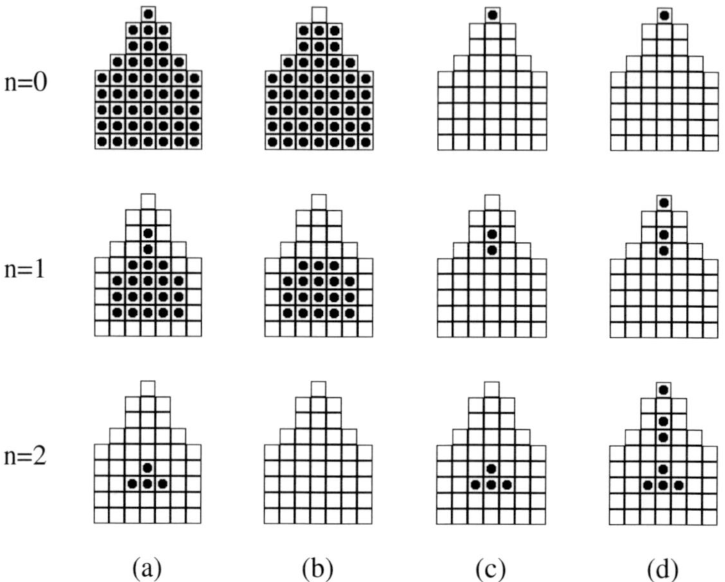

Figure 3.29 Skeletonization by skeletal subsets: (a) $S \ominus nE$, (b) $(S \ominus nE) \circ E$, (c) $Skel(S; n)$, (d) $\bigcup_{k=0}^{n} Skel(S; k)$.

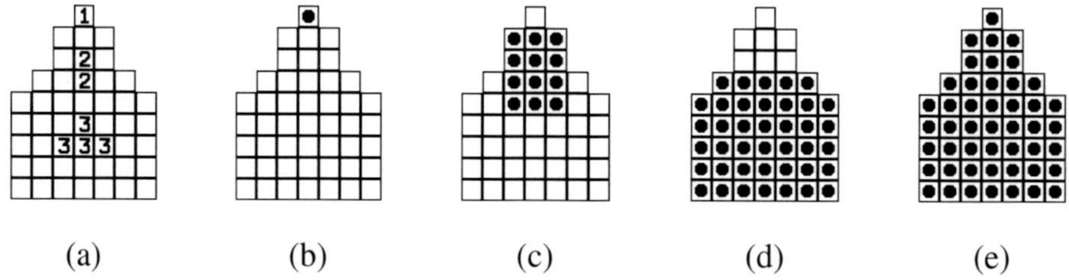

Figure 3.30 Reconstruction from the quench function: (a) quench function q_S, (b) $\Xi_1(q_S) \oplus 0E$, (c) $\Xi_2(q_S) \oplus 1E$, (d) $\Xi_3(q_S) \oplus 2E$, (e) reconstruction is the union.

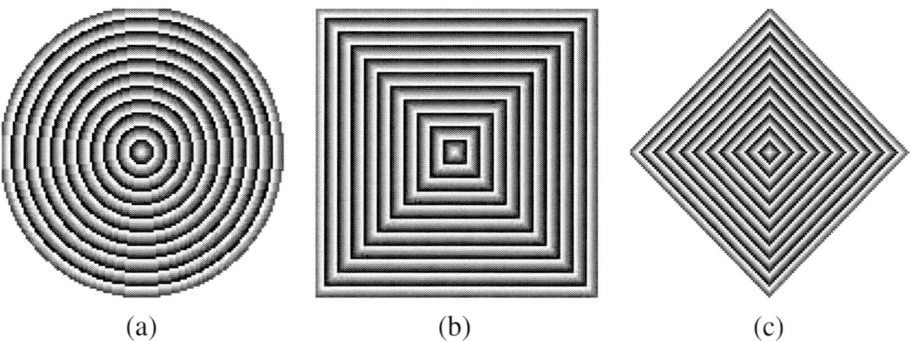

Figure 3.31 Family of disks: (a) Euclidean, (b) chessboard, (c) city-block.

points are the same, ii) the distances from one point to another and vice-versa are the same, and iii) when going from one point to another, if stopping at a third point in between, the overall distance can never be less than going nonstop. In sum:

$$
\begin{aligned}
d(u,v) &\ge 0, \text{ and } d(u,v) = 0 \text{ if and only if } u = v,\\
d(u,v) &= d(v,u),\\
d(u,v) &\le d(u,w) + d(w,v).
\end{aligned}
\tag{3.34}
$$

Three popular metrics are Euclidean (d_E), chessboard (d_8), and city-block (d_4):

$$
\begin{aligned}
d_E[(x_1,y_1),(x_2,y_2)] &= \sqrt{(x_1 - x_2)^2 + (y_1 - y_2)^2},\\
d_8[(x_1,y_1),(x_2,y_2)] &= \max\{|x_1 - x_2|, |y_1 - y_2|\},\\
d_4[(x_1,y_1),(x_2,y_2)] &= |x_1 - x_2| + |y_1 - y_2|.
\end{aligned}
\tag{3.35}
$$

The Euclidean metric is the most natural one as it is rotation invariant, but the city-block and chessboard metrics deserve to be mentioned as they are very suitable for the square discrete grid of digital images.

A **disk** of radius r is the set of points within r of the origin:

$$
D(r) = \{u : d(u, (0,0)) \le r\}.
\tag{3.36}
$$

Figure 3.31 shows three families of disks using the Euclidean, the chessboard, and the city-block metrics. The latter two disks are equivalent to the structuring elements rE_8 and rE_4, where E_8 and E_4 are the 3×3 box and diamond structuring elements.

For a given set A, define the **distance function** or **distance transform** $DT(A)$ in the following manner: for any point x in A, let $DT(A)(x)$ be the distance from x to the complement of A:

$$
DT(A)(x) = \min\{d(x,y), y \in A^c\}.
\tag{3.37}
$$

Figure 3.32 illustrates the distance transform of a synthetic binary image [part (a)] using: (b) the Euclidean, (c) the chessboard, and (d) the city-block metrics.

The iso-contour lines are displayed by computing the distance transform using the modulo 6 function.

The distance transform of A stores the erosions of A by a family of disks. In the discrete case, the pixels that are of a distance below or equal to r from the background are the pixels removed by the erosion of A by the disk of radius r. This property is useful for geometric interpretation of the distance transform and the erosion by disks, but it is also useful in the design of efficient algorithms both for erosion by disks or to compute the distance transform. Specifically,

$$A \ominus D(r) = X_{r+1}(DT(A)), \tag{3.38}$$

where $X_k(f) = \{x : f(x) \geq k\}$ is the threshold set of f at level k. Figure 3.33 shows the erosion of the same image used in the distance transform of Fig. 3.32. The erosions are computed using the Euclidean disk of radius 6 [part (b)], radius 12 [part (c)], and radius 18 [part (d)].

3.12 Geodesic Distance Transform

The foregoing description of the distance concept gives the shortest distance between two points u and v if no obstacles exist between them. When there are restrictions on the path joining the points, we refer to **geodesic distance** restricted to set A, $d_A(u, v)$, the length of the shortest path in A between the points u and v.

A **geodesic disk** at center c of radius r restricted to the set A is the set of points with a geodesic distance from the center less than or equal to r:

$$D_A(c, r) = \{u : d_A(u, c) \leq r\}. \tag{3.39}$$

Figure 3.34 illustrates Euclidean, chessboard and city-block geodesic disks. Similarly to the standard disks, the chessboard and city-block geodesic disks of radius r can be generated by conditionally dilating r times a single point at c restricted to the set A.

3.13 Exercises

1. Find the 4-connected and 8-connected components of the image S in Exercise 17.

2. Find the 4-connected and 8-connected external, internal, and morphological-

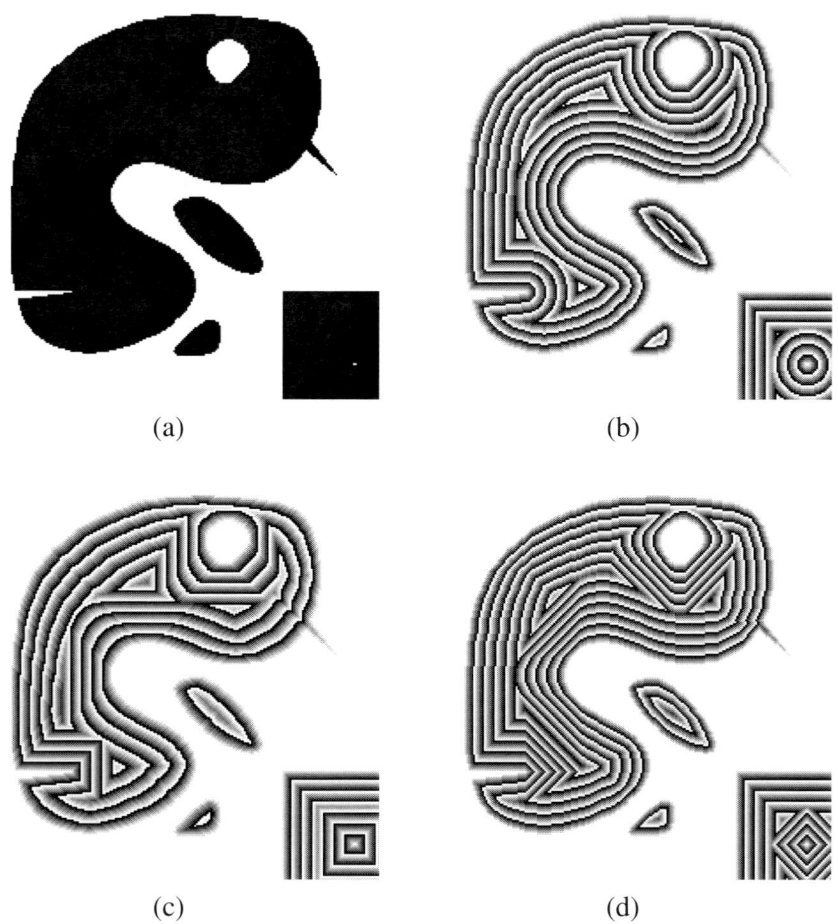

(a)   (b)

(c)   (d)

Figure 3.32 Distance transform. The gray values of the distance transform are illustrated by iso-contour lines highlighted by computing it using the modulo 6 function: (a) input image, (b) Euclidean, (c) chessboard, (d) city-block.

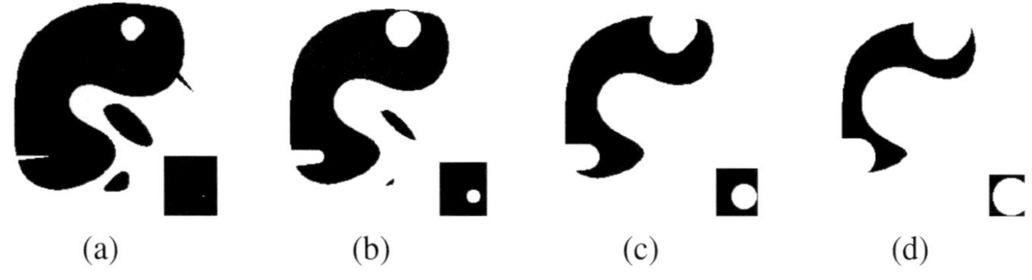

(a)   (b)   (c)   (d)

Figure 3.33 Distance transform and erosion by Euclidean disks: (a) input image, (b) eroded by $D(6)$, (c) eroded by $D(12)$, (d) eroded by $D(18)$.

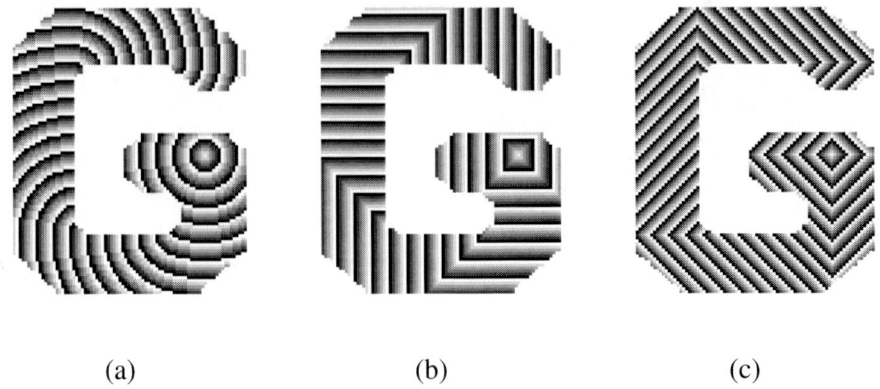

(a)   (b)   (c)

Figure 3.34 Family of geodesic disks restricted to a shaped letter G binary image: (a) Euclidean geodesic disks, (b) chessboard geodesic disks, (c) city-block geodesic disks.

gradient boundaries of the image

$$S = \begin{pmatrix} 0 & 0 & 0 & 0 & 0 & 0 & 0 \\ 0 & 1 & 1 & 1 & 1 & 0 & 0 \\ 0 & 1 & 1 & 1 & 1 & 0 & 0 \\ 0 & 1 & 1 & 1 & 1 & 1 & 0 \\ 0 & 0 & 1 & 1 & 1 & 1 & 0 \\ 0 & 0 & 0 & 1 & 1 & 1 & 0 \\ 0 & 0 & 0 & 0 & 0 & 0 & 0 \end{pmatrix}$$

3. Apply 4-connected and 8-connected reconstruction via geodesic dilation to the image S in Exercise 17 using the pixel in the upper-right corner as the marker.

4. An operator Ψ is **self-dual** if it produces the same result if applied to A or to the complement of A: $\Psi(A) = \Psi(A^c)$. Is the morphological gradient a self-dual operator? What is the relationship between the internal and external gradients?

5. Is the morphological gradient an increasing operator?

6. Note that in Eq. (3.7), T is the first input parameter for the reconstruction, whereas S is the first input for the size-n geodesic dilation. Why is the order of the parameters inverted?

7. Prove that the disjunctive opening in Eq. (3.15), which is a union of reconstructive openings, is equivalently formed by taking the union of openings by the same structuring elements and then reconstructing.

8. Can the same thing be said about conjunctive openings that was said in Exercise 7 about disjunctive openings? If so, prove it; if not, give a counterexample.

9. Prove that a conjunctive opening is antiextensive and idempotent.

10. Prove that a disjunctive opening is antiextensive and idempotent.

11. Comparing Eqs. (2.18) and (3.15), we see that a disjunctive opening is formed in the same manner as a τ-opening except that the openings forming the union are reconstructive, which is why a disjunctive opening is also called a reconstructive τ-opening. Now suppose the intersection of Eq. (3.16) involved openings instead of reconstructive openings. Would the resulting operator be idempotent. If so, prove it; if not, give a counterexample.

12. Prove that a logical opening is antiextensive and idempotent.

13. Give an example of an LSF that is not increasing.

14. Construct an LSF that can pick out the vertical bar in the image

$$S = \begin{pmatrix} 0 & 0 & 0 & 0 & 0 & 0 & 1 & 0 & 0 & 0 & 1 & 0 & 0 \\ 0 & 0 & 0 & 0 & 0 & 0 & 1 & 0 & 0 & 0 & 1 & 0 & 0 \\ 1 & 1 & 1 & 1 & 1 & 0 & 1 & 0 & 1 & 1 & 1 & 1 & 1 \\ 0 & 0 & 0 & 0 & 0 & 0 & 1 & 0 & 0 & 0 & 1 & 0 & 0 \\ 0 & 0 & 0 & 0 & 0 & 0 & 1 & 0 & 0 & 0 & 1 & 0 & 0 \end{pmatrix}$$

15. Find the skeletal subsets and skeleton of the image

$$S = \begin{pmatrix} 1 & 1 & 1 & 1 & 0 & 0 & 0 \\ 1 & 1 & 1 & 0 & 0 & 0 & 0 \\ 1 & 1 & 1 & 0 & 0 & 0 & 0 \\ 1 & 1 & 1 & 0 & 0 & 0 & 0 \\ 1 & 1 & 1 & 1 & 1 & 0 & 0 \\ 1 & 1 & 1 & 1 & 1 & 1 & 1 \\ 1 & 1 & 1 & 1 & 1 & 1 & 1 \end{pmatrix}$$

16. What is the relationship between the geodesic disk and the size-n geodesic dilation?

3.14 Laboratory Experiments

Figure 3.35 lists the **MT** functions introduced in this chapter.

1. Use the concept of closing-of-holes operator and morphological reconstruction to detect the "rings" of Fig. 3.20.

name	description
mmgradm:	morphological gradient (Fig. 3.6)
mmlabel:	labeling [Eq. (3.3)]
mmcdil:	size-n geodesic dilation [Eq. (3.6)]
mminfrec:	inf-reconstruction [Eq. (3.7)]
mmcero:	size-n geodesic erosion [Eq. (3.9)]
mmsuprec:	sup-reconstruction [Eq. (3.10)]
mmclohole:	closing of holes [Eq. (3.11)]
mmopenrec:	reconstructive opening [Eq. (3.12)]
mmopenrecth:	reconstructive opening top-hat [Eq. (3.14)]
mmcloserec:	reconstructive closing [Eq. (3.22)]
mmareaopen:	area open [Eq. (3.26)]
mmskelm:	morph. skeleton [Eqs. (3.29) and (3.31)]
mmskelmrec:	reconstr. from skeleton [Eq. (3.32)]
mmsedisk:	disk structuring element [Eq. (3.36)]
mmdist:	distance transform [Eq. (3.37)]
mmgdist:	geodesic distance transform (Fig. 3.34)

Figure 3.35 MT operators presented in this chapter.

2. Design a LSF to identify just the "=" sign in Fig. 3.17(a).

3. Compute the skeleton of Fig. 3.28 using the 2×2 structuring element and compare the result with the skeleton of Fig. 3.28(b) that was computed using the 3×3 structuring element.

4. Verify that the reconstruction from the quench function using only values above the value n is equivalent to opening the original image by nE, as given by Eq. (3.33).

5. Reproduce Figs. 3.34 (b) and (c) using the function mmcdil instead of the geodesic distance transform mmgdist. Why it is not possible to reproduce the Euclidean metric of Fig. 3.34(c) using the geodesic dilation?

6. Suppose a binary image coded using the quench function is being received through a low-speed connection, such that a progressive image transmission is used where the pixels with higher quench function values are transmitted first. Write a program that reproduces the appearance of a slow reception of a binary image using this method.

References

1. Y. Chen and E. R. Dougherty. Adaptive reconstructive τ-openings: Convergence and steady-state distribution. *Journal of Electronic Imaging*, 5(3):266–282, 1996.

2. E. R. Dougherty and Y. Chen. Logical structural filters. *Optical Engineering*, 37(6):1668–1676, 1998.

3. J. Goutsias. Morphological analysis of discrete random shapes. *Journal of Mathematical Imaging and Vision*, 2(2/3):193–215, 1992.

4. R. M Haralick, P. L. Katz, and E. R. Dougherty. Model-based morphology: the opening spectrum. *Graphical Models and Image Processing*, 57(1):1–12, 1995.

5. H. J. A. M. Heijmans. Connected morphological operators for binary images. *Computer Vision and Image Understanding*, 73(1):99–120, 1999.

6. H. J. A. M. Heijmans. Introduction to connected operators. In E. R. Dougherty and J. T. Astola, editors, *Nonlinear Filters for Image Processing*, pages 207–235. SPIE/IEEE Presses, Bellingham, WA, 1999.

7. H. J. A. M. Heijmans and C. Ronse. Annular filters for binary images. *IEEE Transactions on Image Processing*, 8(10):1330–1340, 1999.

8. P. Maragos. Morphological skeleton representation and coding of binary images. *IEEE Transactions on Acoustics, Speech and Signal Processing*, 34:1228–1244, 1986.

9. C. Ronse and B. Macq. Morphological shape and region description. *Signal Processing*, 25:91–105, 1991.

10. K. Sivakumar and J. Goutsias. Binary random fields, random closed sets, and morphological sampling. *IEEE Transactions on Image Processing*, 5(6):899–912, 1996.

11. A. Tuzikov and H. J. A. M. Heijmans. Minkowski decomposition of convex polygons into their symmetric and asymmetric parts. *Pattern Recognition Letters*, 19(3–4):247–254, 1998.

12. A.V. Tuzikov, J.B.T.M. Roerdink, and H.J.A.M. Heijmans. Similarity measures for convex polyhedra based on Minkowski addition. *Pattern Recognition*, 33(6):979–995, 2000.

13. L. Vincent. Morphological transformations of binary images with arbitrary structuring elements. *Signal Processing*, 22:3–23, 1991.

Chapter 4

Hit-or-Miss Transform

Rather than simply probe the inside or the outside of an image, it can be very fruitful to probe both at the same time in order to study the relation between figure and background. The hit-or-miss transform accomplishes this, and it has been proved useful in the solution of various problems, such as character recognition and thinning. We treat it in the present chapter.

4.1 The Transform

When eroding an image by a structuring element, the erosion acts as a marker locating the structuring element fits. Although the marked points depend on the location of the origin relative to the structuring element, the shape of the output image does not: changes in origin position only result in translated outputs. A similar comment applies to the dual operation, dilation, it being the complement of the eroded complement of the image. The **hit-or-miss transform** captures both inner and outer markings in a single operation. It requires two structuring elements, E and F, which are treated as a **hit-or-miss template** $T = (E, F)$, with the assumption that $E \subset F^c$. E is used to probe the inside and F to probe the outside, and it is defined by

$$A \circledast T = (A \ominus E) \cap (A^c \ominus F). \tag{4.1}$$

A point is in the hit-or-miss output if and only if E translated to the point fits inside A and F translated to the point fits outside A. It is assumed E and F are disjoint, for otherwise it would be impossible for both fits to occur simultaneously.

Because it operates by fitting structuring elements into both the image and its complement, the hit-or-miss transform probes the relationship between the image and its complement relative to the structuring pair. Note that if F is made of all zeros, the hit-or-miss degrades to the erosion by E. The transform is illustrated in Fig. 4.1. Rather than write the structuring pair, it is common practice to write a single template and to mark as foreground (1) pixels in the template used for the hit structuring element E, mark with background (0) those used for the miss structuring element F, and mark with an "×" or simply not mark those pixels used for neither, these being the so-called "don't care" pixels.

4.2 Object Recognition

Use of the hit-or-miss transform for object recognition is straightforward; nonetheless, like most recognition schemes, application to real-world images is highly

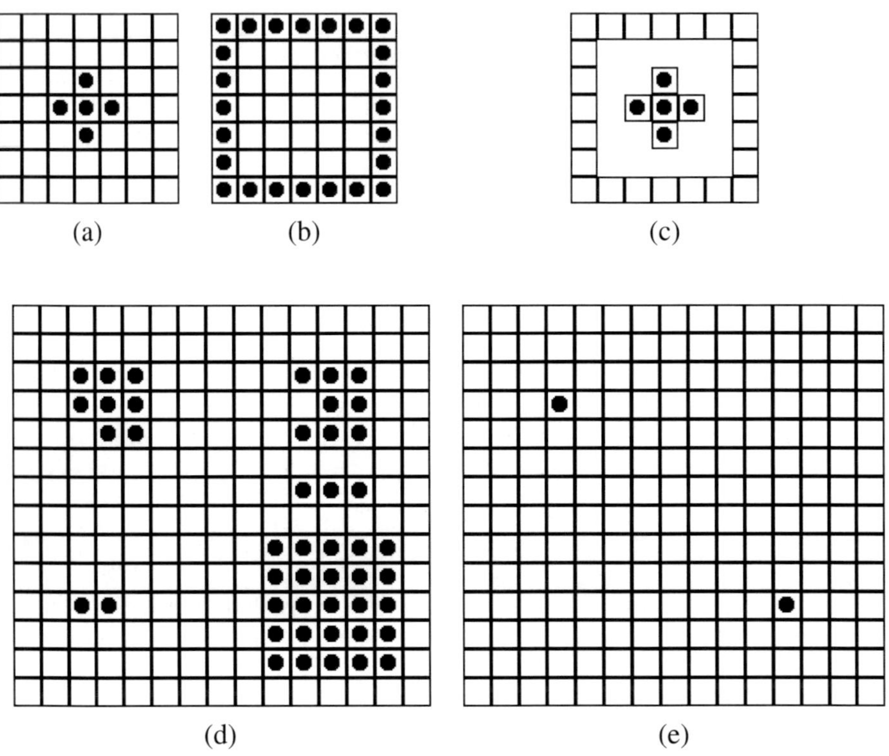

Figure 4.1 Hit-or-miss transform: (a) hit structuring element E, (b) miss structuring element F, (c) hit-or-miss template $T_{E,F}$, (d) input image A, (e) hit-or-miss output $A \circledast T$.

problematic and requires various adaptations of the basic algorithm. We illustrate the approach with reference to a particular example.

In Fig. 4.2 there is an image A composed of six objects: a rectangle, a square, a square with a small extrusion, and also the same three objects with noisy edges. If our task is to mark the location of the clean square-with-extrusion, then we can proceed by forming a structuring element that is an exact copy of the desired shape. Erosion of the image by the square-with-extrusion will yield a single marked point, the point of the origin where the structuring element fits exactly. In the present circumstance erosion accomplishes matching.

The problem is different if we wish to find the clean square. Erosion by an exact copy of the square will mark two locations, one for the square and one for the square-with-extrusion. One solution is to apply the hit-or-miss transform with structuring pair $T = (E, F)$, where E is the square and F is a thin exterior boundary of the square. The representation of this hit-or-miss template is depicted in the figure (T). Applying the hit-or-miss transform will then yield a single point, the center of the desired square. By forcing F to fit into the complement, we distinguish between the square and the square-with-extrusion.

A variation of the same problem is also illustrated in Fig. 4.2, with the noisy objects. The exact hit-or-miss template T does not detect the noisy square. Now consider a new hit-or-miss template T_n where the hit and miss templates are a slightly eroded version of the square and the exterior boundary of a slightly dilated version of the square. This "loose" hit-or-miss template will accomplish the task of recognizing the noisy square and the clean square. However, the resulting marker will not be a single point; it will be a small region, the intersection of the erosion of the noisy square by the eroded square and the erosion of the complement of the noisy square by the exterior boundary of the dilated square. If our aim is to reconstruct the object once it has been found, then the size of the marker, so long as it lies within the object of interest, is not important.

Figure 4.3(a) shows a practical application of template matching using the hit-or-miss transform. The goal is to detect the ground symbol in an electronic circuit drawing. The designed template is shown on part (a) of the figure, and in part (b) the detected symbols are highlighted on the input image. Note that the template has few hit and miss points to fit five quite different ground symbols. The result of the hit-or-miss transform was conditionally dilated by a rectangle of the same size as the hit template relative to the input image to reveal only the detected symbols.

The hit-or-miss recognition procedure has been applied to various problems in machine vision and character recognition. In principle, and in a noiseless environment, if we are given n distinct shapes, then these can be distinguished by using n structuring pairs in which the first element in a pair is identical in shape to one of the given shapes, and the second element in the same pair is a thin exterior boundary of the first element in the pair. If there is noise, the problem may become unsolvable. We might try to use eroded versions of the shapes and exterior boundaries of dilated

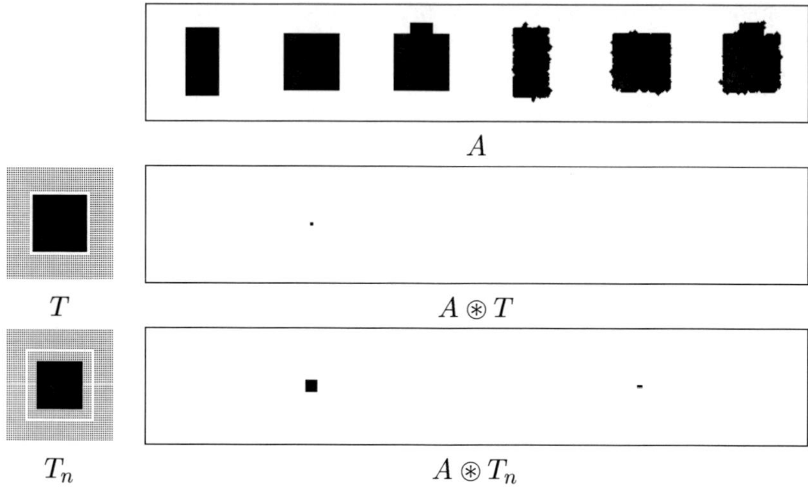

Figure 4.2 Recognition of noisy object. The exact hit-or-miss template T cannot recognize the noisy object, whereas the loose hit-or-miss template T_n can.

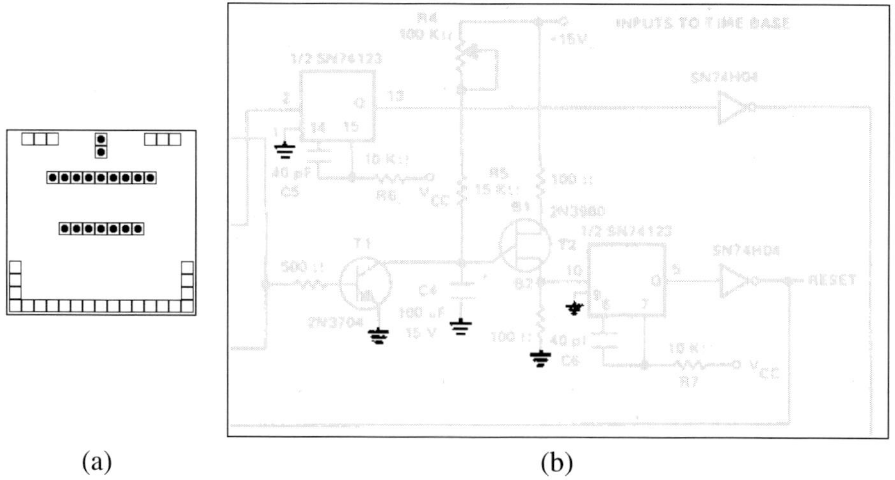

(a)   (b)

Figure 4.3 Object recognition using hit-or-miss, (a) hit-or-miss template, (b) ground symbol detected.

versions of the shapes, but such an approach might result in misclassification.

The design of the hit-or-miss templates can become a very difficult problem in real-life applications. Various methods have been applied to construct structuring pairs that accomplish acceptable recognition rates.

4.3 Thinning

Various morphological algorithms depend on the hit-or-miss transform. One of the most commonly employed thins digital images. For the structuring pair $T = (E, F)$, we define the **thinning** of S by T as

$$S \odot T = S - (S \circledast T), \tag{4.2}$$

that is, $S \odot T$ is the set-theoretic difference between S and $S \circledast T$.

More generally, a sequence of templates $T_1, T_2, \ldots$ is employed iteratively to generate a sequence of outputs:

$$\begin{aligned} S_1 &= S \odot T_1, \\ S_2 &= S_1 \odot T_2, \\ S_3 &= S_2 \odot T_3, \end{aligned} \tag{4.3}$$

(and continuing). As the iteration proceeds, the successive sets are ever-thinner, and (assuming the original input is finite) eventually the process halts, yielding a thin, or skeletal-type image. The choice of structuring pairs is only limited by the requirement that the elements of each pair do not intersect. In fact, each T_k could be the same, which means that the iteration proceeds by repeatedly thinning with the same template. In practice, a set of templates is usually chosen and the iteration repeatedly cycles through them until it halts, halting occurring when a full cycle yields no change.

Figure 4.4 illustrates the procedure for successive iteration with the single template T operating on the image S. Owing to the geometry of the structuring elements, the hit-or-miss transform marks lower-left corners of the image, and at each stage in the iteration these are removed by the thinning.

Because only a single template is employed in Fig. 4.4, the thinning is directional. By cycling through a set of eight "compass" templates, thinning is accomplished in a more symmetric manner. Figure 4.5 shows eight compass templates that are cycled through from top-down and left-to-right. Figure 4.6 shows the effect of thinning when cycling through these eight pairs. We use $(S \odot T)[i; j]$ to denote the output of the jth compass template on the ith cycle and $(S \odot T)[i; j, j+1, \ldots]$ to indicate that a particular image occurs on iteration $[i, j]$ and it remains unchanged on some number of further iterations in the ith cycle.

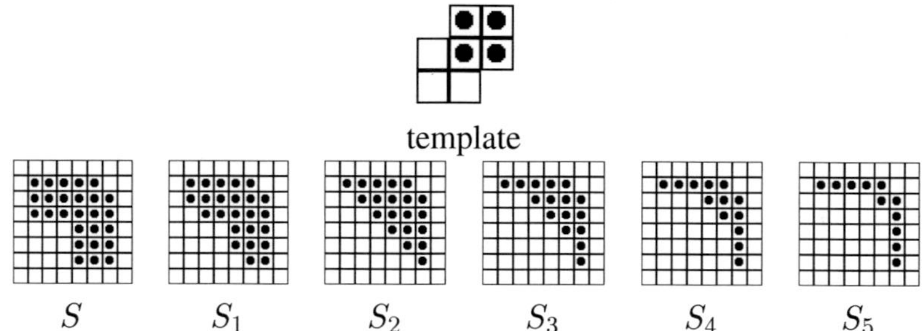

Figure 4.4 Sequential thinning with a single hit-or-miss template.

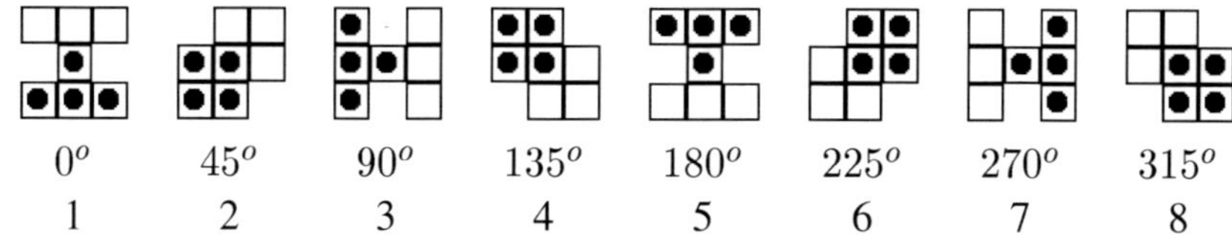

Figure 4.5 Eight compass templates for sequential thinning.

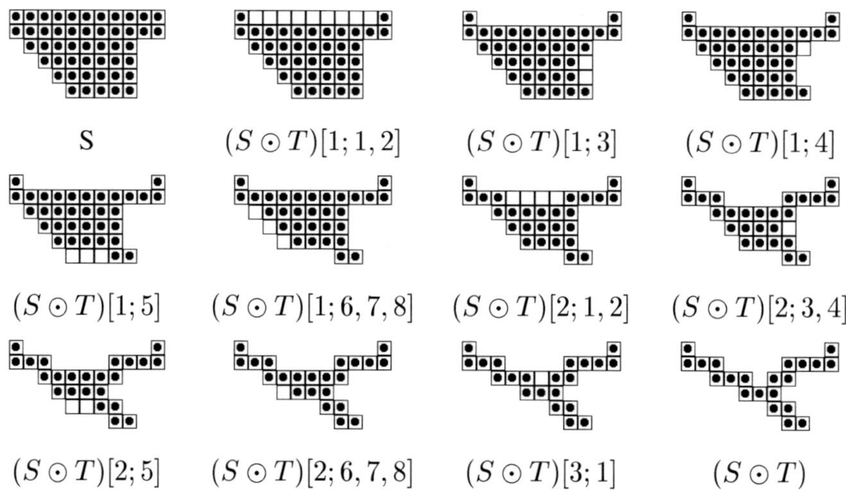

Figure 4.6 Sequential thinning using compass templates.

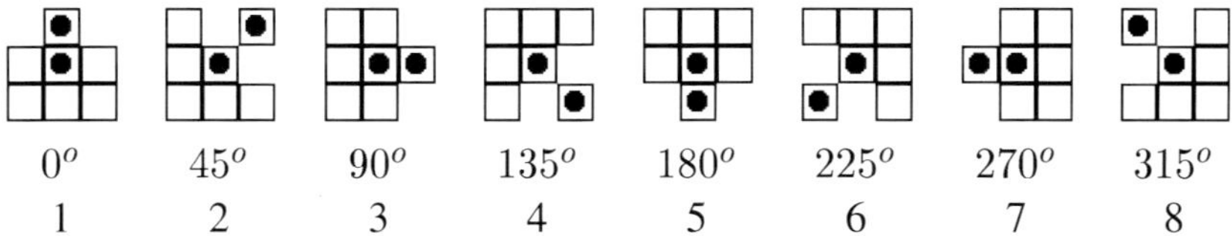

Figure 4.7 Eight compass templates for sequential pruning.

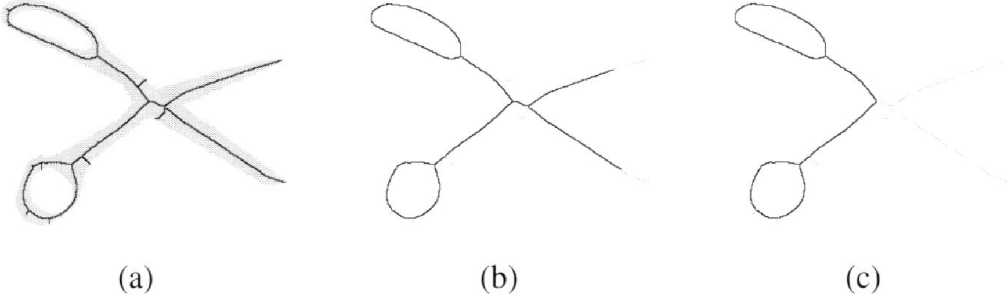

Figure 4.8 Thinning and pruning: (a) sequential thinning using eight compass templates, (b) 15 cycles of sequential pruning using eight compass templates, (c) pruning until stability.

4.4 Pruning

If a shape is relatively thick, it is possible to remove narrow extrusions by opening with a "ball-like" structuring element such as a 3×3 square, the assumption being that whichever pixels do not lie in some fitted ball are spurious to the true shape. If, however, the shape is thin, such as a skeleton, an opening approach cannot be used.

For skeletal shapes a beneficial approach is to employ a variant of thinning known as **pruning**. This is often necessary because thinning-type algorithms such as skeletonization often leave "fuzz." The objective is to eliminate all pixels that are endpoints. This can be accomplished by cycling through a thinning sequence with the eight compass templates of Fig. 4.7. Unlike thinning, where the algorithm cycles through the pairs until it halts of its own accord, for pruning it is also common to have a predetermined number of thinning cycles, else there could be a continuing decreasing to a single point or ring-like structures, one for each connected component in the image.

In Fig. 4.8(a) we see a thinned scissors image with some extraneous pixels attached, a common occurrence in typical thinning transforms. Using 15 pruning cycles eliminates the fuzz [Fig. 4.8(b)]. If the pruning cycles are repeated until no further changes are made, then we get the pruned result of Fig. 4.8(c), where the two loops of the scissors handle are joined together, removing all ending points of the skeleton.

There are problems with the preceding cleaning method. Parts of the lines near the tips of the scissors are erroneously eliminated together with spurious pixels

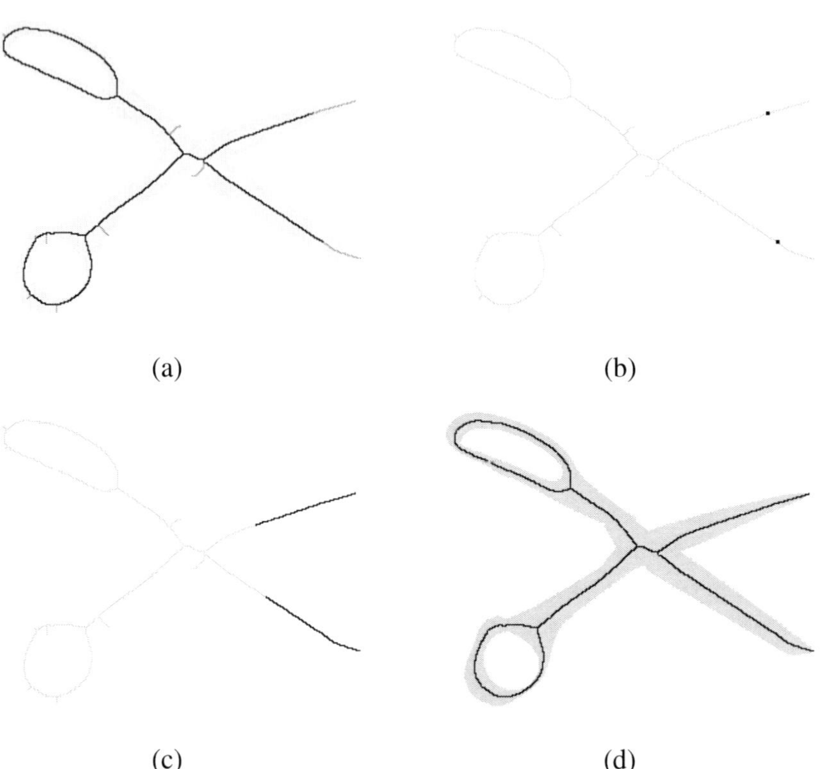

(a)   (b)

(c)   (d)

Figure 4.9 Reconstruction of pruned skeleton by geodesic dilation: (a) skeleton with 15 cycles of pruning, (b) endpoint set, (c) 45 iterations of geodesic dilation of the endpoint set relative to the skeleton, (d) union of the geodesic dilation with pruned skeleton.

[See Fig. 4.9(a)]. One corrective approach is reconstruction by geodesic dilation. Perform one more set of hit-or-miss transformations with the endpoint template collection, but instead of performing thinning with these operations, simply save the eight outputs. This will give the endpoints of the already pruned image. To reconstruct the desired image (hopefully without the undesirable spurs), perform a geodesic dilation of this endpoint set using the 3×3 structuring element relative to the initial image upon which the pruning was performed and with three times the number of iterations as pruning cycles. The reason we need more geodesic dilations than pruning cycles is that each pruning cycle can remove up to three pixels with the eight compass templates. Finally, union this geodesic dilation with the pruned image. Figure 4.9 illustrates this reconstruction: (a) sequentially thinned scissors with 15 cycles of pruning, (b) endpoints detection, (c) 15 cycles of reconstruction applied to the skeleton image, and (d) the union of the reconstruction with the pruned skeleton resulting in a clean skeleton image.

To illustrate the manner in which the methodologies of the present chapter can be employed, we consider a problem involving inspection of printed circuit boards, the specific task being to find open connections on the top layer of the PCB. We outline the steps in the algorithm. In Fig. 4.10(a), we see a defective connection in the upper side of the image. An opening of this image by a disk of radius 6, shown in Fig. 4.10(b), detects the pads of the PCB. Two iterations of geodesic dilation extract the pad regions, as shown in Fig. 4.10(c). This image will be used to discard endpoints at the pads. If instead of opening and dilating the binary image we thin to generate a connected one-pixel-thick skeletal sub-image, then the image of Fig. 4.10(d) is obtained. This image is pruned for two cycles to remove spurious endpoints, shown in Fig. 4.11(a). Fig. 4.11(b) shows the endpoints of the pruned skeleton. A subtraction of the pad image from the endpoint image [Fig. 4.11(c)] leaves endpoint pixels marking breaks in the circuit and in the boundary of the image. A dilation of these endpoints, shown in Fig. 4.11(d), allows the elimination of the endpoints detected near the image frame using the edgeoff function as illustrated in Fig. 3.15. The final endpoint detection is shown in Fig. 4.11(e).

4.5 Exercises

1. Construct a hit-or-miss transform containing a minimal number of pixels in its structuring elements that can pick out the vertical bar in the image of Exercise 14.

2. Apply thinning to the image of Exercise 15 using the compass templates of Fig. 4.5.

3. Apply two pruning cycles to the image

$$
S = \begin{pmatrix}
1 & 1 & 1 & 1 & 0 & 0 & 0 \\
1 & 1 & 0 & 0 & 1 & 0 & 0 \\
1 & 1 & 0 & 0 & 0 & 0 & 0 \\
1 & 1 & 1 & 0 & 0 & 1 & 0 \\
1 & 1 & 0 & 0 & 0 & 1 & 0 \\
1 & 1 & 1 & 1 & 1 & 1 & 1 \\
1 & 1 & 1 & 1 & 1 & 1 & 1
\end{pmatrix}
$$

4. An external boundary for S is given by $(S \oplus E) - S$, where E consists of the 3×3 diamond structuring element. Express this external boundary operator as a union of hit-or-miss transforms.

5. Using the 3×3 diamond structuring element, express the opening top-hat as a union of hit-or-miss transforms.

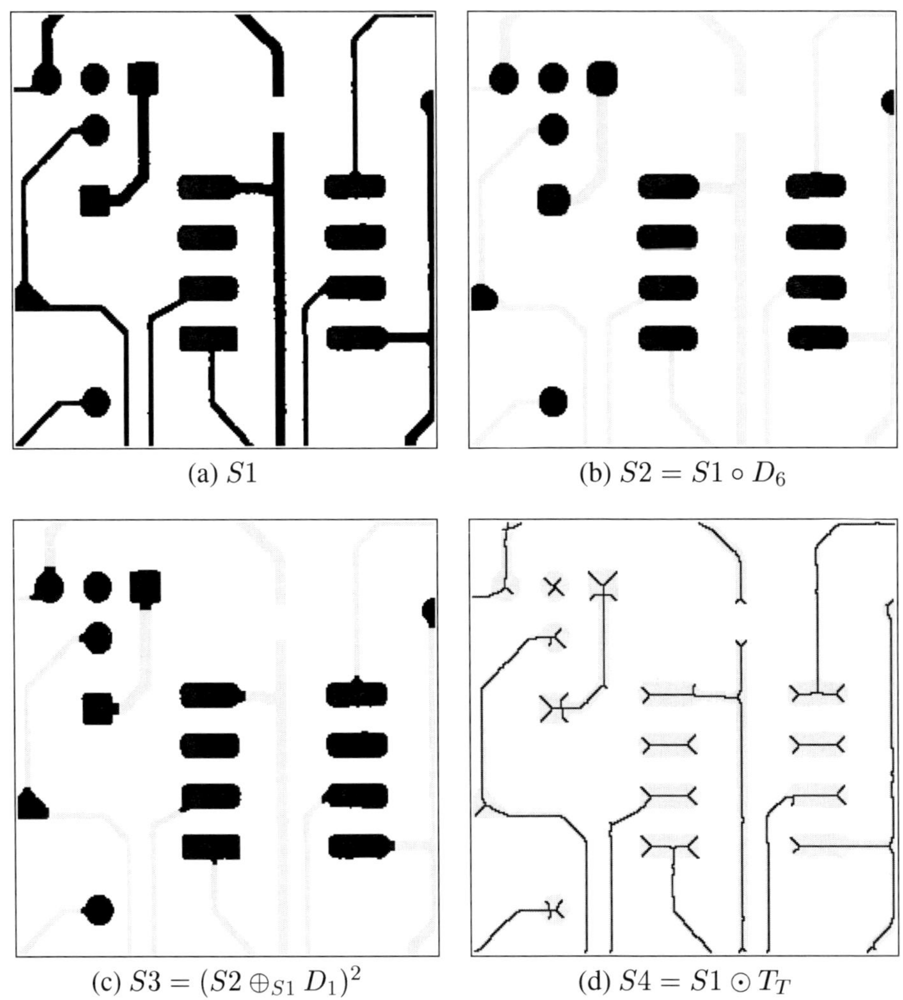

(a) $S1$

(b) $S2 = S1 \circ D_6$

(c) $S3 = (S2 \oplus_{S1} D_1)^2$

(d) $S4 = S1 \odot T_T$

Figure 4.10 Detection of open connection in printed circuit: (a) input image, (b) pad detection, (c) pad extraction, (d) thinning of input image.

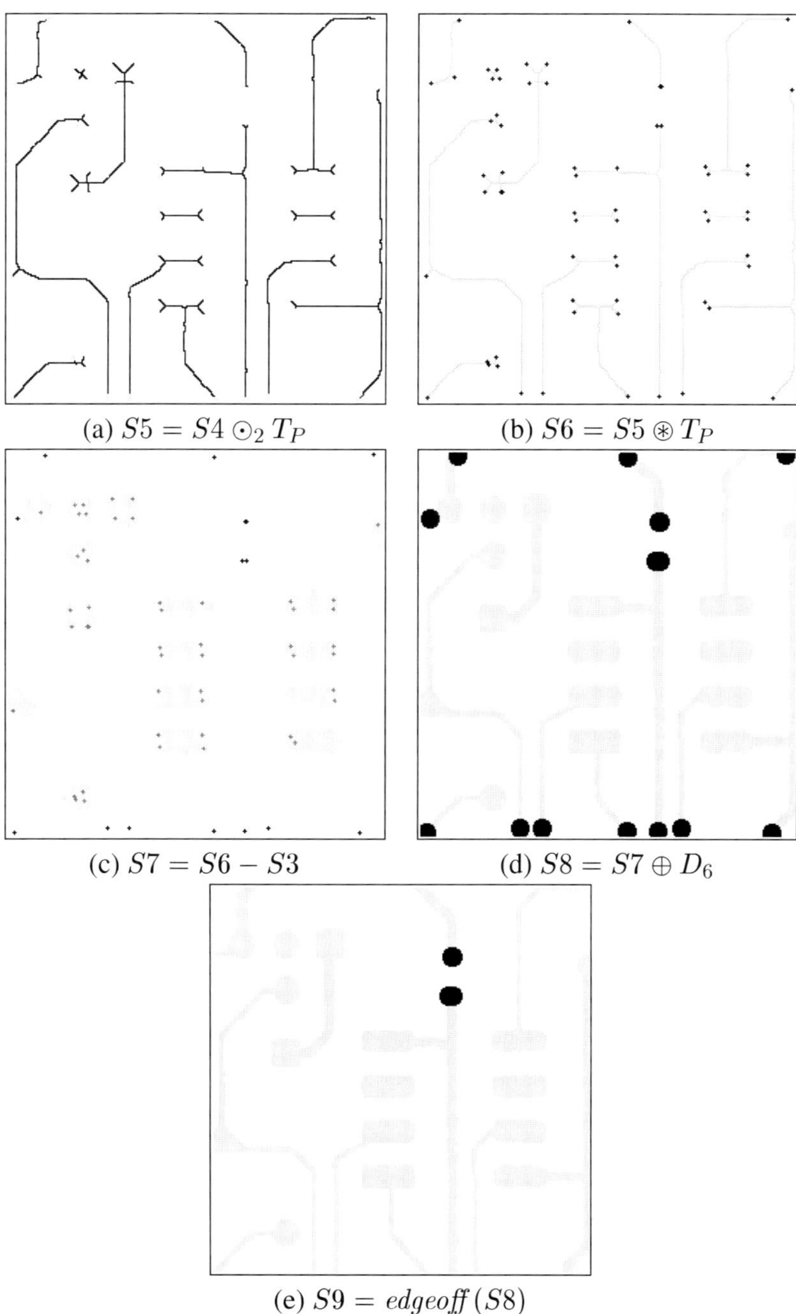

(a) $S5 = S4 \odot_2 T_P$

(b) $S6 = S5 \circledast T_P$

(c) $S7 = S6 - S3$

(d) $S8 = S7 \oplus D_6$

(e) $S9 = \mathit{edgeoff}\,(S8)$

Figure 4.11 Detection of open connection in printed circuit: (a) pruning of the skeleton, (b) endpoints, (c) endpoints not at the pads, (d) mark endpoints, (e) open connection detection, removed points near image frame.

name	description
`mmsupgen:`	`hit-or-miss transform [Eq. (4.1)]`
`mmse2hmt:`	`hit-or-miss template [Fig. 4.1(c)]`
`mmthin:`	`sequential thinning [Fig. 4.4]`
`mmhomothin:`	`compass templates for thinning (Fig. 4.5)`
`mmendpoints:`	`compass templates for pruning (Fig. 4.7)`

Figure 4.12 MT operators presented in this chapter.

6. Find an operator consisting of a union of hit-or-miss transforms that can perfectly reconstruct the 3-pixel-thick cross in the image

$$S = \begin{pmatrix} 0 & 1 & 1 & 1 & 1 & 0 & 0 \\ 0 & 0 & 1 & 1 & 1 & 0 & 0 \\ 1 & 1 & 1 & 1 & 0 & 1 & 1 \\ 1 & 1 & 1 & 1 & 1 & 1 & 1 \\ 1 & 0 & 1 & 1 & 1 & 1 & 1 \\ 0 & 0 & 1 & 1 & 1 & 0 & 1 \\ 0 & 0 & 1 & 1 & 1 & 0 & 0 \end{pmatrix}$$

4.6   Laboratory Experiments

Figure 4.12 lists the **MT** functions introduced in this chapter.

1. Detect the electronic components of the circuit diagram shown in Fig. 3.25 in the same way as the PCB image was decomposed in its parts in Sec. 2.8. Use hit-or-miss templates and openings.

2. Compare the use of morphological skeleton (`mmskelm`) and the sequential thinning (`mmthin`) on the "scissors" image.

3. Design an operator to mark a single middle point in each connected component using sequential thinning and pruning. Consider the case where the components may have holes.

References

1. D. Zhao and D. Daut. Morphological hit-or-miss transformation for shape recognition. *Journal of Visual Communication and Image Representation*, 2(3):230–243, 1991.

Chapter 5

Gray-Scale Morphology

Our attention now turns to gray-scale morphology, which means that the morphological operators act on real-valued functions defined on n-dimensional Euclidean space or the n-dimensional Cartesian grid. For signals, $n = 1$, and for images, $n = 2$. It is useful to look at a gray-scale image as a surface. Figure 5.1 shows a gray-scale image made of three Gaussian-shape peaks of different heights and variances. The image is depicted in four different graphical representations: (a) the pixel values mapped in gray scale: low values are dark and high values are bright gray tones; (b) the pixel values also mapped in gray scale but in a reverse order: low values are bright and high values are dark gray tones; (c) the same image but as a top-view shading surface; and (d) a mesh plot of the same surface.

Although our main concern is with image processing, we will develop the gray-scale theory for signals, our aim being to keep notation as simple as possible and to facilitate straightforward illustrative figures. Once the underlying gray-scale theory has been presented for signals, one need only recognize that by treating points on the line as spatial points in the plane the theory at once goes over into the imaging domain, the fundamental point being that the theory itself is independent of domain dimensionality.

5.1 Mathematical Preliminaries

A Euclidean signal is a function f defined on some domain, $D[f]$, which is a subset of the real line; a digital signal is defined on a domain within the set of integers. If the variable is denoted by x, then $f(x)$ denotes the functional value of the signal at x. There are fundamental theoretical differences between Euclidean and digital gray-scale morphology; however, to a great extent these are grist for the theoretician and we do not wish to make them central issues in the present text. Rather, we will develop gray-scale morphology from the Euclidean perspective, so as to provide strong geometric intuition, and we will do so making certain implicit assumptions that need not concern the non-theoretician. We are safe on these grounds because, for the digital case, which is the case one actually implements, these implicit assumptions always hold. To keep ourselves honest and to not mislead the person who might someday go deeply into mathematical morphology, we will point out key assumptions when we first make use of them.

Before defining gray-scale erosion, we need to provide counterparts for the binary building blocks of translation, subset, and union. As is typically done, we view a signal in terms of its graph, and the graph can be translated in two ways, horizontally or vertically. A horizontal translation, or shift, of a signal f to the right by x, to be called **translation** by x, is defined by $f_x(z) = f(z - x)$. A

91

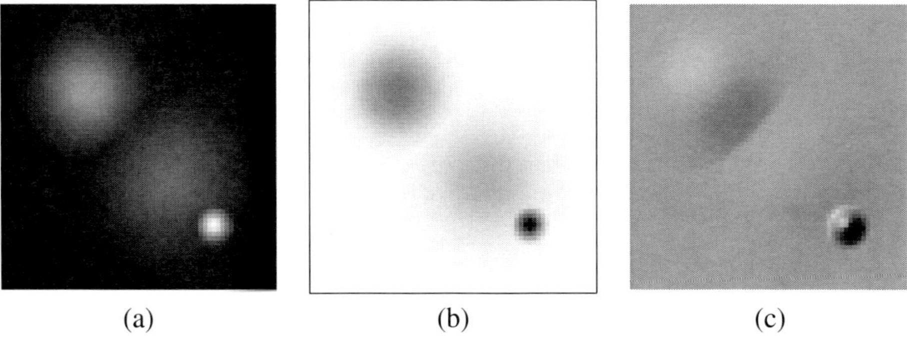

(a)   (b)   (c)

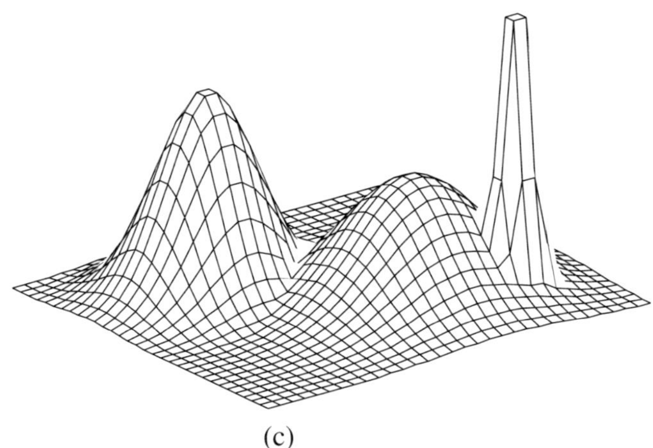

(c)

Figure 5.1 Four graphical representations of a gray-scale image: (a) gray-scale mapping: zero is dark and 255 is bright gray tone, (b) reverse gray-scale mapping: zero is bright and 255 is dark, (c) top-view shading surface, (d) surface mesh plot.

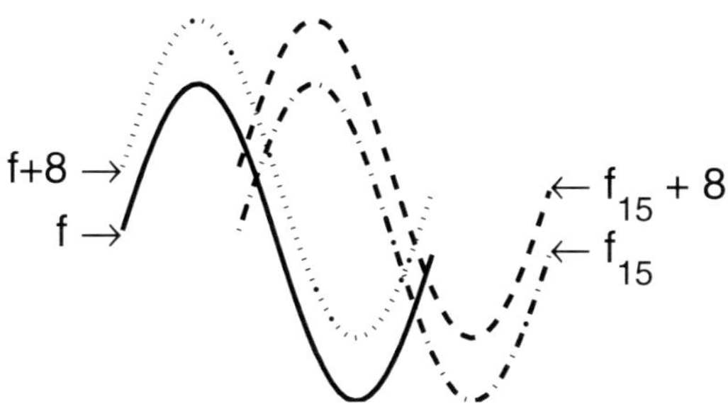

Figure 5.2 Signal translations.

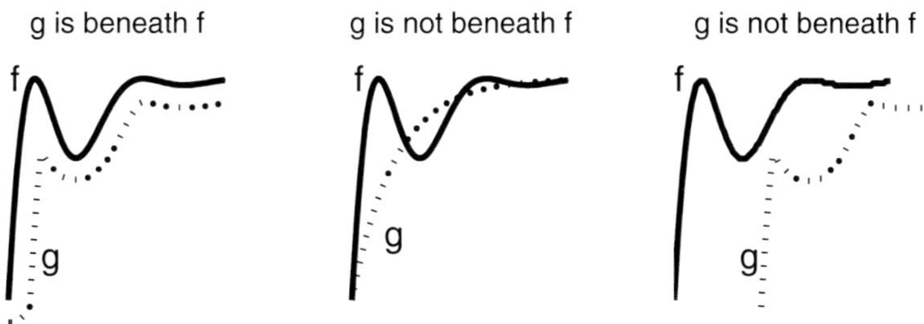

Figure 5.3 Signal ordering.

vertical translation of f by y, to be called an **offset** by y, is defined by $(f + y)(z) = f(z) + y$. When both translation and offsetting are applied together, we obtain a **morphological translation** $f_x + y$, given by

$$(f_x + y)(z) = f(z - x) + y. \tag{5.1}$$

Figure 5.2 illustrates a signal f, its translation by x, its offset by y, and the corresponding morphological translation.

In the binary setting the notion of subset provides an order relation between images; an appropriate analog is required for the gray scale. If g and f are signals with domains $D[g]$ and $D[f]$, respectively, we say that g is **beneath** f if (1) the domain of g is a subset of the domain of f, and (2) for any x in the domain of g, which must be the common domain of both, $g(x) \leq f(x)$. We write $g \leq f$. The following situations occur in Fig. 5.3: (a) $g \leq f$; (b) g is not beneath f because there is a point x in the domain of g at which $g(x) > f(x)$; and (c) g is not beneath f because the domain of g is not a subset of the domain of f.

The genesis of the order relation $\leq$ is in the manner in which we treat a function off its domain. In linear processing, where integration plays the key role, it is common to treat a signal or function as being identically zero off its domain, since such a convention maintains the integral. In morphological processing, an image is treated as negative infinity off its domain (frame). If the domain of g is not a subset of the domain of f, then there must be some point x in the set-theoretic

Figure 5.4 Maximum and minimum operations.

difference $D[g] - D[f]$, and at this point $g(x)$ is finite, whereas $f(x)$ is negative infinity. Hence, g is not beneath f.

Intersection and union play central roles in binary morphology. We now address their counterparts in gray-scale morphology, minimum and maximum. Given two signals f and g, we must define their minimum in a manner consistent with their being treated as negative infinity off their domains. Since negative infinity is less than any other value, at each point x, if either $f(x)$ or $g(x)$ is negative infinity, then their minimum will be negative infinity. Thus, we define the **minimum** of f and g in the following manner: if x lies in the intersection of the domains, $D[f] \cap D[g]$, then

$$(f \wedge g)(x) = \min\{f(x), g(x)\}; \qquad (5.2)$$

otherwise, x does not lie in the domain of $f \wedge g$. Actually, Eq. (5.2) gives the definition for any x, so long as we allow the value negative infinity.

Similar reasoning shows that the **maximum** $f \vee g$ is defined pointwise by

$$(f \vee g)(x) = \max\{f(x), g(x)\}, \qquad (5.3)$$

for x lying in $D[f] \cap D[g]$. It is defined by $(f \vee g)(x) = f(x)$ if x lies in the domain of f but not in the domain of g, and by $(f \vee g)(x) = g(x)$ if x lies in the domain of g but not in the domain of f. Finally, $f \vee g$ is not defined at x if x lies in neither domain, that is, if x is not a point of $D[f] \cup D[g]$. The full definition is in accord with the negative-infinity stipulation regarding points outside a signal's domain. For instance, if x lies in $D[f]$ but not $D[g]$, then $g(x) = -\infty$, so that taking the maximum of $f(x)$ and $g(x)$ yields $f(x)$. Figure 5.4 shows two signals f and g, their maximum, and their minimum.

The foregoing maximum and minimum considerations extend to more than two signals; however, if we were to consider an infinite collection of signals, for a full mathematical rigor we would have to replace the maximum and minimum by their continuous counterparts, supremum and infimum. Since our ultimate purpose is digital processing we will forego such an approach, merely accepting the fact that it is possible to make the necessary adjustments to the theory.

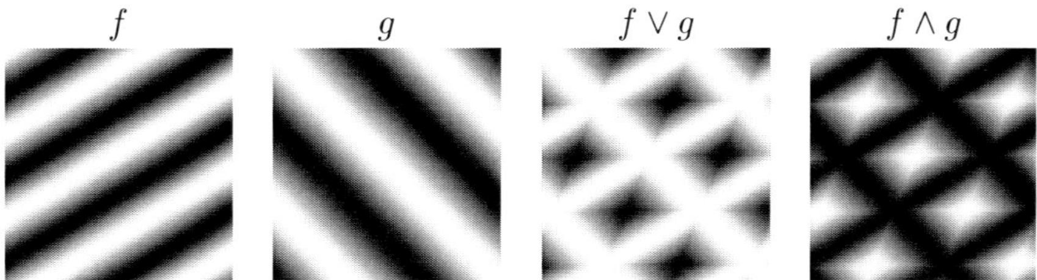

Figure 5.5 Maximum and minimum between two gray-scale images.

Figure 5.6 Signal reflection.

Figure 5.5 illustrates the maximum and minimum operations for gray-scale images. These operations create the maximum and minimum of their corresponding surfaces, respectively.

The gray-scale analog to complement of a set is the negation. The **negation** of f, denoted f^c, is given by $-f(x)$.

We need to consider one more operation: the reflection of a gray-scale signal. If h is a signal with domain $D[h]$, the **reflection** of h is defined by

$$\breve{h}(x) = h(-x). \tag{5.4}$$

As illustrated in Fig. 5.6, reflection is accomplished by reflecting the signal through the vertical axis.

5.2 Gray-Scale Erosion

Because erosion and dilation satisfy a number of algebraic identities, there are a number of equivalent ways of defining them. Since the genesis of morphology is fitting and gray-scale morphological processing is concerned with the topography of a signal's (image's) graph, we define the gray-scale operations directly in terms of fitting, as we did with the binary operations. The **erosion** of signal f by structuring element g (also a signal) is defined pointwise by

$$(f \ominus g)(x) = \max\{y : g_x + y \leq f\}. \tag{5.5}$$

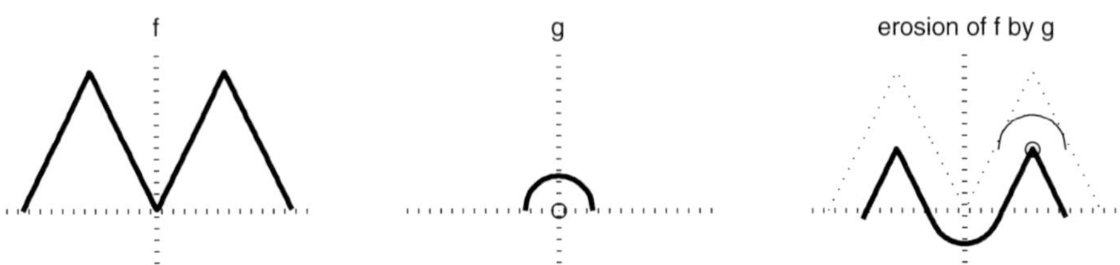

Figure 5.7 Signal erosion by a nonflat structuring element.

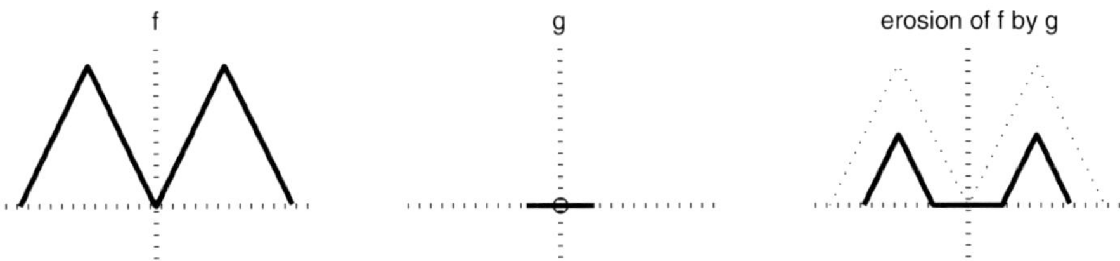

Figure 5.8 Signal erosion by a flat structuring element.

Geometrically speaking, to find the erosion of a signal by a structuring element at a point x, we slide the structuring element spatially so that its origin (which for signals is the Euclidean-plane origin relative to the structuring element) is located at x, and then we find the maximum amount we can offset ("push-up") the structuring element and have it be beneath the signal. Since the structuring element must be beneath the signal, the domain of the spatially translated structuring element must be a subset of the signal domain; otherwise the erosion is not defined (is $-\infty$) at the point.

Equation (5.5) is illustrated in Fig. 5.7. The effect is as if the semicircular structuring element were "rolled along" under the signal and the origin traced, there being the restriction that the element can never be translated so that it is not beneath the signal. As a disk filters a binary image from the inside, in the present circumstance the semicircular element filters the signal from beneath.

In Fig. 5.8, a flat structuring element is employed for the signal of Fig. 5.7. Note the filtering effect of the flat element. Flat structuring elements play an important role in many applications.

In Figs. 5.7 and 5.8, one can observe a fundamental property of gray-scale erosion relative to binary erosion: the domain of the gray-scale eroded signal equals the binary erosion of the signal domain by the structuring element domain.

Rather than finding the maximum of Eq. (5.5) by pushing up the structuring element, we can instead find the minimum difference between the signal values and the translated-structuring element values over the domain of the translated structuring element, since this minimum is the same as the maximum we can push up. This leads to the following formulation of erosion:

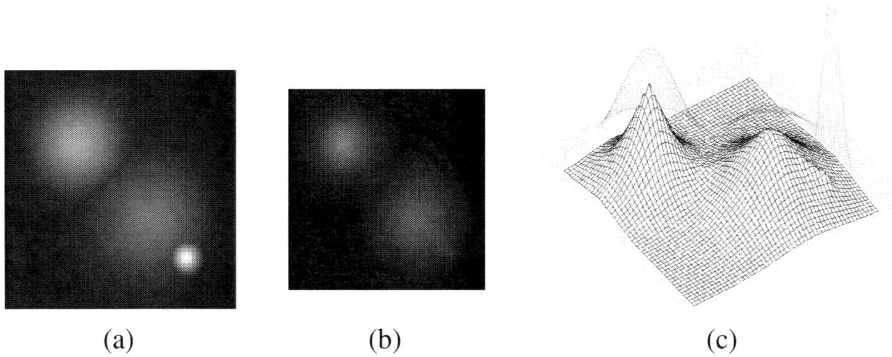

(a)   (b)   (c)

Figure 5.9 2D Gray-scale erosion by a semi-sphere structuring element: (a) input image, (b) eroded image, (c) surface mesh plot of input (light-gray) and eroded (dark-gray) images.

$$(f \ominus g)(x) = \min\{f(z) - (g_x)(z) : z \in D[g_x]\}, \tag{5.6}$$

where it is understood that the erosion is not defined at any point where the translated structuring element does not lie beneath the signal.

Figure 5.9 illustrates 2D gray-scale erosion by a semi-sphere structuring element. The erosion turned the image darker and removed entirely the bottom-right white dot as the structuring element cannot fit from below. Note the image domain of the eroded image is shrunk by the radius of the structuring element.

Turning to signals defined on the integers, we must differentiate two cases. First, we might consider **sampled signals**, these being defined on the integers but possessing real gray values. Second, we can consider **digital signals**, these being defined on integers and taking their gray values amongst the integers. It is the latter kind that we process digitally; in fact, for digital processing we are restricted to an integer gray-scale range between some bounds, such as between 0 and 255 (8 bits). In order to avoid digressions into numerical computation, we will presume our signals to be sampled over the integers, so that the gray range is the real line; however, our examples will always be strictly digital, so that fractions do not occur. Thus, a typical signal is of the form

$$f = \begin{pmatrix} -\infty & 0 & 2 & 1 & \mathbf{5} & 9 & 6 & 1 & 0 \end{pmatrix}, \tag{5.7}$$

where negative infinity $(-\infty)$ indicates that the signal is undefined at the point and the bold character indicates the origin position relative to the signal.

Suppose we wish to erode f by the structuring element

$$g = \begin{pmatrix} -\infty & -\infty & \mathbf{5} & 5 & 4 \end{pmatrix}. \tag{5.8}$$

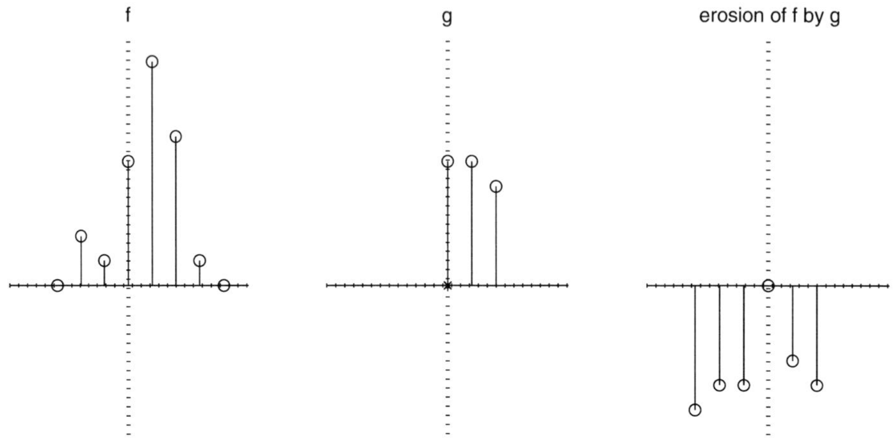

Figure 5.10 Digital signal erosion.

Translating g from left to the right, the first time it lies beneath f is when it is translated -3 units, so that we are considering the translation

$$g_{-3} = \begin{pmatrix} 5 & 5 & 4 & -\infty & -\infty & -\infty & -\infty \end{pmatrix}. \tag{5.9}$$

Applying Eq. (5.6),

$$(f \ominus g)(-3) = \min\{0 - 5, 2 - 5, 1 - 4\} = -5. \tag{5.10}$$

Successively translating g rightward and taking the minima yields the eroded signal

$$f \ominus g = \begin{pmatrix} -5 & -4 & -4 & \mathbf{0} & -3 & -4 & -\infty \end{pmatrix}. \tag{5.11}$$

The final point in the domain of the erosion is $x = 2$, because beyond that point the translated structuring element no longer lies beneath the signal. The signal, structuring element, and eroded signal are depicted in Fig. 5.10.

Rather than characterize erosion pointwise, there is a global Minkowski-subtraction formulation:

$$f \ominus g = \bigwedge_{x \in D[\breve{g}]} f_x - \breve{g}(x), \tag{5.12}$$

where $\breve{g}$ is the reflection of g through the vertical axis. To make this formulation useful we write it in an equivalent form by applying the definition of $\breve{g}$:

$$f \ominus g = \bigwedge_{x \in D[g]} f_{-x} - g(x). \tag{5.13}$$

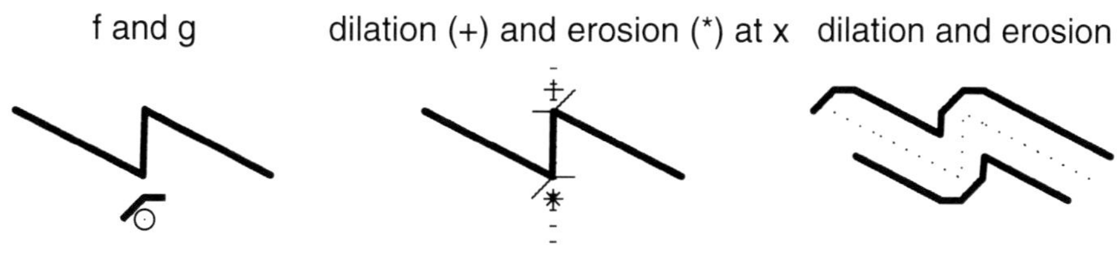

Figure 5.11 Gray-scale erosion and dilation fittings.

According to this formulation, for each point x in the domain of the structuring element g, the signal f is translated by $-x$ and $g(x)$ is subtracted from each translated-signal value. Having one signal for each point in the structuring element domain, we then take the pointwise minimum of these, keeping in mind the negative-infinity interpretation of values outside the domain.

Applying Eq. (5.13) to the signal of Eq. (5.7) and the structuring element of Eq. (5.8) yields three signals after translation and subtraction:

$$
\begin{array}{llllllllllll}
f_0 - 5 & = (& & -\infty & -5 & -3 & -4 & \mathbf{0} & 4 & 1 & -4 & -5 & & &), \\
f_{-1} - 5 & = (& & -5 & -3 & -4 & 0 & \mathbf{4} & 1 & -4 & -5 & -\infty & &), \\
f_{-2} - 4 & = (& -4 & -2 & -3 & 1 & 5 & \mathbf{2} & -3 & -4 & -\infty & -\infty & -\infty &).
\end{array}
$$
$$(5.14)$$

Taking the minimum down each "column" of the array in Eq. (5.14) yields the erosion of Eq. (5.11). Note the ease of computation with this approach.

5.3 Gray-Scale Dilation

As in the binary setting, dilation is defined in a dual manner to erosion. Rather than translating the structuring element and finding the maximum the translated element can be pushed up and still be beneath the signal, we instead take the negation of the reflection of the structuring element and find the minimum it needs to be pushed up to be above the signal when the signal is restricted to the domain of the translated structuring element. This last proviso is necessary because otherwise the domain of the signal would likely extend outside the domain of the translated reflected structuring element and the signal would never lie beneath the morphological translation of the reflected structuring element. Mathematically, the **dilation** of f by g is defined pointwise by

$$(f \oplus g)(x) = \min\{y : -\breve{g}_x + y \geq f\}, \qquad (5.15)$$

where the domain proviso applies. This "fitting" formulation of dilation, along with the dual formulation for erosion, is illustrated in Fig. 5.11 for an asymmetric structuring element.

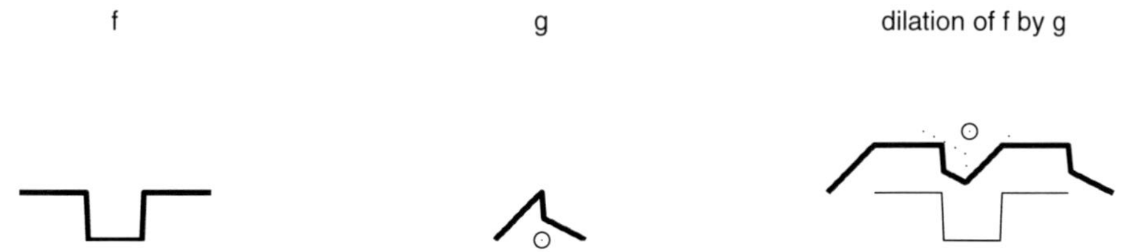

Figure 5.12 Gray-scale dilation: fitting from above with the negated and reflected structuring element, according to Eq. (5.15).

Another illustration of dilation is given in Fig. 5.12, where the conical structuring element is asymmetric with respect to the vertical axis. Note how the negation of the reflection of the structuring element is fit into the signal graph from the top. There is also a global Minkowski-addition formulation of dilation:

$$f \oplus g = \bigvee_{x \in D[g]} f_x + g(x). \tag{5.16}$$

Here, the signal is spatially translated by each point in the domain of the structuring element and then offset by the corresponding structuring element value. Like its counterpart for erosion, the Minkowski-addition formulation facilitates computation. Consider the signal

$$f = \left(\; 7 \quad 9 \quad 8 \quad \mathbf{3} \quad 8 \quad 9 \quad 9 \;\right), \tag{5.17}$$

and structuring element

$$g = \left(\; -3 \quad \mathbf{0} \quad -1 \;\right). \tag{5.18}$$

Equation (5.16) involves a pointwise maximum of three signals:

$$
\begin{aligned}
f_{-1} - 3 &= (\quad 4 \quad\;\; 6 \quad\;\; 5 \quad 0 \quad \mathbf{5} \quad 6 \quad 6 \quad -\infty \quad -\infty \quad), \\
f_0 - 0 &= (\;\; -\infty \quad 7 \quad\;\; 9 \quad 8 \quad \mathbf{3} \quad 8 \quad 9 \quad\;\; 9 \quad\;\; -\infty \quad), \\
f_1 - 1 &= (\;\; -\infty \quad -\infty \quad 6 \quad 8 \quad \mathbf{7} \quad 2 \quad 7 \quad\;\; 8 \quad\;\;\; 8 \quad).
\end{aligned}
\tag{5.19}
$$

Taking the maximum down each column of the array yields

$$f \oplus g = \left(\; 4 \quad 7 \quad 9 \quad 8 \quad \mathbf{7} \quad 8 \quad 9 \quad 9 \quad 8 \;\right). \tag{5.20}$$

Notice how the domain has been expanded, the domain of the dilation being the binary dilation of the domains, and notice how the amount of the drop-out at the origin has been lessened by fitting the "conical" element $-\breve{g}$ from the top.

As in the binary setting, dilation is both commutative and associative. By commutativity we can interchange the roles of signal and structuring element in Eq. (5.16) to obtain

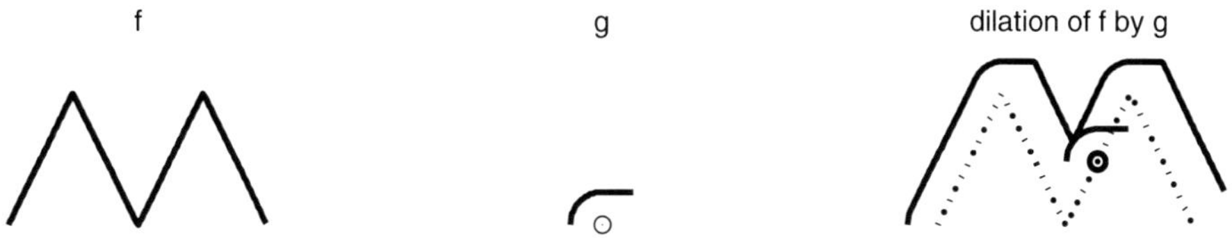

Figure 5.13 Signal dilation by a nonflat structuring element.

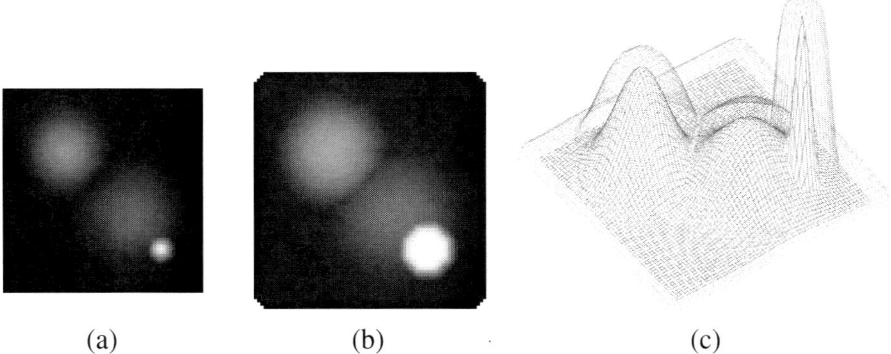

(a)　　　　　(b)　　　　　(c)

Figure 5.14 Two-dimensional gray-scale dilation by a semi-sphere structuring element: (a) input image, (b) dilated image, (c) surface mesh plot of input (dark-gray) and dilated (light-gray) images.

$$f \oplus g = \bigvee_{x \in D[f]} g_x + f(x). \tag{5.21}$$

Computationally, this reformulation is less practical, since instead of requiring a small number of translations—the number of points in the structuring element—it now requires a translation for each point in the signal. Nonetheless, it does provide an important geometric interpretation. As illustrated in Fig. 5.13, the dilation is accomplished by translating and offsetting the structuring element so that its origin lies on the signal graph, doing this for each point on the graph, and then taking the maximum over all such copies of the structuring element. This is analogous to the Minkowski-addition formulation of binary dilation, where the structuring element is translated to all points in the binary image and the union is then taken. Pointwise, Eq. (5.21) takes the form

$$(f \oplus g)(x) = \max\{g(x - z) + f(z) : z \in D[f]\}. \tag{5.22}$$

Figure 5.14 illustrates 2D gray-scale dilation by a semi-sphere structuring element. The dilation turns the image lighter and dilates the bottom-right white dot. Note the image domain of the dilated image is also dilated by the structuring element domain. This means that $D[f \oplus g] = D[f] \oplus D[g]$.

5.4 Algebraic Properties

Corresponding to the binary Minkowski algebra there is a gray-scale algebra. We have already mentioned commutativity and associativity. For the most part we leave discussion of this algebra to the literature; however, we mention some useful propositions. If the structuring element can be represented by the maximum of two signals, $g \vee h$, then

$$f \oplus (g \vee h) = (f \oplus g) \vee (f \oplus h), \tag{5.23}$$

$$f \ominus (g \vee h) = (f \ominus g) \wedge (f \ominus h). \tag{5.24}$$

If the structuring element can be represented by the dilation of two signals, $g \oplus h$, then

$$f \oplus (g \oplus h) = (f \oplus g) \oplus h, \tag{5.25}$$

$$f \ominus (g \oplus h) = (f \ominus g) \ominus h. \tag{5.26}$$

The gray-scale dilation is an algebraic dilation and the gray-scale erosion is an algebraic erosion, meaning that they commute with maximum and minimum respectively:

$$(f \vee g) \oplus h = (f \oplus h) \vee (g \oplus h), \tag{5.27}$$

$$(f \wedge g) \ominus h = (f \ominus h) \wedge (g \ominus h). \tag{5.28}$$

There is also a duality relation between erosion and dilation. It takes the form

$$f \oplus g = [(f^c) \ominus \breve{g}]^c. \tag{5.29}$$

However, here one must be careful. For the duality relation to hold in full, one must allow signal values to be positive infinity, which would be the negation of negative infinity. This means that negating a signal results in a signal that is infinity off its frame. In practice this is often not done; instead, when a signal is undefined $(-\infty)$, the negation is also taken to be undefined. Under these circumstances Eq. (5.29) only holds on the domain of $f \ominus \breve{g}$. If we explicitly pad the signal with $-\infty$, the duality relation holds in full.

For an illustration of duality under the convention that the signal is explicitly expanded with $-\infty$ and that the infinite negation is applied properly, $[\,(-\infty)^c = +\infty$ and $(+\infty)^c = -\infty]$, consider the extended signal f and structuring element g given by

$$f = \begin{pmatrix} -\infty & -\infty & -\infty & 2 & 3 & \mathbf{7} & 2 & 3 & -\infty & -\infty & -\infty \end{pmatrix}, \quad (5.30)$$

$$g = \begin{pmatrix} 0 & 1 & 2 \end{pmatrix}. \quad (5.31)$$

Then $\breve{g} = \begin{pmatrix} 2 & 1 & 0 \end{pmatrix}$ and

$$f \oplus g = \begin{pmatrix} -\infty & -\infty & -\infty & 2 & 3 & 7 & \mathbf{8} & 9 & 4 & 5 & -\infty & -\infty \end{pmatrix}, \quad (5.32)$$

$$f^c = \begin{pmatrix} \infty & \infty & \infty & -2 & -3 & \mathbf{-7} & -2 & -3 & \infty & \infty & \infty \end{pmatrix}, \quad (5.33)$$

$$(f^c) \ominus (\breve{g}) = \begin{pmatrix} \infty & \infty & \infty & -2 & -3 & -7 & \mathbf{-8} & -9 & -4 & -5 & \infty & \infty & \infty \end{pmatrix}. \quad (5.34)$$

We reiterate that a more rigorous mathematical approach in which we were to pay careful attention to the role of infinity is required. There are strict theoretical reasons for such a treatment of infinity, and these have to do with the appropriate abstract mathematical setting in which to place mathematical morphology, which happens to be a complete lattice. We will not pursue these theoretical questions and suggest that the interested reader pursue the literature.

5.5 Filter Properties

Discussion of filter properties for gray-scale morphology entails a reformulation of the basic filter properties. For binary erosion and dilation these are mainly concerned with translation invariance and increasing monotonicity, and they are formulated relative to image filters Ψ that take an input image A and yield an output image $\Psi(A)$. In the present setting we are concerned with gray-scale signals, so a filter Ψ will take an input signal f and yield an output signal $\Psi(f)$.

When treating gray-scale signals (images) morphologically, the topography of the graph as a subset of the plane (space) plays the central role. Consequently, translation invariance is defined relative to morphological translation, Ψ being **translation invariant** if

$$\Psi(f_x + y) = (\Psi(f))_x + y, \quad (5.35)$$

for any signal f and any parameters x and y. In words, morphologically translating and then filtering is equivalent to filtering and then morphologically translating. If

Ψ is translation invariant, then ipso facto it is both spatially translation invariant and offset invariant:

$$\Psi(f_x) = (\Psi(f))_x, \tag{5.36}$$

$$\Psi(f + y) = \Psi(f) + y, \tag{5.37}$$

the first relation following from Eq. (5.35) by letting $y = 0$, and the second following by letting $x = 0$. On the other hand, it is certainly possible for a filter to satisfy Eq. (5.36) and not Eq. (5.37), or vice versa.

Both gray-scale erosion and dilation are translation invariant:

$$(f_x + y) \ominus g = (f \ominus g)_x + y, \tag{5.38}$$

$$(f_x + y) \oplus g = (f \oplus g)_x + y. \tag{5.39}$$

Whereas Eqs. (5.38) and (5.39) refer to erosion and dilation as signal operators, so that it is translation of the input signal that is of concern, as we did in the binary case we might ask what happens if we translate the structuring element prior to filtering. The results in the gray-scale setting are analogous to those in the binary [Eqs. (1.24) and (1.25)]. Because dilation is commutative, Eq. (5.39) at once applies to structuring element translation; namely,

$$f \oplus (g_x + y) = (f \oplus g)_x + y. \tag{5.40}$$

Looking at the fitting characterization of erosion, we see that structuring element translation results in an opposite effect on the output; namely,

$$f \ominus (g_x + y) = (f \ominus g)_{-x} - y. \tag{5.41}$$

Turning to monotonicity, the key point is that a filter should preserve the underlying ordering of signals (images). Thus, Ψ is said to be **increasing** if whenever f is beneath h, then $\Psi(f)$ is beneath $\Psi(h) : f \leq h$ implies $\Psi(f) \leq \Psi(h)$. Erosion and dilation are increasing filters: $f \leq h$ implies $f \ominus g \leq h \ominus g$ and $f \oplus g \leq h \oplus g$.

From the perspective of structuring elements, the order relation is preserved for dilation and inverted for erosion. If $g \leq k$, then $f \oplus g \leq f \oplus k$ and $f \ominus g \geq f \ominus k$. The latter relation is very important to morphological filter theory. Its genesis is straightforward. If g is beneath k, then g can be pushed up at least as much as k and still lie beneath f. Consequently, erosion by g is at least as great as erosion by k.

As in the binary case there is a duality theory for gray-scale operations. The **dual** of filter Ψ is defined by $\Psi^*(f) = \Psi(f^c)^c$. According to Eq. (5.29), gray-scale erosion and dilation are dual, so long as we treat the negation of negative infinity as positive infinity and vice versa. Such careful handling of infinity is necessary if

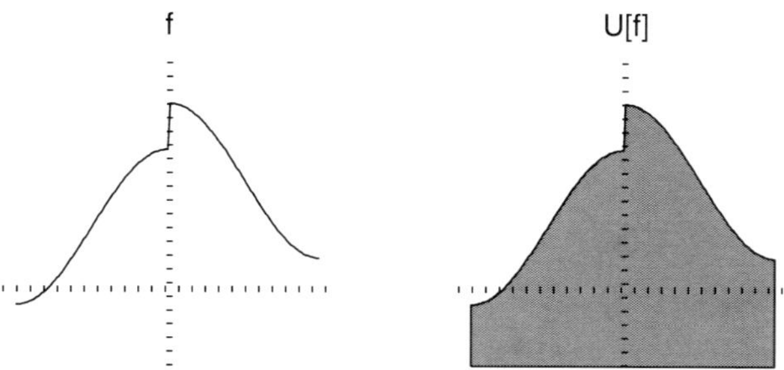

Figure 5.15 Umbra of a signal.

gray-scale duality is to be meaningful. In computer implementations there is often no facility for handling the conundrums when dealing with negations and infinity, so, as mentioned when we discussed erosion-dilation duality, one must proceed with care.

A final point: properties such as translation invariance and increasing monotonicity are not solely the concern of morphological filtering; rather, they are general properties and play important roles in both linear and nonlinear filtering.

5.6 Umbra Transform

As evidenced thus far, there is a close relationship between gray-scale and binary morphology. This relationship is formalized by the umbra transform. While umbra techniques are less-used today, it should be recognized that the umbra played a key role in the early formulation of gray-scale morphology, and it therefore appears frequently in the literature.

For any signal f, let $G[f]$ denote the usual **graph** of f,

$$G[f] = \{(x, f(x)) : x \in D[f]\}, \tag{5.42}$$

which is a subset of the plane. The **umbra** of f consists of all points in the plane lying beneath the graph of f, including the points on the graph itself. It is denoted $U[f]$ and is formally defined by

$$U[f] = \{(x, y) : x \in D[f] \text{ and } y \leq f(x)\}. \tag{5.43}$$

The umbra is illustrated in Fig. 5.15.

Given a set A in the plane, we wish to formalize the notion of its **surface**. For instance, the surface of a signal's umbra is the graph of the signal. If we assume set A is topologically closed, which means it contains its boundary, then defining a surface is straightforward. Rigorously, we define the surface of set A by

$$S[A] = \{(x, y) \in A : y \geq z \text{ for any } (x, z) \in A\}. \tag{5.44}$$

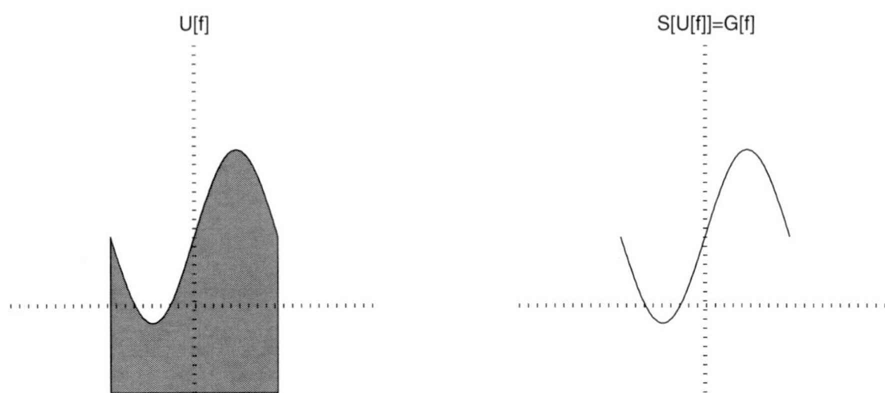

Figure 5.16 Surface operator as the inverse of the umbra transform.

The surface operator "peels off the top" of a set. We can consider $S[A]$ as either a set in the plane or as defining the graph of a signal.

For any signal f, the surface of its umbra is its graph, namely,

$$S[U[f]] = G[f]. \tag{5.45}$$

Thus, as indicated in Fig. 5.16, the surface operator acts as an inverse operator for the umbra transform applied to signals.

The importance of the umbra in mathematical morphology is that it provides a mechanism for expressing gray-scale operations in terms of binary operations. This correspondence facilitates intuition and can be theoretically useful. Although use of the umbra is not logically necessary for the development of gray-scale morphology, it has played a key role in the historical evolution of the subject and still is employed to gain insight into fundamental relationships.

Regarding erosion and dilation, there are two fundamental umbra propositions relating binary and gray-scale morphology. For erosion,

$$f \ominus g = S[U[f] \ominus U[g]], \tag{5.46}$$

so that gray-scale erosion can be performed by taking the umbrae of the signal and structuring element, performing a binary erosion using these umbrae, and then taking the surface of the result. For dilation,

$$f \oplus g = S[U[f] \oplus U[g]], \tag{5.47}$$

so that gray-scale dilation can be performed by dilating the respective umbrae and then taking the surface. The erosion and dilation by the umbra transform is illustrated in Fig. 5.17.

Because we have only defined the surface operator for topologically closed sets, a mathematical question arises as to whether Eqs. (5.46) and (5.47) are restricted in

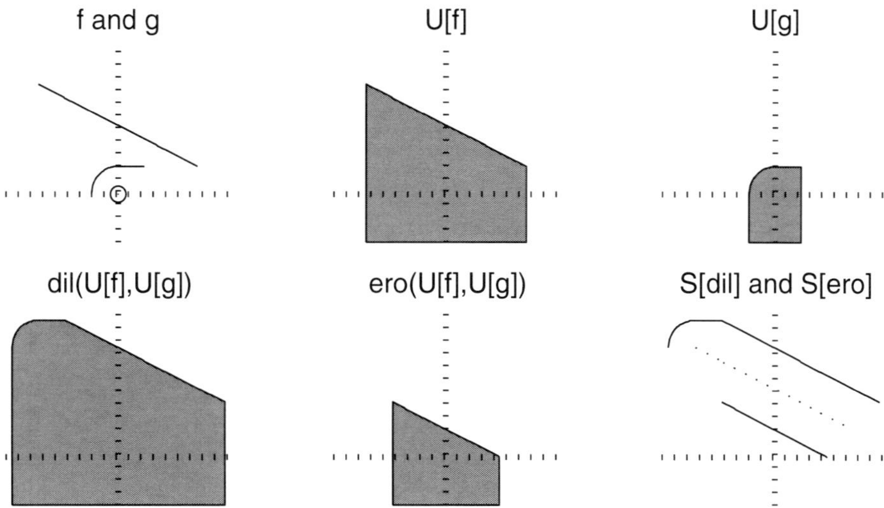

Figure 5.17 Dilation and erosion by means of the umbra transform.

any way. In fact, they are not. Although we have restricted our definition to avoid mathematical difficulties, even if the dilation of the two umbrae does not contain its boundary, the surface operator can still be given meaning by defining it in a limiting fashion, and the proposition still holds.

Why, then, are we making such a fuss about the matter? Simply because we wish to avoid some historical confusion in the literature regarding the use of umbrae. Put as simply as possible, it is quite apparent that, for any signal f, $U[S[U[f]]] = U[f]$; however, this relationship does not necessarily hold when an arbitrary set A is used in place of f, unless we mantain the assumption that A is topologically closed. It is known that if $U[f]$ and $U[g]$ are topologically closed umbrae, then so too is their dilation. Thus, we can apply the umbra transform to both sides of Eq. (5.47) to obtain

$$U[f \oplus g] = U[f] \oplus U[g]. \tag{5.48}$$

But are we assured that the umbrae of f and g are both topologically closed? In fact, we are not, unless some restriction is placed on the kind of signals allowed. Because this matter is of concern to Euclidean morphology but not to digital processing, we will content ourselves with Eqs. (5.46) and (5.47) and leave the matter at that.

5.7 Flat Structuring Elements

Historically, flat structuring elements have played a key role in gray-scale morphology, where by a flat element we mean one that is constant over its domain. Because of the translation property of Eq. (5.41), when dealing with a flat structuring element we might as well assume that it is zero on its domain, because if it is not we can always offset it, operate, and then re-offset in the opposite direction. Thus, the class of flat structuring elements can be viewed as the class of structuring elements defined by their domains and zero thereon. But, this class is identical to the set of subsets of the line (plane for images). Consequently, we may consider erosion or dilation by a flat structuring element as erosion or dilation of a signal by a set. Hence, if g is zero on its domain D, it is common to write $f \ominus D$ and $f \oplus D$ to denote the erosion and dilation of f by g, respectively. Throughout the remainder of the present section, when we refer to operation by a flat structuring element we implicitly mean it is zero on its domain, so that we do not differentiate between erosion by a flat structuring element and erosion by a set.

If we apply Eq. (5.6) to erosion by set D, it becomes

$$(f \ominus D)(x) = \min\{f(z) : z \in D_x\}, \tag{5.49}$$

so that $f \ominus D$ is simply a moving-minimum filter over the window D.

Correspondingly, applying the pointwise form of Eq. (5.15) to dilation by a set D yields

$$(f \oplus D)(x) = \max\{f(z) : z \in D_x\}, \tag{5.50}$$

so that $f \oplus D$ is a moving-maximum filter over the window D. Thus, erosion and dilation by flat structuring elements are simply special cases of order-statistic filters. Of course, in a real sense, such filters are not "true" gray-scale morphological filters, since there is no gray-scale variation in the probes.

An example of gray-scale flat erosion by a disk on a gray-scale image is shown in Fig. 5.18. The two images on the left, input and eroded, are represented in gray-scale shades and the two on the right are the same images represented by their top-view surfaces. Note how well the term "erosion" applies to this illustration. The eroded surface appears as being created by a pantograph engraving machine equipped with a flat disk milling cutter. The pantograph is guided to follow the original surface while shaping the eroded surface using the flat disk milling cutter.

The binary nature of erosion and dilation of signals by sets can be seen in another way. The **threshold decomposition** of a gray-scale signal f is the collection of all the **threshold sets** $X_t(f)$ at gray level t:

$$X_t(f) = \{z : f(z) \geq t\}. \tag{5.51}$$

The signal can be characterized uniquely by its threshold decomposition collection. The signal can be recovered from its threshold sets by **stack reconstruction**:

$$f(x) = \max\{t : x \in X_t(f)\}. \tag{5.52}$$

For any signal, as t increases, $X_t(f)$ decreases; i.e., if $t_1 \geq t_2$, then $X_{t_1}(f) \subset X_{t_2}(f)$. This is known as the **stack property**.

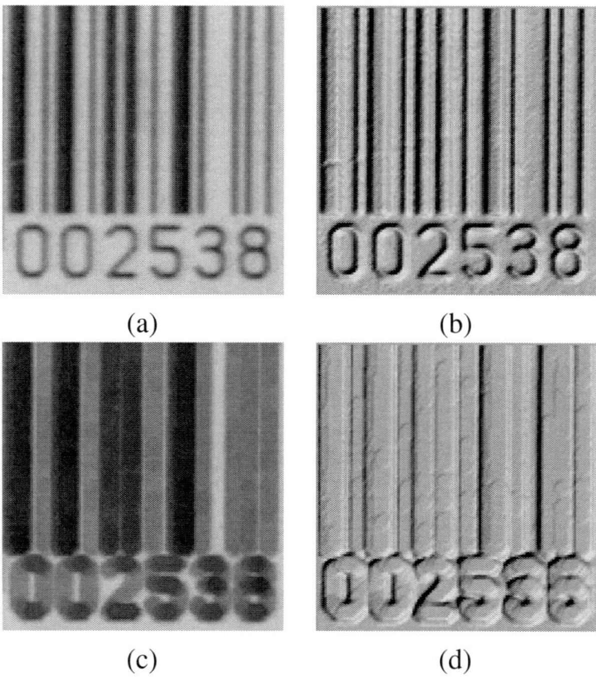

(a)   (b)

(c)   (d)

Figure 5.18 Illustration of gray-scale erosion by a flat disk. (a) Input image, (b) surface view of the input image, (c) erosion by Euclidean disk, (d) surface view of the eroded image.

To erode f by a set D we desire at each x the minimum function value over the translation D_x, this value being $(f \ominus D)(x)$. By the definition of a moving minimum, this value is greater than or equal to t if and only if $f(z) \geq t$ for any $z \in D_x$. Hence, $(f \ominus D)(x) \geq t$ if and only if D_x is a subset of the slice $X_t(f)$, which means that $(f \ominus D)(x) \geq t$ if and only if x lies in the binary erosion $X_t(f) \ominus D$. Consequently, the set erosion $f \ominus D$ can be found by means of the binary erosions $X_t(f) \ominus D$ using the stack reconstruction:

$$(f \ominus D)(x) = \max\{t : x \in X_t(f) \ominus D\}. \tag{5.53}$$

In fact, Eq. (5.53) is simply a restatement of the classical umbra formulation of Eq. (5.46) for flat structuring elements. To see this, suppose g is flat with domain D. Then the umbra $U[g]$ is a column and a pair (x, y) lies in the binary erosion $U[f] \ominus U[g]$ if and only if the translation $D_{(x,y)}$ fits inside $U[f]$, which in turn means that $x \in X_y(f) \ominus D$. Since S in Eq. (5.46) means to take the maximum over all such y, Eqs. (5.46) and 5.53 are identical (for flat structuring elements).

Figure 5.19 illustrates the discrete gray-scale erosion by means of threshold decomposition. The input signal is $(3 \ -\infty \ 5\ 4\ 5\ 6\ 3\ 1\ 4)$ and the structuring element is $(0\ \mathbf{0}\ 0)$. The double arrows show the gray-scale operation and the single arrows show the same operation using (1) threshold decomposition, (2) binary erosions,

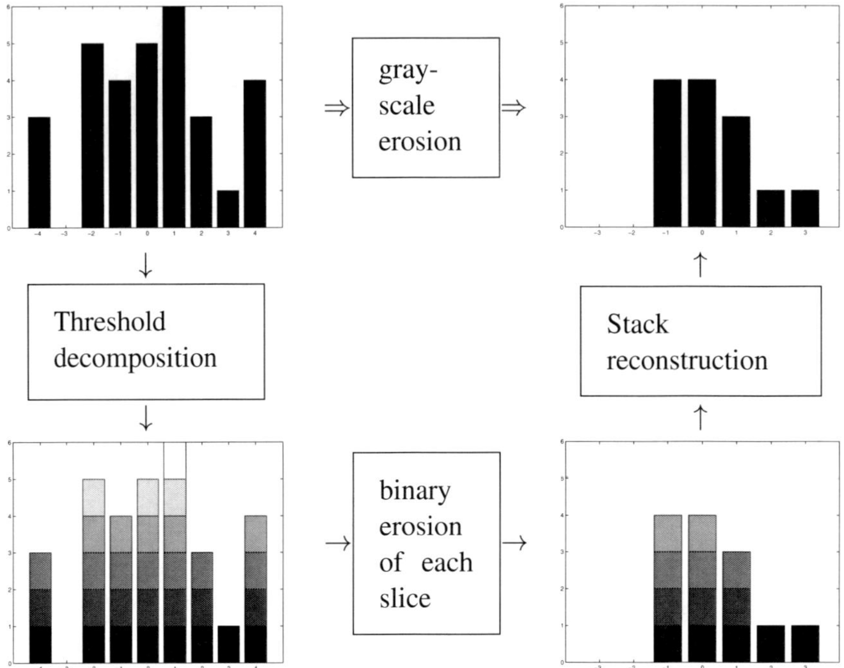

Figure 5.19 Illustration of gray-scale flat erosion by means of the threshold decomposition.

and (3) stack reconstruction.

The filters that can be implemented by threshold decomposition are called **stack filters**. A stack filter can be built from any binary filter as long as it does not break the stack property required for the stack reconstruction. Any monotonically increasing binary filter satisfies this condition. Dilation and erosion by flat structuring elements are in this sense stack filters. So too is the median filter.

Figure 5.20 illustrates the gray-scale flat dilation by a circular disk. At the right of the gray-scale images (original and dilated), there are three threshold sets, at gray levels 50, 100, and 200, respectively. A practical characteristic of a stack filter is that it stores all results of the filtering of the input thresholded images.

5.8 Gray-Scale Morphology for Discrete Images

To this point we have stayed with gray-scale signals to keep notation under control and to be able to provide understandable illustrations; we now provide some examples to demonstrate the computations necessary for digital processing. We employ bound matrix notation, with entries now being gray values. As with digital signals, the minus infinity appears at points where the image is undefined and also outside the bound matrix.

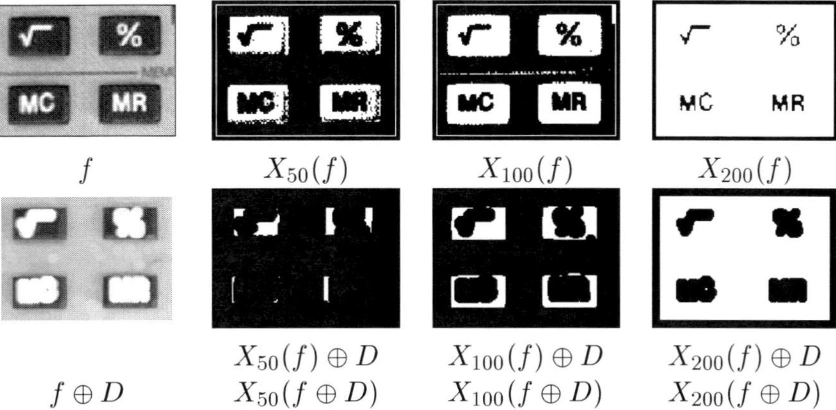

$$f \qquad X_{50}(f) \qquad X_{100}(f) \qquad X_{200}(f)$$

$$
\begin{array}{cccc}
& X_{50}(f) \oplus D & X_{100}(f) \oplus D & X_{200}(f) \oplus D \\
f \oplus D & X_{50}(f \oplus D) & X_{100}(f \oplus D) & X_{200}(f \oplus D)
\end{array}
$$

Figure 5.20 Illustration of gray-scale flat dilation by means of the threshold decomposition.

Consider the image

$$f = \begin{pmatrix} 0 & 2 & 2 & 2 & 1 \\ 1 & 2 & 6 & 2 & 1 \\ 0 & 6 & \mathbf{7} & 2 & 1 \\ 1 & 1 & 6 & 1 & -\infty \\ 1 & 0 & 2 & 2 & 1 \end{pmatrix}, \tag{5.54}$$

and structuring element

$$g = \begin{pmatrix} -\infty & 0 & -1 \\ -\infty & \mathbf{3} & 4 \\ -\infty & -\infty & -\infty \end{pmatrix}. \tag{5.55}$$

The domain of the structuring element, using the $(row, column)$ matrix notation, is

$$D[g] = \{(-1,0),(0,0),(-1,1),(0,1)\}; \tag{5.56}$$

and for the points of the domain, the corresponding gray values of the structuring element are $0, 3, -1$, and 4, respectively. Thus, according to the Minkowski-addition formulation, Eq. (5.16), dilation of f by g is given by

$$f \oplus g = \max\{f_{(-1,0)} + 0, f_{(0,0)} + 3, f_{(-1,1)} - 1, f_{(0,1)} + 4\}, \tag{5.57}$$

where $f_{(i,j)}$ is the spatial translation of the image f along the vector (i,j). The four images composing the maximum and the resulting dilation are given in Fig. 5.21, where boldface values indicate origin positions.

$$f_{(-1,0)} + 0 = \begin{pmatrix} 0 & 2 & 2 & 2 & 1 \\ 1 & 2 & 6 & 2 & 1 \\ 0 & 6 & 7 & 2 & 1 \\ 1 & 1 & \mathbf{6} & 1 & -\infty \\ 1 & 0 & 2 & 2 & 1 \\ -\infty & -\infty & -\infty & -\infty & -\infty \\ -\infty & -\infty & -\infty & -\infty & -\infty \end{pmatrix}$$

$$f_{(0,0)} + 3 = \begin{pmatrix} 3 & 5 & 5 & 5 & 4 \\ 4 & 5 & 9 & 5 & 4 \\ 3 & 9 & \mathbf{10} & 5 & 4 \\ 4 & 4 & 9 & 4 & -\infty \\ 4 & 3 & 5 & 5 & 4 \end{pmatrix}$$

$$f_{(-1,1)} - 1 = \begin{pmatrix} -\infty & -\infty & -1 & 1 & 1 & 1 & 0 \\ -\infty & -\infty & 0 & 1 & 5 & 1 & 0 \\ -\infty & -\infty & -1 & 5 & 6 & 1 & 0 \\ -\infty & -\infty & 0 & \mathbf{0} & 5 & 0 & -\infty \\ -\infty & -\infty & 0 & -1 & 1 & 1 & 0 \\ -\infty & -\infty & -\infty & -\infty & -\infty & -\infty & -\infty \\ -\infty & -\infty & -\infty & -\infty & -\infty & -\infty & -\infty \end{pmatrix}$$

$$f_{(0,1)} + 4 = \begin{pmatrix} -\infty & -\infty & 4 & 6 & 6 & 6 & 5 \\ -\infty & -\infty & 5 & 6 & 10 & 6 & 5 \\ -\infty & -\infty & 4 & \mathbf{10} & 11 & 6 & 5 \\ -\infty & -\infty & 5 & 5 & 10 & 5 & -\infty \\ -\infty & -\infty & 5 & 4 & 6 & 6 & 5 \end{pmatrix}$$

$$f \oplus g = \begin{pmatrix} -\infty & 0 & 2 & 2 & 2 & 1 & 0 \\ -\infty & 3 & 5 & 6 & 6 & 6 & 5 \\ -\infty & 4 & 6 & 9 & 10 & 6 & 5 \\ -\infty & 3 & 9 & \mathbf{10} & 11 & 6 & 5 \\ -\infty & 4 & 5 & 9 & 10 & 5 & 0 \\ -\infty & 4 & 5 & 5 & 6 & 6 & 5 \\ -\infty & -\infty & -\infty & -\infty & -\infty & -\infty & -\infty \end{pmatrix}$$

Figure 5.21 Image dilation.

To erode f by g according to the Minkowski-subtraction formulation of Eq. (5.13), the erosion is given by

$$f \ominus g = \min\{f_{(1,0)} - 0, f_{(0,0)} - 3, f_{(1,-1)} + 1, f_{(0,-1)} - 4\}, \tag{5.58}$$

and is computed in Fig. 5.22.

5.9 Gray-Scale Morphology for Discrete Bounded Signals

As we did in the binary case, we have covered in detail the gray-scale morphology as it applies to unbounded domain and range. It is in this venue that the theory and intuition of morphological image processing are understood. Nevertheless, when dealing with real-life discrete images it is convenient to have the input and output images with the same domain (size). For this we need to introduce the bounded gray-scale operations. For these, the digital signal has a limited domain $\{0, 1, 2, \ldots, n\}$ and is restricted to an integer gray-scale range between the bounds k_1 and k_2. In practical implementation these bounds often depend on the pixel data type. For unsigned 8-bit images, the range is from $k_1 = 0$ to $k_2 = 255$, for unsigned 16-bit images the range is from $k_1 = 0$ to $k_2 = 65535$, and for signed 32-bit images the range is from $k_1 = -(2^{31} - 1)$ to $k_2 = 2^{31} - 1$. In the examples shown in this section, we will use smaller ranges to facilitate the illustration.

The **negation** applied to bounded signals between k_1 and k_2 is given by

$$f^{\dot{c}}(x) = k_1 + k_2 - f(x). \tag{5.59}$$

When $k_1 = -k_2$, the negation is $f^{\dot{c}} = -f$, which is consistent with the negation seen so far. For signed 32-bit images the range is chosen such that $k_1 = -k_2$. For unsigned-byte gray-scale images, the negation is given by $f^{\dot{c}} = 255 - f(x)$.

The **addition** and **subtraction** of bounded signals f and g, both limited between k_1 and k_2, is given by

$$(f +^* g)(z) = \begin{cases} k_1 & \text{if } f(z) + g(z) < k_1, \\ f(z) + g(z) & \text{if } k_1 \leq f(z) + g(z) \leq k_2, \\ k_2 & \text{if } f(z) + g(z) > k_2. \end{cases} \tag{5.60}$$

and

$$(f -^* g)(z) = \begin{cases} k_1 & \text{if } f(z) - g(z) < k_1, \\ f(z) - g(z) & \text{if } k_1 \leq f(z) - g(z) \leq k_2, \\ k_2 & \text{if } f(z) - g(z) > k_2. \end{cases} \tag{5.61}$$

$$f_{(1,0)} - 0 = \begin{pmatrix} -\infty & -\infty & -\infty & -\infty & -\infty \\ -\infty & -\infty & -\infty & -\infty & -\infty \\ 0 & 2 & 2 & 2 & 1 \\ 1 & 2 & \mathbf{6} & 2 & 1 \\ 0 & 6 & 7 & 2 & 1 \\ 1 & 1 & 6 & 1 & -\infty \\ 1 & 0 & 2 & 2 & 1 \end{pmatrix}$$

$$f_{(0,0)} - 3 = \begin{pmatrix} -3 & -1 & -1 & -1 & -2 \\ -2 & -1 & 3 & -1 & -2 \\ -3 & 3 & \mathbf{4} & -1 & -2 \\ -2 & -2 & 3 & -2 & -\infty \\ -2 & -3 & -1 & -1 & -2 \end{pmatrix}$$

$$f_{(1,-1)} + 1 = \begin{pmatrix} -\infty & -\infty & -\infty & -\infty & -\infty & -\infty & -\infty \\ -\infty & -\infty & -\infty & -\infty & -\infty & -\infty & -\infty \\ 1 & 3 & 3 & 3 & 2 & -\infty & -\infty \\ 2 & 3 & 7 & \mathbf{3} & 2 & -\infty & -\infty \\ 1 & 7 & 8 & 3 & 2 & -\infty & -\infty \\ 2 & 2 & 7 & 2 & -\infty & -\infty & -\infty \\ 2 & 1 & 3 & 3 & 2 & -\infty & -\infty \end{pmatrix}$$

$$f_{(0,-1)} - 4 = \begin{pmatrix} -4 & -2 & -2 & -2 & -3 & -\infty & -\infty \\ -3 & -2 & 2 & -2 & -3 & -\infty & -\infty \\ -4 & 2 & 3 & \mathbf{-2} & -3 & -\infty & -\infty \\ -3 & -3 & 2 & -3 & -\infty & -\infty & -\infty \\ -3 & -4 & -2 & -2 & -3 & -\infty & -\infty \end{pmatrix}$$

$$f \ominus g = \begin{pmatrix} -\infty & -\infty & -\infty & -\infty & -\infty \\ -2 & -1 & -2 & -3 & -\infty \\ -3 & 2 & \mathbf{-2} & -3 & -\infty \\ -3 & -2 & -3 & -\infty & -\infty \\ -4 & -3 & -2 & -\infty & -\infty \end{pmatrix}$$

Figure 5.22 Image erosion.

These operations are used in the **MT**, where they are `mmaddm` and `mmsubm`, respectively. They are important for digital implementation, so that no overflow or underflow occurs.

We will introduce the bounded gray-scale morphology using the umbra transform for illustrative purposes. It will be restricted to the limited domain and limited range.

The **bounded gray-scale erosion** of a bounded signal f and structuring element h is defined by

$$(f \dot{\ominus} h)(x) = \min\{f(z) \dot{-} h_x(z) : z \in D[h_x] \cap D[f]\}. \tag{5.62}$$

There are three main differences between this definition and Eq. (5.6) defining the ordinary gray-scale erosion: (i) for each point x, z is scanned in the domain of the translated structuring element, but restricted to the domain of f; (ii) the translation of the structuring element h_x is the one described in Sec. 5.1; and (iii) the subtraction ($\dot{-}$) is defined differently from normal subtraction, namely,

$$t \dot{-} v = \begin{cases} k_1 & \text{if } t < k_2 \text{ and } t - v \leq k_1, \\ t - v & \text{if } t < k_2 \text{ and } k_1 \leq t - v \leq k_2, \\ k_2 & \text{if } t < k_2 \text{ and } t - v > k_2, \\ k_2 & \text{if } t = k_2, \end{cases} \tag{5.63}$$

where k_1 and k_2 are the limits of the finite range of t. One can observe that this rule simulates the behavior of the subtraction when the signal range is from $-\infty$ and ∞: $-\infty - v = -\infty$, $+\infty - v = +\infty$.

The **bounded gray-scale dilation** of a bounded signal f and structuring element h is defined by

$$(f \dot{\oplus} h)(x) = \max\{f(z) \dot{+} h_x(z) : z \in D[h_x] \cap D[f]\}, \tag{5.64}$$

where the addition $\dot{+}$ is defined by

$$t \dot{+} v = \begin{cases} k_1 & \text{if } t = k_1, \\ k_1 & \text{if } t > k_1 \text{ and } t + v \leq k_1, \\ t + v & \text{if } t > k_1 \text{ and } k_1 \leq t + v \leq k_2, \\ k_2 & \text{if } t > k_1 \text{ and } t + v \geq k_2. \end{cases} \tag{5.65}$$

Figure 5.23 illustrates bounded erosion and dilation where the domain of the signal f is over the integers 0 to 300 and its values range from 0 to 255. Note that the erosion did not make the signal negative nor decrease the wide portion touching 255. In the bounded dilation the signal is not expanded from 0 in the regions around coordinate 100. A key advantage of bounded operations is that there is no overflow or underflow on the pixel values. The domain and image of the output signal are exactly the same as for the input signal.

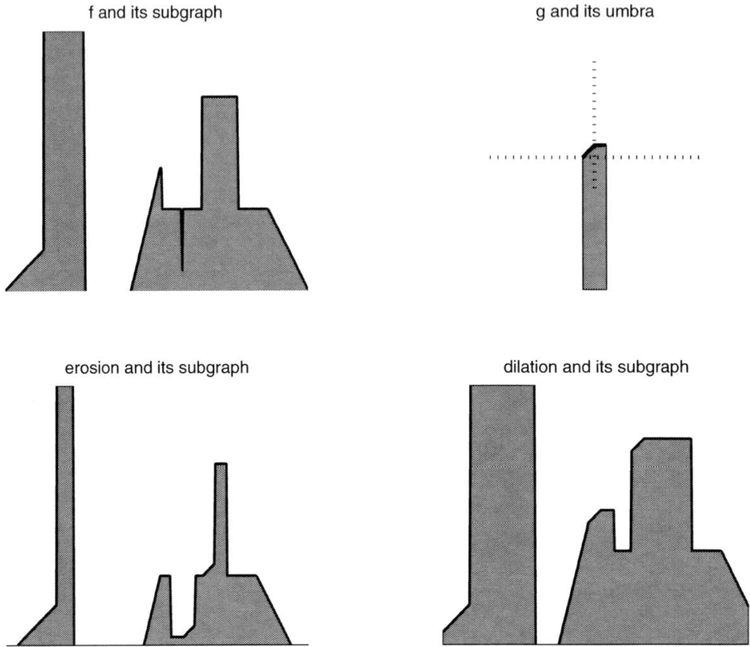

Figure 5.23 Bounded erosion and dilation.

The bounded dilation and erosion preserve most of the algebraic properties of Sec. 5.4 with the exception to translation invariance and commutativity of dilation. The duality of bounded dilation and erosion is illustrated in Fig. 5.24.

There are several advantages of using the bounded operations: (i) the negation works without any assumptions outside the image domain; (ii) there is no overflow or underflow on the output pixel values; (iii) there is no change on the output image domain, the output image size being the same as the input image size.

5.10   Gray-Scale Opening and Closing

Given the primary gray-scale morphological operations of erosion and dilation, we can define the secondary operations, gray-scale opening and closing. As in the binary case these are dual operations, and both can be characterized directly in terms of fitting.

Gray-scale **opening** can be defined analogously to its definition in the binary setting, namely, as a composition of erosion followed by dilation:

$$f \circ g = (f \ominus g) \oplus g. \tag{5.66}$$

As in the binary setting, it is usually better to view opening in terms of fitting, the analog to Eq. (2.2) being

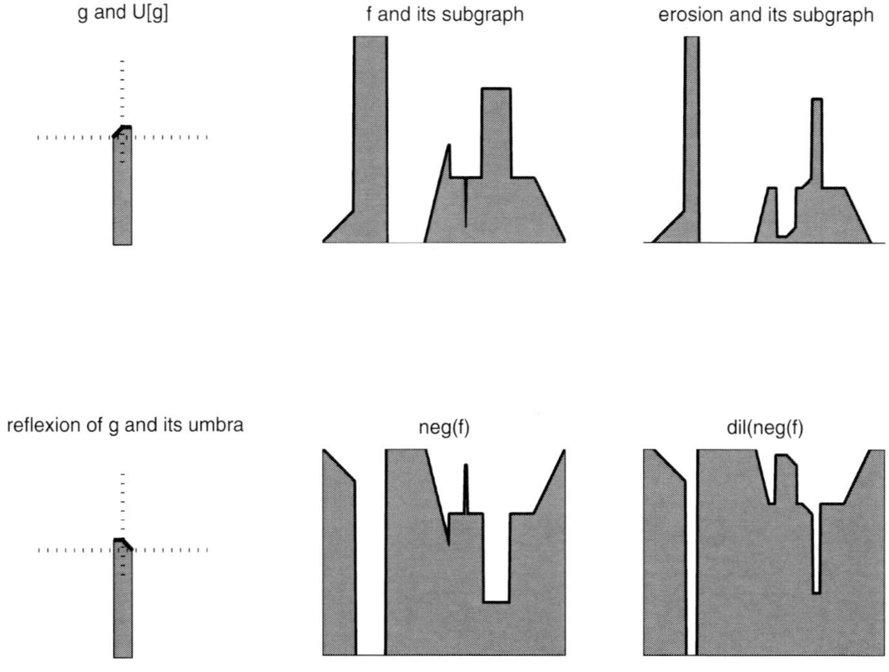

Figure 5.24 Duality between bounded erosion and dilation.

$$f \circ g = \bigvee \{g_x + y : g_x + y \leq f\}. \tag{5.67}$$

According to Eq. (5.67), the opening is found by taking the maximum over all morphological translations of the structuring element that fit beneath the input signal. The fitting formulation gives the geometric intuition for opening: slide the structuring element along beneath the signal and at each point record the point on the structuring element translation that is highest at that point. The position of the origin relative to the structuring element is irrelevant.

There is also an umbra characterization of opening:

$$f \circ g = S[U[f] \circ U[g]]. \tag{5.68}$$

Opening examples are given in Figs. 5.25, 5.26 and 5.27. Note the manner in which the curves are filtered from the bottom in accordance with the shape of the structuring element. In Fig. 5.26 there is a "rolling ball" effect, just as there would be when opening a binary set by a disk. Note also in Fig. 5.26 the boundary effects owing to the requirement for the structuring element to be beneath the signal. Specifically, there is a rolling-off at the edges.

A gray-scale image opening example is shown in Fig. 5.28. Note the differences from the erosion example of Fig. 5.9. The open image is much closer to the

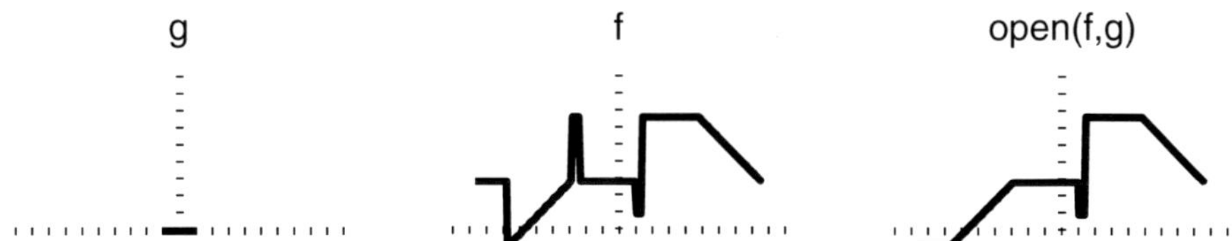

Figure 5.25 Opening by a flat structuring element.

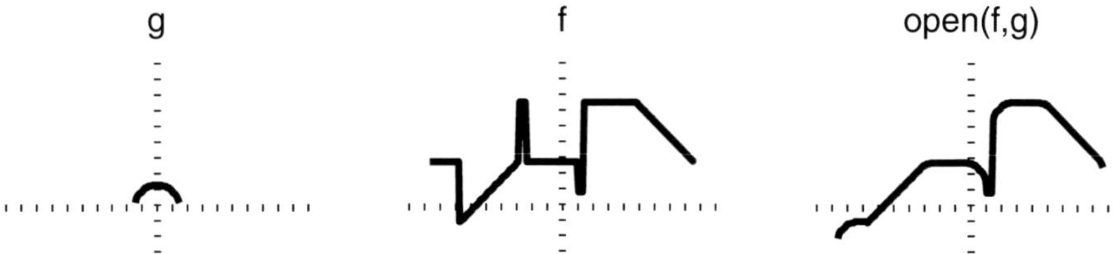

Figure 5.26 Opening by a semicircle.

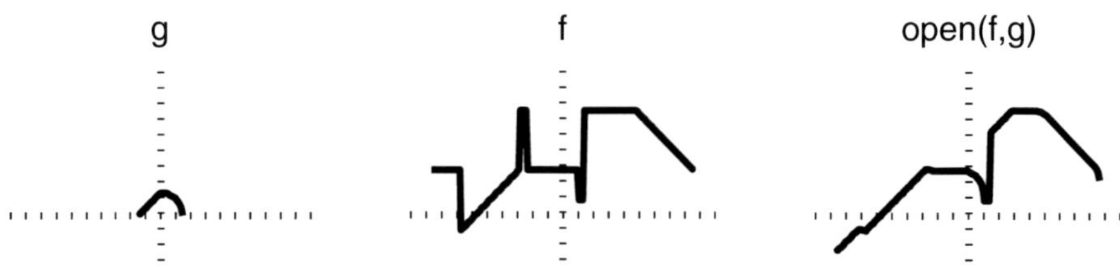

Figure 5.27 Opening by a nonsymmetrical nonflat structuring element.

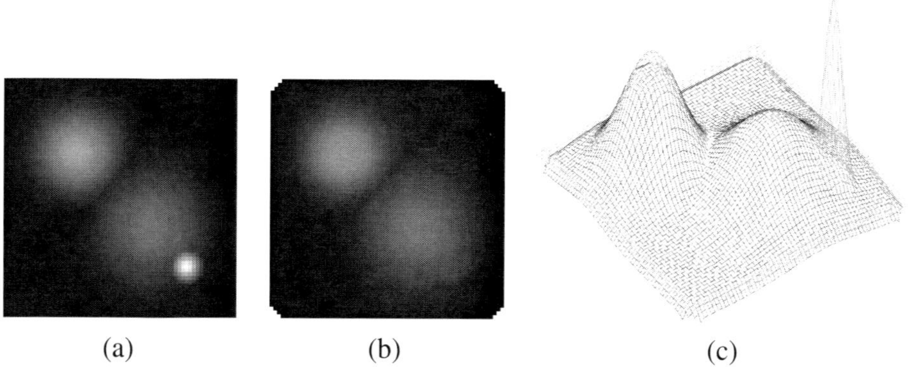

(a)   (b)   (c)

Figure 5.28 Two-dimensional gray-scale opening by a semi-sphere structuring element: (a) input image, (b) opened image, (c) surface mesh plot of opened (dark-gray) and input (light-gray) images.

input image, just with narrow peaks removed, where the meaning of narrowness is relative to the structuring element shape.

Figure 5.29 illustrates opening by a flat disk of a real gray-scale image. The two left pictures of Fig. 5.29 are the shade representations of the input and opened images. The two pictures on the right are their equivalent surface representations. Note how the round seeds are highlighted while other bright narrow structures are suppressed. Note how the thin black shadow of the twig has resulted in a wider black region following the twig.

As a filter, opening is translation invariant:

$$(f_x + y) \circ g = (f \circ g)_x + y. \tag{5.69}$$

This follows from the fact that opening is a composition of erosion and dilation, each of these is translation invariant, and a composition of translation invariant filters is translation invariant. It is also increasing: if $f \leq h$, then $f \circ g \leq h \circ g$.

The binary opening is antiextensive and idempotent. The extension of these filter properties to the gray-scale is immediate. A filter Ψ is **antiextensive** if $\Psi(f)$ lies beneath f, $\Psi(f) \leq f$. It is **idempotent** if operating twice in succession by the filter is equivalent to operating once, $\Psi(\Psi(f)) = \Psi(f)$. The gray-scale opening is both antiextensive and idempotent: $f \circ g \leq f$ and

$$(f \circ g) \circ g = f \circ g. \tag{5.70}$$

A gray-scale filter that is translation invariant, monotonically increasing, antiextensive, and idempotent is called a τ-**opening**, and there exists a gray-scale extension of the representation for binary τ-openings. Specifically, any gray-scale τ-opening is of the form

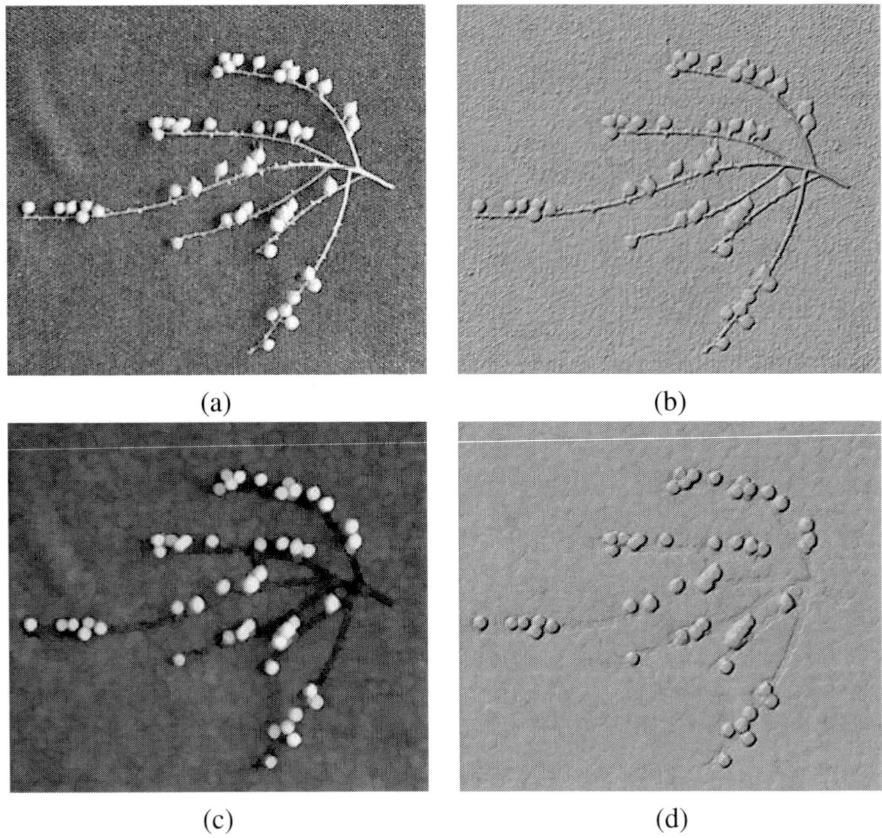

(a)   (b)

(c)   (d)

Figure 5.29 Illustration of gray-scale opening by a flat disk: (a) input image f, (b) surface view of f, (c) opened $f \circ D$, (d) surface view of $f \circ D$.

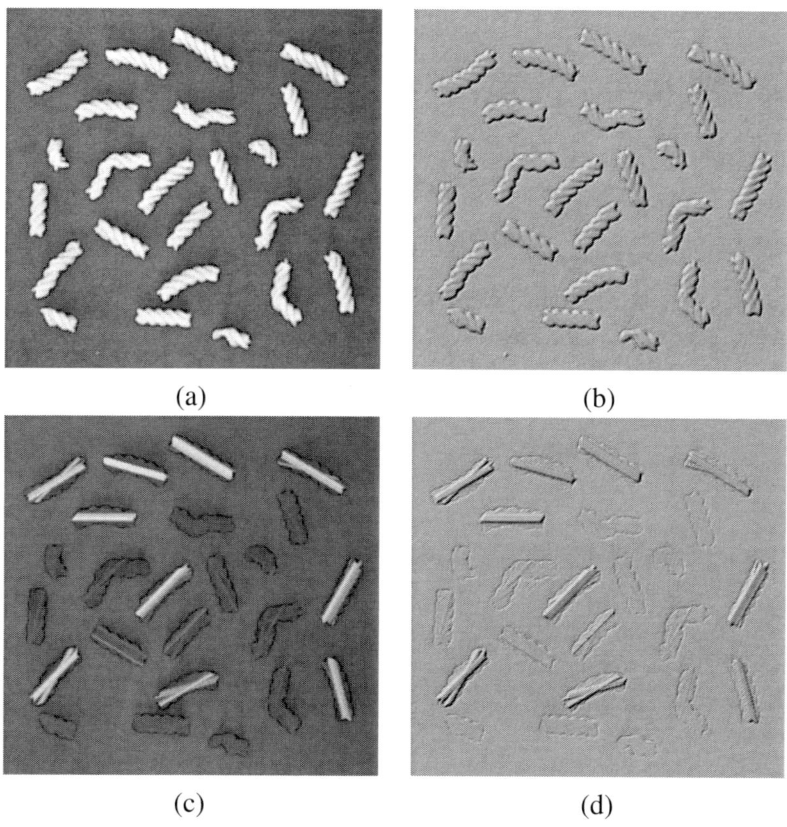

(a) (b)

(c) (d)

Figure 5.30 Radial opening: (a) input image, (b) surface view of the input image, (c) radial opening, (d) top-view surface of the radial opening.

$$\Psi(f) = \bigvee_{g \in \mathbf{B}} f \circ g, \tag{5.71}$$

where $\mathbf{B}$ is a **base** for Ψ. In the gray-scale case, the invariant class of Ψ consists of all maxima of morphological translations of signals in $\mathbf{B}$.

Like in the binary case, the most popular gray-scale τ-opening is the flat radial opening. Figure 5.30 shows a radial opening using rotated linear structuring elements from 0-deg to 180-deg in steps of 15-deg. Note how the opening has marked only straight elongated pasta chips.

Like opening, the gray-scale closing is defined analogously to the binary case:

$$f \bullet g = (f \oplus g) \ominus g. \tag{5.72}$$

Duality takes the form

$$f \bullet g = (f^c \circ \breve{g})^c. \tag{5.73}$$

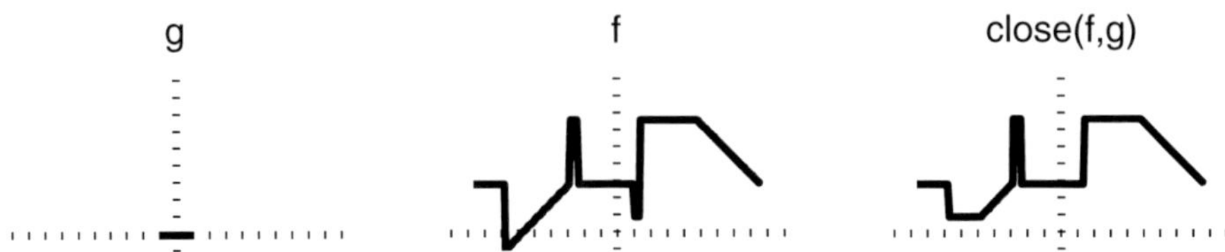

Figure 5.31 Closing by a flat structuring element.

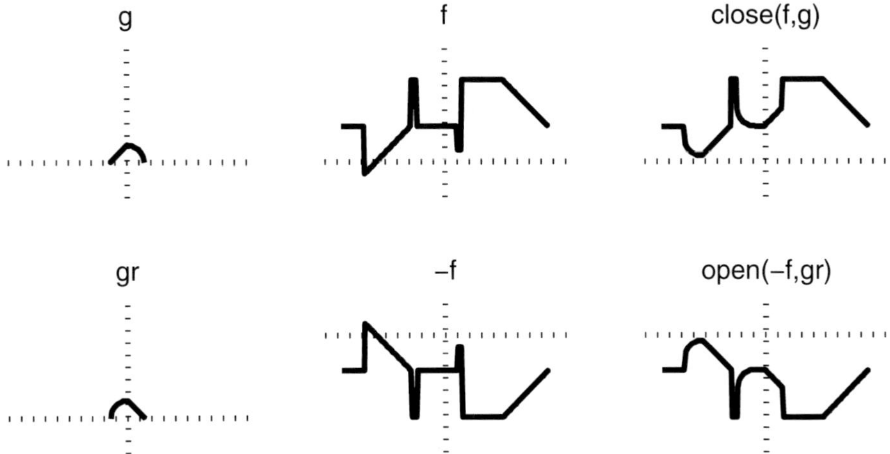

Figure 5.32 Closing by a nonsymmetrical nonflat structuring element, and the duality to the opening.

Geometric intuition regarding closing can be obtained from the duality relation. Whereas opening filters from below the signal, closing filters from above; indeed, by duality closing is an opening of f^c. Hence, to apply closing, simply flip the signal across the horizontal axis, filter by the opening, and then reflip. However, there is a subtlety: the choice of the structuring element. Choose the structuring element h with which you wish to open f^c. Since the duality relation involves opening by the reflection, the desired closing is with structuring element $\breve{h}$. The procedure is illustrated in Figs. 5.31 and 5.32. Here one must be careful. As with dilation and erosion, for the duality to hold in full, f^c must have positive infinity values outside the signal frame.

A gray-scale image closing example is shown in Fig. 5.33. Note the differences from the dilation example of Fig. 5.14. The closed image in this case is almost the same as the input image. The differences are in the filling of the valleys between the peaks. The strongest filling is between the narrow high peak (white dot) and the lower wider hill. The shapes of these fillings are dependent of the structuring element, which in this case is a semi-sphere.

A useful geometric intuition of the closing filter when applied to a gray-scale

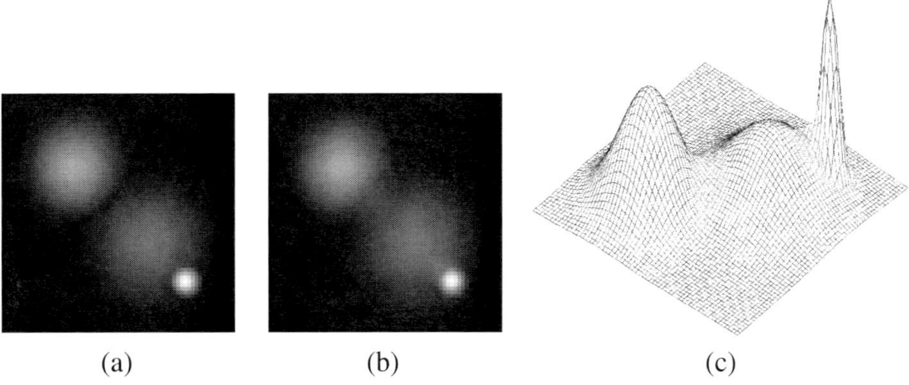

(a)   (b)   (c)

Figure 5.33 Two-dimensional gray-scale closing by a semi-sphere structuring element: (a) input image, (b) closed image, (c) surface mesh plot of input (dark-gray) and closed (light-gray) images.

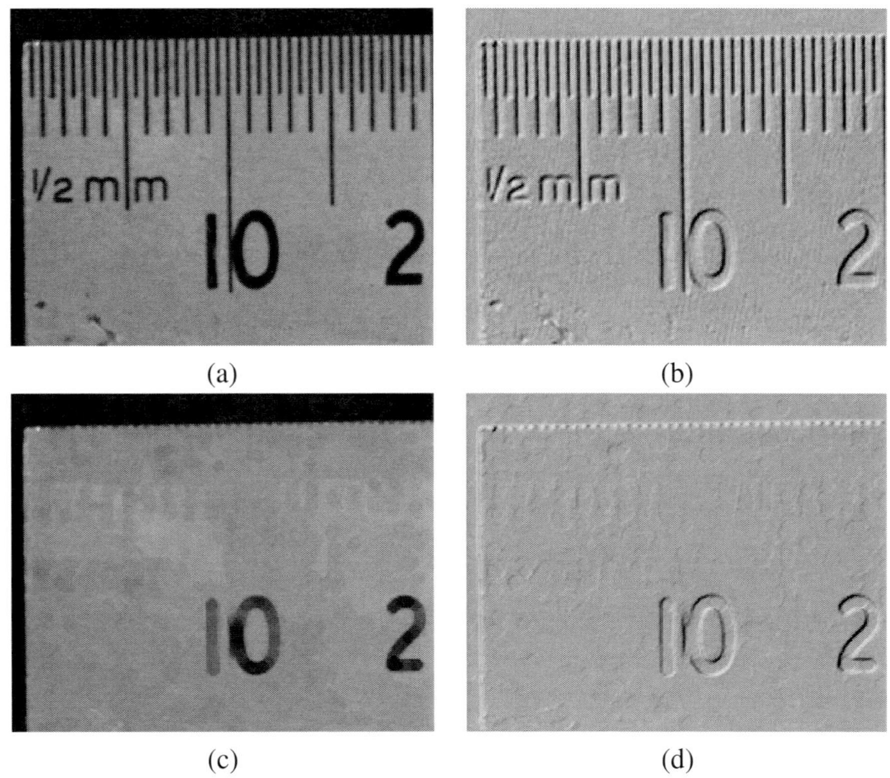

(a)   (b)

(c)   (d)

Figure 5.34 Illustration of gray-scale closing by a flat disk: (a) input image f, (b) surface view of f, (c) closing $f \bullet D$, (d) surface view of $f \bullet D$.

image can be seen in Fig. 5.34, from the image surface representation. Closing can be seen as a process where plaster is coated on the surface and the excess is removed by a tool shaped by the structuring element (in this case a flat disk) which can translate but not rotate. The two left-side pictures of Fig. 5.34 are the shade representations of the input and closed images. The two pictures on the right are their equivalent surface representations. Note how the narrow tick markers of the closed ruler are coated, whereas the thick numbers are recovered because the tool tip can fit inside them to remove the plaster excess.

5.11 Exercises

1. For the signals

$$
\begin{aligned}
f &= (\quad 0 \quad 1 \quad 2 \quad \mathbf{1} \quad 0 \quad -1 \quad -2 \quad), \\
g &= (\quad 1 \quad \mathbf{0} \quad -1 \quad), \\
h &= (\quad -\infty \quad \mathbf{0} \quad 0 \quad),
\end{aligned}
\tag{5.74}
$$

find each of the following expressions:

<table>
<tr><td>(a) $f \oplus g$</td><td>(g) $(f \oplus 2h) \vee f$</td></tr>
<tr><td>(b) $g \oplus h$</td><td>(h) $f \dot{\oplus} g$</td></tr>
<tr><td>(c) $(f \oplus g) \oplus h$</td><td rowspan="2">(i) $f \dot{\ominus} g$</td></tr>
<tr><td>(d) $f \ominus g$</td></tr>
<tr><td>(e) $f \oplus \breve{g}$</td><td>(j) $f \circ g$</td></tr>
<tr><td>(f) $2g$</td><td>(k) $f \bullet g$</td></tr>
</table>

2. Prove Eq. (5.23).

3. Prove Eq. (5.26).

4. Erode f by g using Eq. (5.46), where $f = (5\ 6\ \mathbf{9}\ 4\ 3\ 4)$ and $g = (0\ \mathbf{1}\ 0)$.

5. Dilate f by g using Eq. (5.47), where f and g are given in the preceding exercise.

6. Erode f by g using Eq. (5.53), where $f = (-\infty\ 6\ \mathbf{9}\ 4\ 3\ 4)$ and $g = (0\ \mathbf{0}\ 0)$.

7. Let

$$
f = \begin{pmatrix}
1 & 0 & 1 & 0 & 1 & 0 & 1 \\
1 & 7 & 7 & 2 & 0 & 0 & 0 \\
2 & 6 & 7 & 0 & 1 & 0 & 3 \\
1 & 1 & 0 & 1 & 1 & -6 & 1 \\
-\infty & 1 & 0 & 1 & 1 & 1 & 1
\end{pmatrix}
$$

name	description
mmsurf:	top-view shaded surface (Fig. 5.1)
mmis:	relationship between images (Fig. 5.3)
mmintersec:	minimum [Eq. (5.2)]
mmunion:	maximum [Eq. (5.3)]
mmsereflect:	structuring element reflection [Eq. (5.4)]
mmseero:	gray-scale erosion [Eq. (5.5)]
mmsedil:	gray-scale dilation [Eq. (5.12)]
mmbinary:	threshold decomposition [Eq. (5.51)]
mmneg:	bounded negation [Eq. (5.59)]
mmaddm:	bounded addition [Eq. (5.60)]
mmsubm:	bounded subtraction [Eq. (5.61)]
mmero:	bounded gray-scale erosion [Eq. (5.62)]
mmdil:	bounded gray-scale erosion [Eq. (5.64)]
mmopen:	gray-scale opening [Eq. (5.67)]
mmclose:	gray-scale closing [Eq. (5.10)]

Figure 5.35 MT operators presented in this chapter.

and

$$g = \begin{pmatrix} -\infty & 1 & 2 \\ -\infty & 0 & 1 \\ 0 & -1 & -\infty \end{pmatrix}.$$

Find $f \oplus g$, $f \ominus g$, $f \circ g$, and $f \bullet g$.

5.12 Laboratory Experiments

Figure 5.35 lists the **MT** functions introduced in this chapter.

1. Write a program to illustrate the duality between the bounded opening and closing. Use a 9×9 image and a 3×3 structuring element generated randomly several times to confirm if the implementation is correct.

References

1. B. De Baets, E. Kerre, and M. Gupta. The fundamentals of fuzzy mathematical morphology. Part 1: Basic concepts. *International Journal of General Systems*, 23:155–171, 1994.

2. B. De Baets, E. Kerre, and M. Gupta. The fundamentals of fuzzy mathematical morphology. Part 2: Idempotence, convexity and decomposition. *International Journal of General Systems*, 23:307–322, 1995.

3. I. Bloch. Geodesic balls in a fuzzy set and fuzzy geodesic mathematical morphology. *Pattern Recognition*, 33(6):897–905, 2000.

4. I. Bloch. On links between mathematical morphology and rough sets. *Pattern Recognition*, 33(9):1487–1496, 2000.

5. I. Bloch and H. Mâitre. Fuzzy mathematical morphology. *Annals of Mathematics and Artificial Intelligence*, 10:55–84, 1994.

6. I. Bloch and H. Mâitre. Fuzzy mathematical morphologies: a comparative study. *Pattern Recognition*, 28(9):1341–1387, 1995.

7. T. Q. Deng and H. J. A. M. Heijmans. Grey-scale morphology based on fuzzy logic. *Journal of Mathematical Imaging and Vision*, 16(2):155–171, 2002.

8. E. R. Dougherty and J. Barrera. Computational gray-scale image operators. In E. R. Dougherty and J. T. Astola, editors, *Nonlinear Filters for Image Processing*, pages 61–98. SPIE/IEEE Presses, Bellingham, WA, 1999.

9. E. R. Dougherty and R. M. Haralick. Unification of nonlinear filtering in the context of binary logical calculus – part I: Binary filters. *Journal of Mathematical Imaging and Vision*, 2(2):173–183, 1992.

10. E. R. Dougherty and R. M. Haralick. Unification of nonlinear filtering in the context of binary logical calculus – part II: Gray-scale filters. *Journal of Mathematical Imaging and Vision*, 2(2):185–192, 1992.

11. E. R. Dougherty and D. Sinha. Computational gray-scale mathematical morphology on lattices (a comparator-based image algebra) part 1: Architecture. *Real-Time Imaging*, 1(1):69–85, 1995.

12. E. R. Dougherty and D. Sinha. Computational gray-scale mathematical morphology on lattices (a comparator-based image algebra) part 2: Image operators. *Real-Time Imaging*, 1:283–295, 1995.

13. E.R. Dougherty and D. Sinha. Computational mathematical morphology. *Signal Processing*, 38:21–29, 1994.

14. M. Gabbouj, E. Coyle, and N. Gallagher, Jr. An overview of median and stack filtering. *IEEE Transactions on Circuits, Systems and Signal Processing*, 11(1):7–45, 1992.

15. P. D. Gader. Separable decompositions and approximations of greyscale morphological templates. *Computer Vision, Graphics and Image Processing*, 53(3):288–296, 1991.

16. J. Goutsias and H. J. A. M. Heijmans. Nonlinear multiresolution signal decomposition schemes. Part I: Linear and morphological pyramids. *IEEE Transactions on Image Processing*, 9(11):1862–1876, 2000.

17. R. M. Haralick, S. R. Sternberg, and X. Zhuang. Image analysis using mathematical morphology. *IEEE Transactions on Pattern Analysis and Machine Intelligence*, 9:532–550, 1987.

18. H. J. A. M. Heijmans. Theoretical aspects of gray-level morphology. *IEEE Transactions on Pattern Analysis and Machine Intelligence*, 13:568–582, 1991.

19. H. J. A. M. Heijmans. On the construction of morphological operators which are self-dual and activity-extensive. *Signal Processing*, 38:13–19, 1994.

20. H. J. A. M. Heijmans. Self-dual morphological operators and filters. *Journal of Mathematical Imaging and Vision*, 6(1):15–36, 1996.

21. H. J. A. M. Heijmans. Composing morphological filters. *IEEE Transactions on Image Processing*, 6(5):713–723, 1997.

22. H. J. A. M. Heijmans and J. Goutsias. Nonlinear multiresolution signal decomposition schemes. Part II: Morphological wavelets. *IEEE Transactions on Image Processing*, 9(11):1897–1913, 2000.

23. H. J. A. M. Heijmans and P. Maragos. Lattice calculus of the morphological slope transform. *Signal Processing*, 59(1):17–42, 1997.

24. H. J. A. M. Heijmans, P. Nacken, A. Toet, and L. Vincent. Graph morphology. *Journal of Visual Communication and Image Representation*, 3:24–38, 1992.

25. H. J. A. M. Heijmans and J. Serra. Convergence, continuity and iteration in mathematical morphology. *Journal of Visual Communication and Image Representation*, 3:84–102, 1992.

26. H. J. A. M. Heijmans and R. van den Boomgaard. Algebraic framework for linear and morphological scale-spaces. *Journal of Visual Communication and Image Representation*, 13:269–301, 2002.

27. P. Maragos and R. W. Schafer. Morphological filters—part I: Their set-theoretic analysis and relations to linear shift-invariant filters. *IEEE Transactions on Acoustics, Speech and Signal Processing*, 35:1153–1169, 1987.

28. P. Maragos and R. W. Schafer. Morphological filters—part II: Their relations to median, order-statistics, and stack filters. *IEEE Transactions on Acoustics, Speech and Signal Processing*, 35:1170–1184, 1987.

29. G. Matheron. Filters and lattices. In J. Serra, editor, *Image Analysis and Mathematical Morphology, Vol. 2: Theoretical Advances*, chapter 6. Academic Press, London, 1988.

30. J.B.T.M. Roerdink. Group morphology. *Pattern Recognition*, 33(6):877–895, 2000.

31. C. Ronse. Why mathematical morphology needs complete lattices. *Signal Processing*, 21:129–154, 1990.

32. C. Ronse. Lattice-theoretical fixpoint theorems in morphological image filtering. *Journal of Mathematical Imaging and Vision*, 4:19–41, 1994.

33. J. Serra. Dilation and filtering for numerical functions. In J. Serra, editor, *Image Analysis and Mathematical Morphology, Vol. 2: Theoretical Advances*. Academic Press, New York, 1988.

34. J. Serra. Introduction to morphological filters. In J. Serra, editor, *Image Analysis and Mathematical Morphology, Vol. 2: Theoretical Advances*, chapter 6. Academic Press, New York, 1988.

35. J. Serra. Mathematical morphology for complete lattices. In J. Serra, editor, *Image Analysis and Mathematical Morphology, Vol. 2: Theoretical Advances*. Academic Press, New York, 1988.

36. J. Serra and L. Vincent. An overview of morphological filtering. *IEEE Transactions on Circuits, Systems and Signal Processing*, 11:47–108, 1992.

37. D. Sinha and E. R. Dougherty. Fuzzy mathematical morphology. *Journal of Visual Communication and Image Representation*, 3(3):286–302, 1992.

38. D. Sinha and E. R. Dougherty. A general axiomatic theory of intrinsically fuzzy mathematical morphologies. *IEEE Transactions on Fuzzy Systems*, 3(4):389–403, 1995.

39. D. Sinha, P. Sinha, E. R. Dougherty, and S. Batman. Design and analysis of fuzzy morphological algorithms for image processing. *IEEE Transactions on Fuzzy Systems*, 5(4):570–584, 1997.

40. S. R. Sternberg. Grayscale morphology. *Computer Vision, Graphics and Image Processing*, 35:333–355, 1986.

41. L. Vincent. Graphs and mathematical morphology. *Signal Processing*, 16:365–388, 1989.

42. P. D. Wendt, E. J. Coyle, and N. C. Callagher Jr. Stack filters. *IEEE Transactions on Acoustics, Speech and Signal Processing*, 34:898–911, 1986.

43. O. Yli-Harja, J. Astola, and Y. Neuvo. Analysis of the properties of median and weighted median filters using threshold logic and stack filter representation. *IEEE Transactions on Signal Processing*, 39(2):395–410, 1991.

Chapter 6

Morphological Processing of Gray-Scale Images

Having discussed the theory of gray-scale mathematical morphology, we now present some procedures that can be employed for various purposes, including filtering, marking, and segmentation.

6.1 Morphological Gradient

Various gradients are used in image processing to detect edges, the basic principle being that large gradients represent points where there is a rapid light-to-dark (or dark-to-light) change. Very often these gradients are digital versions of differential gradients. In morphological image processing there are several gradients that have been developed; however, we will concentrate on only one class of these, typically referred to as the **morphological gradient**. It is defined by

$$grad_{g_e, g_i}(f) = (f \oplus g_e) - (f \ominus g_i), \tag{6.1}$$

where g_e and g_i are structuring elements centered about the origin. The morphological gradient is the sum of two partial gradients, the **external gradient**,

$$grad_{g_e}^{ext}(f) = (f \oplus g_e) - f, \tag{6.2}$$

and the **internal gradient**,

$$grad_{g_i}^{int}(f) = f - (f \ominus g_i). \tag{6.3}$$

From Eq. (6.1) we see that the gradient depends on the size and shape of the structuring element. The gray-scale gradient is the direct generalization of the binary gradient defined in Sec. 3.2, and its action is illustrated in Fig. 6.1. It is common to use a flat structuring element for the gradient. Because dilation and erosion by flat structuring elements yield maximum and minimum filters, respectively, at each point the morphological gradient yields the difference between the maximum and minimum values over the neighborhood at the point determined by the flat structuring element (which is really a set). Figure 6.2 shows the morphological gradient of a gray-scale image.

As is the case with differential gradients, the morphological gradient can be used in conjunction with thresholding to perform gray-scale edge detection. The histogram of the gradient image is used to determine a threshold value and the thresholded gradient is the edge image. As with differential gradients, the procedure is problematic owing to the nonuniformity of the gradient intensity relative to

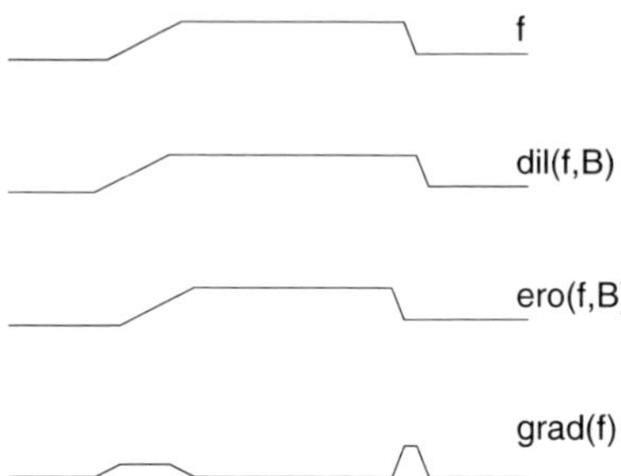

Figure 6.1 Morphological gradient: signal, dilation, erosion, gradient.

edges. This problem will be further explored when we employ the morphological gradient as part of the watershed methodology for gray-scale image segmentation.

6.2   Top-Hat Transform

As mentioned in Sec. 2.1, subtracting an opened image from the original image can leave key markers, such as points of high curvature, that can be employed in recognition algorithms. In the gray scale, this approach can be beneficial in finding small pixel clusters that are dark surrounded by a relatively light background or light pixel clusters surrounded by a relatively dark background. It can also be used to find edges in images possessing little noise.

The operator, known as the **open top-hat transform**, is defined by

$$f \hat{\circ} g = f - (f \circ g), \tag{6.4}$$

where g is an appropriately chosen structuring element. Since opening is an antiextensive operation, the opening lies beneath the original image, so that $f \hat{\circ} g$ is always nonnegative. The operator is illustrated in Fig. 6.3, with B being a flat structuring element with length somewhat longer than the bases of the peaks jutting above the original signal. Notice how the peaks have been detected and can be marked by applying a threshold to the top-hat image.

By choosing an appropriately sized structuring element, one can use the top-hat transform to mark narrow peaks while not marking wide illumination bubbles in the image. In some applications it is impossible to separate desirable and undesirable bright spots simply by using an appropriately sized structuring element, but it is possible to separate them by an appropriately chosen threshold. For successful application, one typically needs to accumulate statistics on peak heights.

Figure 6.4 illustrates the use of the open top-hat to detect the narrow structures in the image. This same picture was used to illustrate the gray-scale opening (Fig. 5.29) detecting the round seeds. The corresponding open top-hat can detect the fine branches where the seeds are connected.

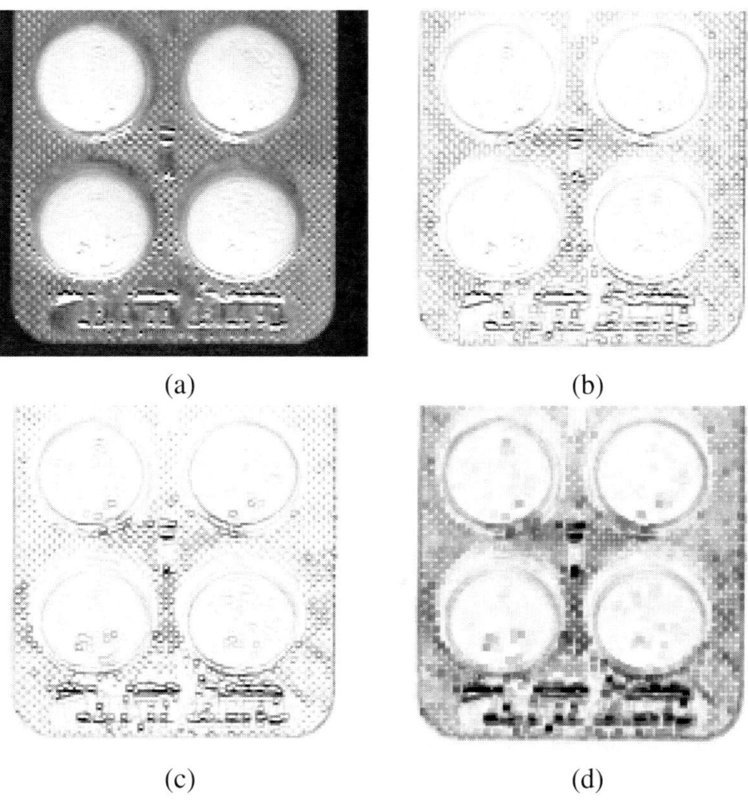

(a)   (b)

(c)   (d)

Figure 6.2 Morphological gradients: (a) input image f, (b) internal gradient $f - (f \ominus B)$, (c) external gradient $(f \oplus B) - f$, (d) morphological gradient $(f \oplus B) - (f \ominus B)$.

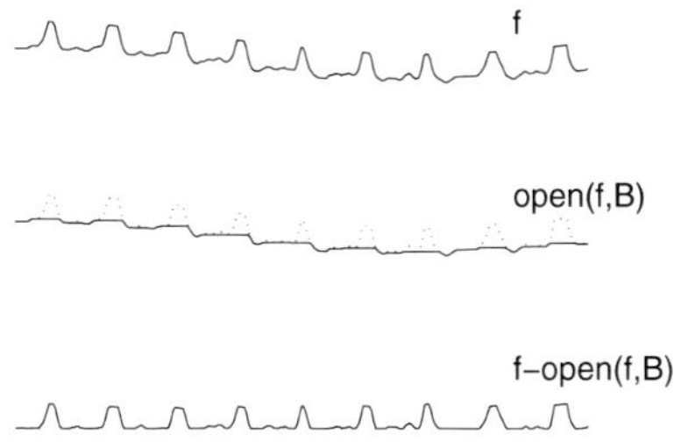

Figure 6.3 Open top-hat transform: signal, opening, top-hat as the residue.

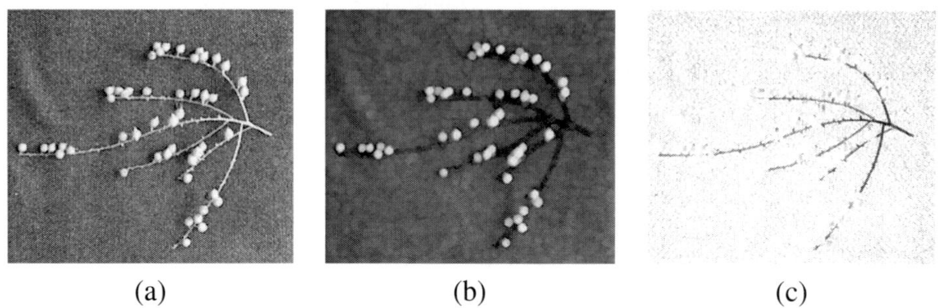

(a)   (b)   (c)

Figure 6.4 Gray-scale top-hat: (a) input image, (b) opened image, (c) open top-hat (negated for visualization).

Open top-hat is very useful as a preprocessing step to correct uneven illumination before applying a threshold, thereby acting as an adaptive thresholding technique. Figure 6.5 illustrates this feature: (a) input image; (b) opened image; (c) top-hat image; and (d) thresholded top-hat image.

For a real-world biological application, we consider fluorescent in-situ hybridization (FISH) imaging. A key technique for molecular diagnosis is in-situ hybridization in which labeled hybridizing agents (such as DNA or RNA probes) are exposed to intact tissue sections. When fluorescent dyes are used as labels, the technique is referred to as fluorescent in-situ hybridization (FISH). The probe hybridizes to a defined target nucleotide sequence of DNA in the cell, and the dye fluoresces to some particular color when excited by a mercury arc lamp or argon laser, so that the labeled probe can be detected when the probed tissue is viewed through a microscope. Each chromosome containing the target DNA sequence will produce a fluorescent signal (spot) in every cell when the specimen is illuminated with suitable excitation. FISH is an excellent method for detection of gene copy number alterations in cancer and other diseases. Figure 6.6 shows the open top-hat transform applied to a FISH image taken from a stack of images on multiple focal planes: (a) FISH image; (b) top-hat transform of the FISH image; (c) binary image resulting from thresholding the top-hat image; and (d) detail of the final result with an arrow indicating the position of each detected spot. Owing to noise, the top-hat methodology typically yields very small spots in the binary image. These can be eliminated by area open. Using the **MT**, the arrows were overlaid automatically in the final image by unioning a dilation of the centroids of the detected spots by an arrow-shape structuring element, with the origin slightly translated from the arrow tip so as to not disturb the visualization of the original spots in the image.

The complementary operator to the open top-hat operator as defined in Eq. (6.4) is the **close top-hat transform**,

$$f \hat{\bullet} g = (f \bullet g) - f. \tag{6.5}$$

The output is again nonnegative because closing is an extensive operation, the clos-

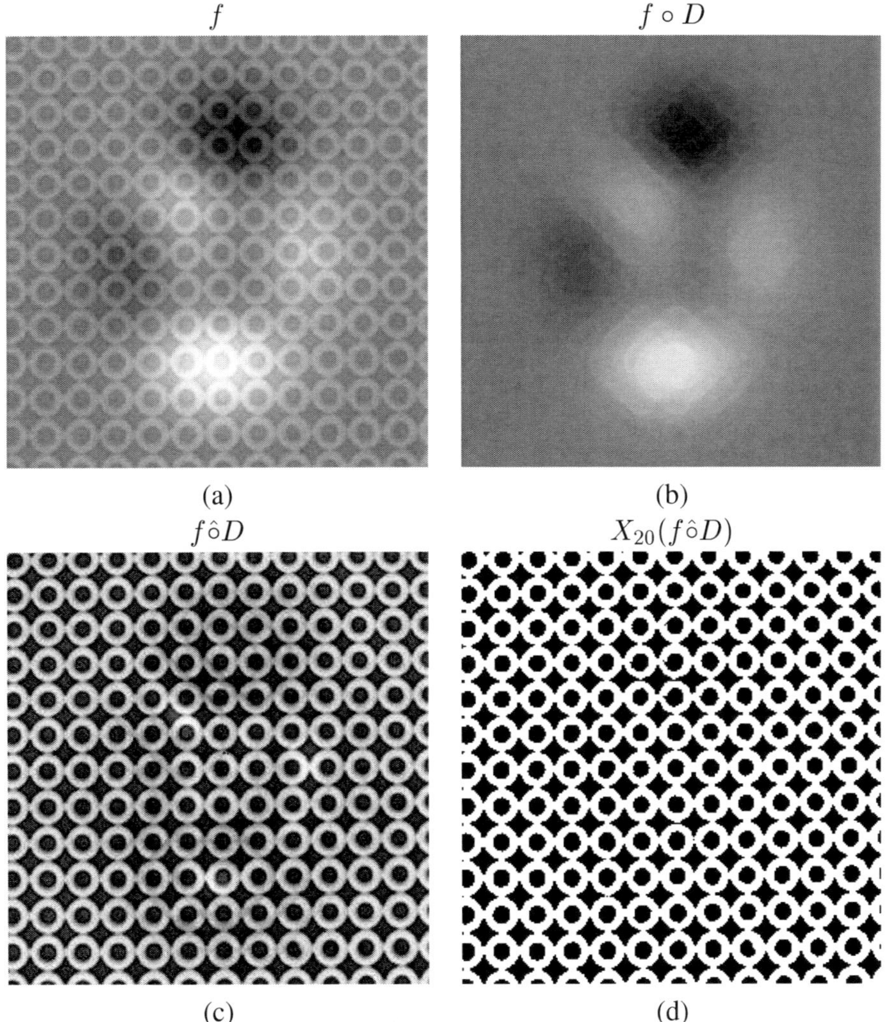

Figure 6.5 Image segmentation using open top-hat to compensate for uneven illumination.

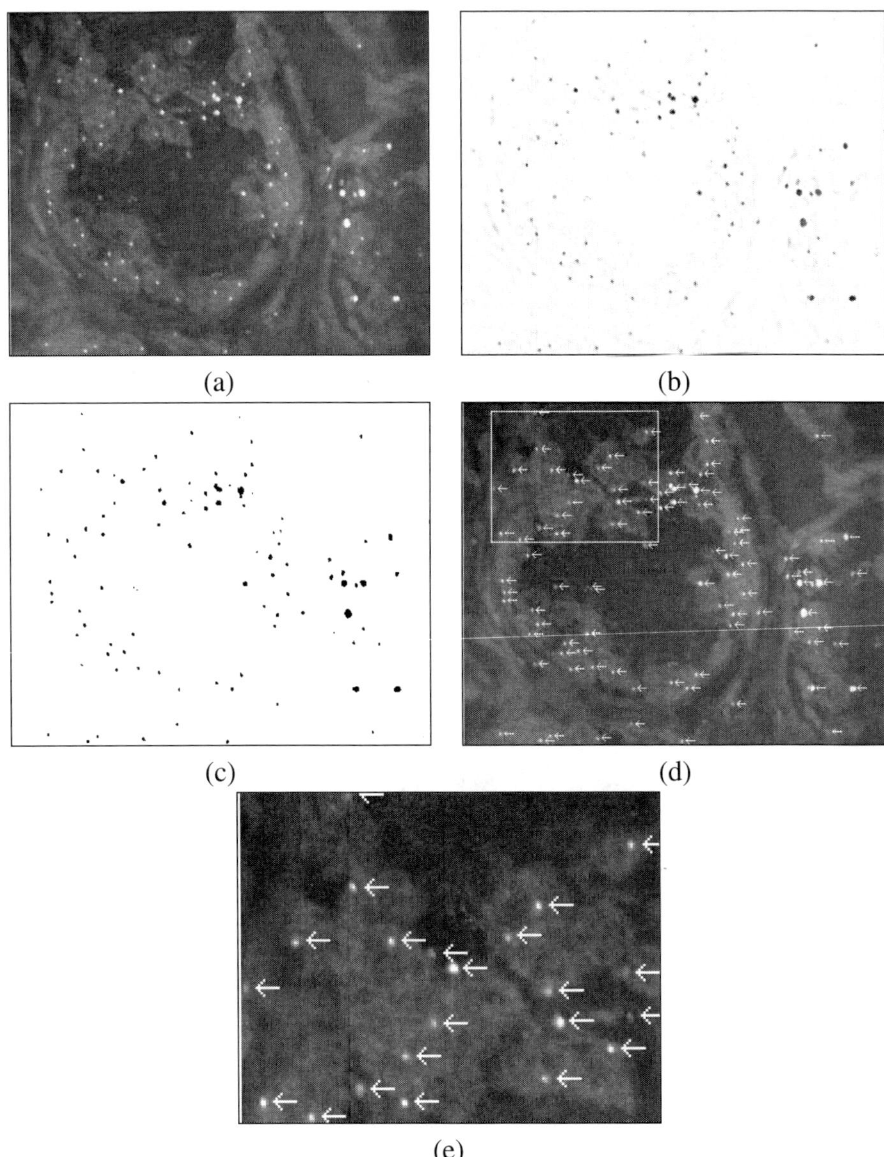

Figure 6.6 Gray-scale top-hat: (a) FISH input image, (b) area open by 2 pixels of the open top-hat by a disk of radius 4, (c) thresholding at gray level 50, (d) dilation of centroids by arrow, for illustration, (e) detail of the rectangle marked on the image of part (d).

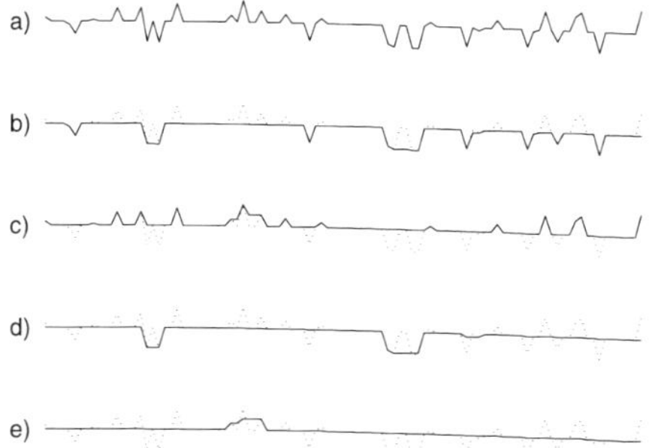

Figure 6.7 ASF filtering: (a) input signal, (b) opening, (c) closing, (d) open-close, and (e) close-open.

ing lying above the input image. The duality relation between opening and closing yields the complementary relation between their respective top-hat transforms:

$$f \hat{\bullet} g = f^c \hat{\circ} \breve{g}. \tag{6.6}$$

To detect both peaks and valleys, one can apply the open top-hat transform, threshold to find peak markers, apply the close top-hat transform, threshold to find valley markers, and then union the two marker images.

6.3 Gray-Scale Alternating Sequential Filters

Binary openings (closings) are useful for filtering images that have been corrupted by union (subtractive) noise. Analogous approaches have been taken in the gray scale. In particular, the gray-scale opening can be employed to filter maximum noise because the noise to be filtered lies above the signal. If the peaks lying above the signal in Fig. 6.7(a) are viewed as noise, then the opened signal in part (b) has been fairly successfully filtered. In a dual manner, and as depicted in Fig. 6.7(c), closing can be used to filter noise spikes beneath the image.

Typically noise is mixed, there being noise spikes both above and below the signal. So long as these noise spikes are sufficiently separated, they can be suppressed by application of either a composition of opening and closing or of closing and opening. Open-close and close-open take the forms

$$CO_g(f) = (f \circ g) \bullet g, \tag{6.7}$$

and

$$OC_g(f) = (f \bullet g) \circ g, \tag{6.8}$$

respectively. Note that the open-close function is written as CO_g in accordance with standard operator notation, in which the inner operator, in this case O, is

(a)   (b)   (c)

Figure 6.8 Two-dimensional gray-scale alternating sequential filtering: (a) input image, (b) close-open by a 3×3 diamond structuring element, (c) ASF close-open with three stages.

applied before the outer operator, in this case C. An analogous statement applies to close-open. Open-close and close-open are illustrated in Fig. 6.7 (d) and (e), respectively, with a semicircular structuring element.

From a statistical perspective, the problem with opening as a filter is that, unless the noisy image lies above the uncorrupted image, as with maximum noise, the filter will suffer bias because the opened noisy image will always lie beneath the noisy image. A dual comment applies to closing bias. One purpose of composition is to mitigate these biases.

As in the binary setting, selection of an appropriately sized structuring element is crucial. In addition, if there is a mixture of spatially sized noise spikes and they are not sufficiently dispersed, one can employ an alternating sequential filter, which is a composition of alternating openings and closings with increasingly wide structuring elements:

$$ASF_{oc,g}^{n}(f) = (((((f \bullet g) \circ g) \bullet 2g) \circ 2g) \ldots \bullet ng) \circ ng, \qquad (6.9)$$

$$ASF_{co,g}^{n}(f) = (((((f \circ g) \bullet g) \circ 2g) \bullet 2g) \ldots \circ ng) \bullet ng. \qquad (6.10)$$

Figure 6.8 shows an example of gray-scale image filtering using alternating sequential filtering.

6.4 Gray-Scale Morphological Reconstruction

Equations (3.5), (3.6), (3.7) and most concepts about connectivity using conditional dilation, morphological reconstruction, and reconstructive filters discussed in Chapter 3 are naturally extended to the gray scale. For **conditional dilation**, minimum replaces intersection:

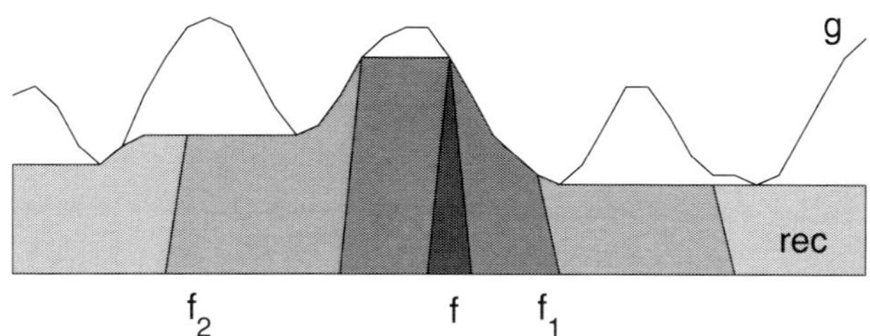

Figure 6.9 Signal morphological reconstruction: g: conditioning signal, f: marker, $f_1 = (f \oplus_g D)^4$, $f_2 = (f \oplus_g D)^{12}$, $rec = g \; \triangle_D \; f$.

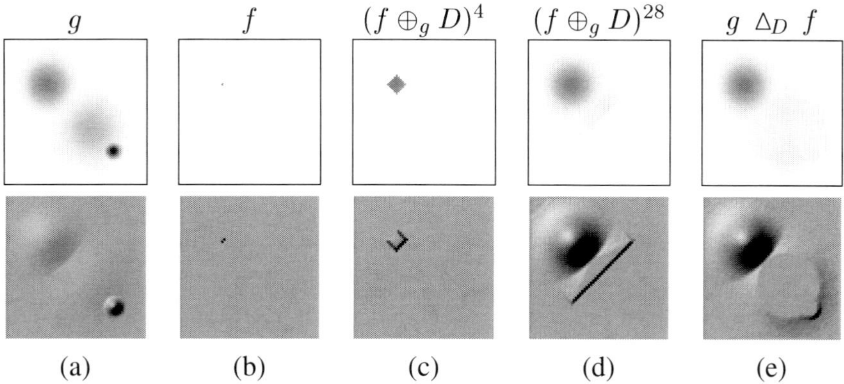

Figure 6.10 Gray-scale morphological reconstruction, gray-tone and top-view surface illustrations.

$$f \oplus_g D = (f \oplus D) \wedge g. \tag{6.11}$$

From this, one can build a sequence of n conditional dilations,

$$(f \oplus_g D)^n = (((f \oplus_g D) \oplus_g D) \oplus_g \ldots \oplus_g D), \tag{6.12}$$

and thereby obtain the **morphological reconstruction** of g from the marker f:

$$g \; \triangle_D \; f = (f \oplus_g D)^\infty. \tag{6.13}$$

Signal reconstruction is illustrated in Fig. 6.9 using a short flat structuring element D, with g and f being the conditioning signal and marker, respectively. Notice how reconstruction by the set D spreads the marker under the local minima of the conditioning signal. If D were longer, then it might raise the reconstructed signal above the minima. Reconstruction for images is illustrated for the same image in Fig. 6.10, which shows a gray-tone and top-view surface, and in Fig. 6.11, which shows a surface view.

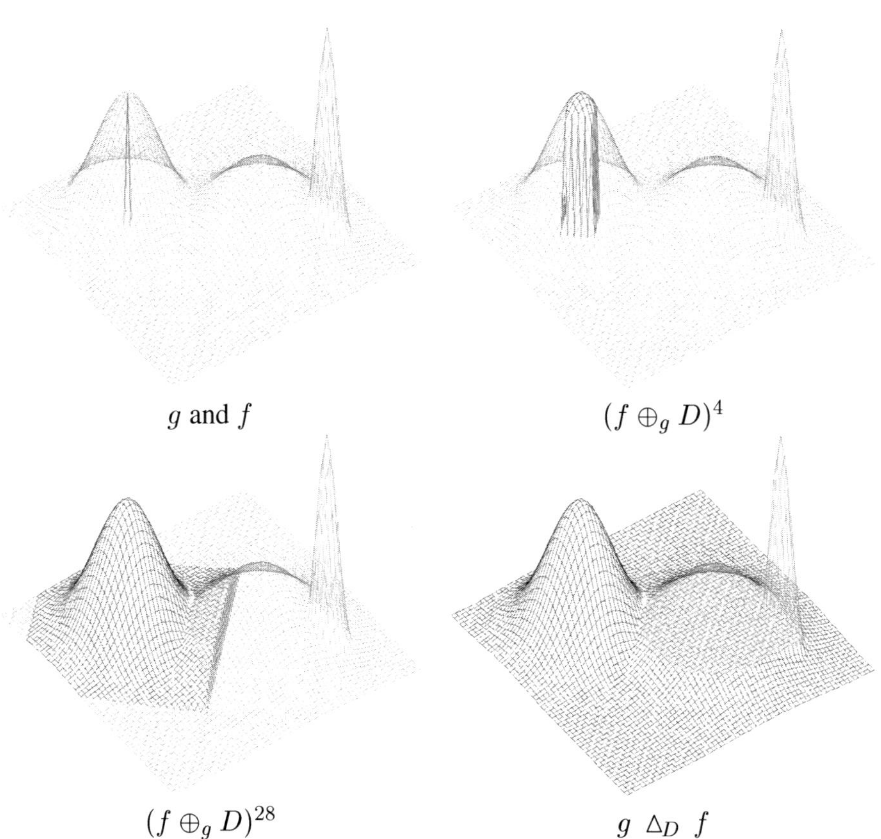

Figure 6.11 Gray-scale morphological reconstruction, surface illustration.

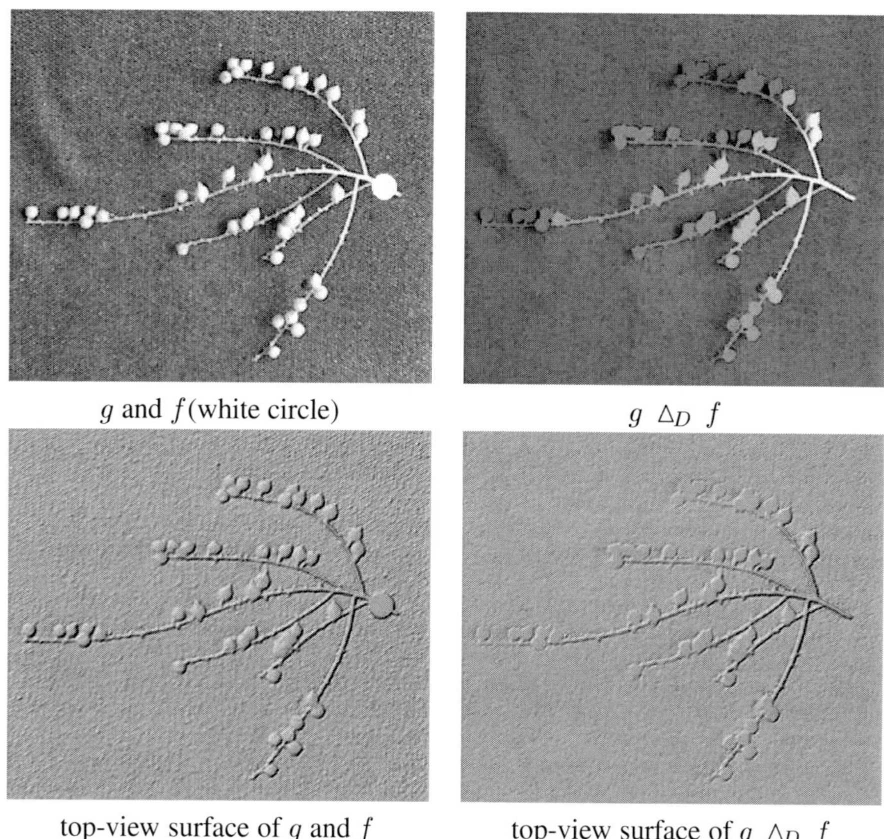

Figure 6.12 Gray-scale morphological reconstruction, marker at a pre-defined location.

As discussed in Chapter 3, there are mainly three ways to design the marker used in reconstruction: *a-priori* selection, selection from the opening, and determination from complex processing. Figure 6.12 illustrates the first case for a real image. Here, a round marker is placed at the right tip of the branch. The idea is to identify pixels that are "gray-scale connected" to the branch tip.

Since the binary morphological reconstruction is monotonically increasing, it can be used in the stack filter framework. In this sense, gray-scale morphological reconstruction using flat structuring elements can be implemented by means of threshold decomposition. Specifically, thresholding commutes with reconstruction:

$$X_t(g) \, \triangle_D \, X_t(f) = X_t((g \, \triangle_D \, f)). \tag{6.14}$$

This property is useful to better visualize the mechanism of the gray-scale reconstruction. Figure 6.13 illustrates this with the same example of Fig. 6.12. The figure shows several threshold sets and their reconstructions.

In analogy to the binary case, the reconstruction of Eq. (6.13) gives the **inf-**

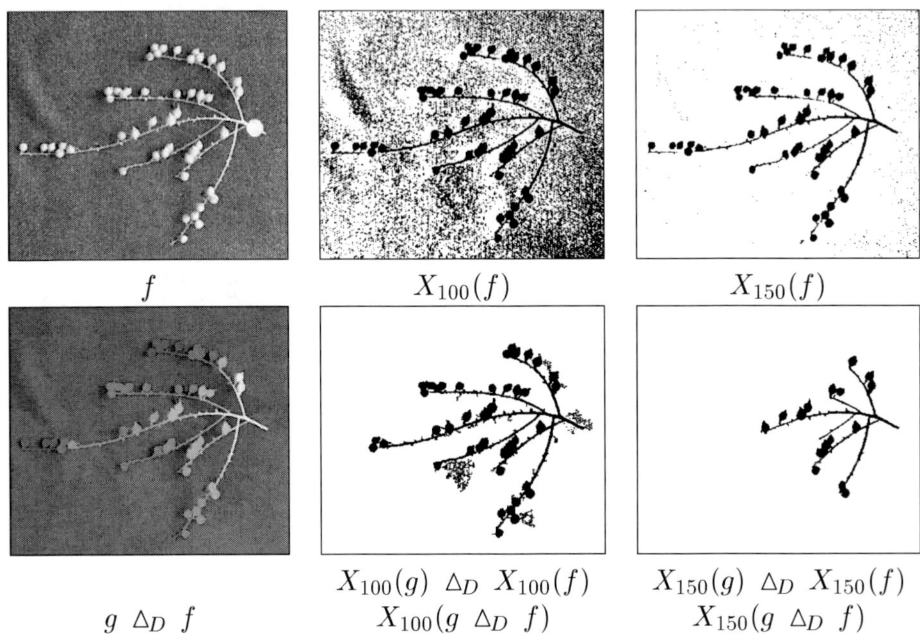

f $X_{100}(f)$ $X_{150}(f)$

$g \; \triangle_D \; f$

$X_{100}(g) \; \triangle_D \; X_{100}(f)$ $X_{150}(g) \; \triangle_D \; X_{150}(f)$
$X_{100}(g \; \triangle_D \; f)$ $X_{150}(g \; \triangle_D \; f)$

Figure 6.13 Illustration of gray-scale morphological reconstruction by means of the threshold decomposition.

reconstruction; that is, reconstruction from dilation. One can also consider reconstruction from erosion. Corresponding to Eqs. (3.8), (3.9), and (3.10), we have **conditional erosion,**

$$f \ominus_g D = (f \ominus D) \vee g, \tag{6.15}$$

a sequence of n conditional erosions,

$$(f \ominus_g D)^n = (((f \ominus_g D) \ominus_g D) \ominus_g \ldots \ominus_g D), \tag{6.16}$$

and the **morphological sup-reconstruction** of g from the marker f,

$$g \, \nabla_D \, f = (f \ominus_G D)^\infty. \tag{6.17}$$

Whenever used without the prefix, reconstruction is assumed to refer to inf-reconstruction.

The hole-filling example presented in Fig. 3.16 is shown in Fig. 6.14 with a gray-scale image: (a) input image, and (b) sup-reconstruction of the input image from a dual marker, which is an image of maximum pixel values inside and a zero in a one-pixel-width frame. Note that three holes in the bottom of the picture are retained, owing to their touching the image frame.

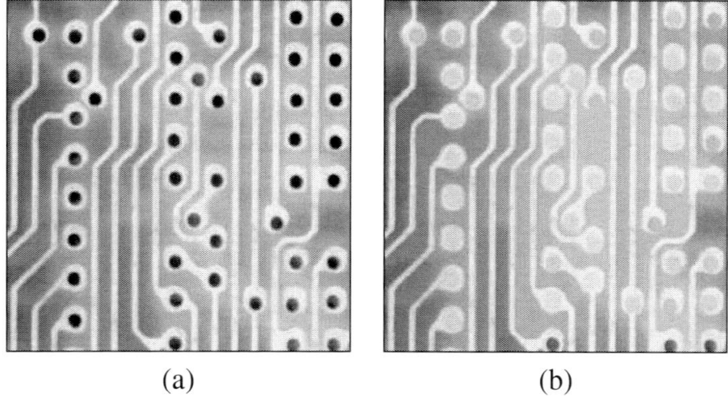

(a)   (b)

Figure 6.14 Gray-scale filling holes: (a) input image, (b) sup-reconstruction of the input image from the dual marker. The marker is an image of maximum pixel values inside and zero in a one-pixel-width frame.

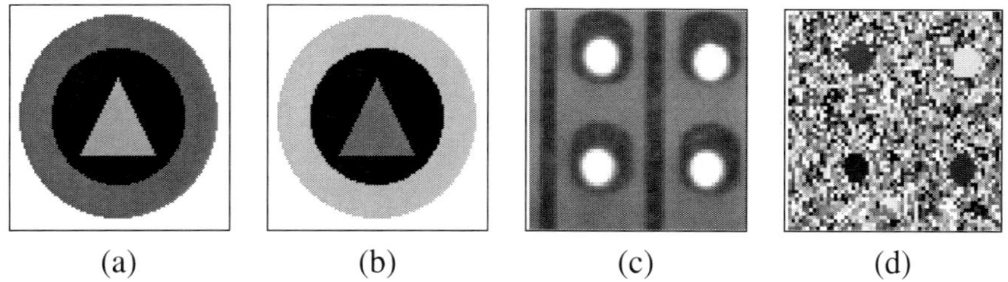

(a)   (b)   (c)   (d)

Figure 6.15 Flat zones: (a) synthetic image, (b) synthetic image has four flat zones, (c) real image, (d) real image has 2,471 flat zones.

6.5 Flat Zones and Connected Filters

A maximal connected component of a gray-scale image with the same pixel values is called a **flat zone**. Flat zones are connected regions of constant gray value. A flat zone can be a single pixel or can be the entire image. Figure 6.15 illustrates the flat zones of a synthetic gray-scale image and a real image. The synthetic image has four flat zones: a triangle, a circle around the triangle, an annular ring around the circle and a square background. A typical real image has a large number of flat zones, depending on the number of gray levels used in the digitalization of the image. The one shown in Fig. 6.15(c) has 2,471 flat zones.

A **connected filter** is an increasing operator that only merges flat zones. The flat zones of the filtered image contain the flat zones of the input image. An advantage of a connected filter is that if an edge exists in the output image, then the edge also exists in the input image. A connected filter never introduces a false edge. An illustration of this is shown in Fig. 6.16 with a binary image and a gray-scale image. Note how the outputs of nonconnected filters introduce artifacts with the structuring element shape.

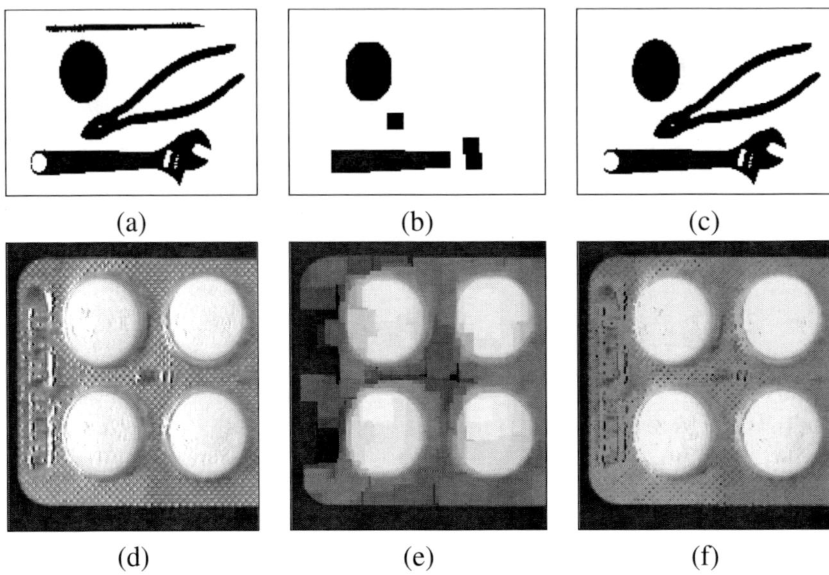

Figure 6.16 Connected filter versus nonconnected filter: (a) binary image, (b) regular opening by a square (nonconnected), (c) reconstructive opening (connected), (d) gray-scale image, (e) regular opening by a square (nonconnected), (f) reconstructive opening (connected).

6.6   Gray-Scale Reconstructive Opening

Morphological reconstruction is one of the most used tools to build connected filters. The **gray-scale reconstructive opening** is a natural extension of Eq. (3.13) and is defined by

$$f \circ_D B = f \; \triangle_D \; (f \ominus B), \tag{6.18}$$

or equivalently by

$$f \circ_D B = f \; \triangle_D \; (f \circ B). \tag{6.19}$$

It is also called **connected opening**, as reconstructive operators are connected operators.

Reconstruction can also be applied from a radial opening. This is illustrated in Fig. 6.17 using the input image and radial opening from Fig. 5.30.

The binary definitions also extend directly to define **disjunctive opening (reconstructive τ-opening)**,

$$\Psi(f) = \bigvee_{k=1}^{n} f \circ_D B_k, \tag{6.20}$$

and **conjunctive opening**,

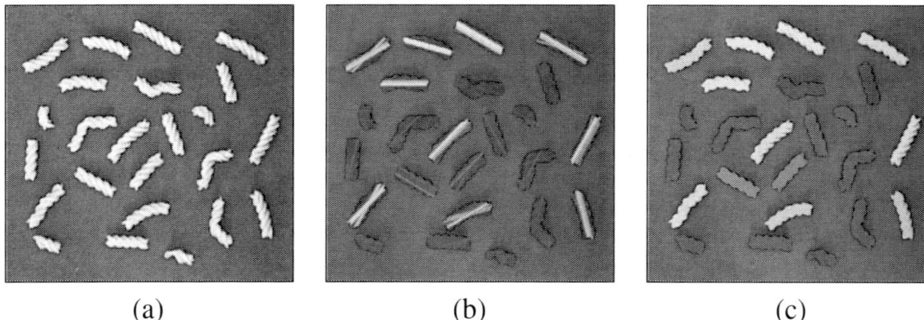

(a)   (b)   (c)

Figure 6.17 Reconstructive radial opening: (a) input image, (b) radial opening as shown in Fig. 5.30, (c) reconstructive radial opening (connected).

$$\Psi(f) = \bigwedge_{k=1}^{n} f \circ_D B_k. \tag{6.21}$$

These accomplish similar effects to the binary setting, except for gray-scale images. They are illustrated in Fig. 6.18 using vertical and diagonal structuring elements: (a) input image; (b) reconstructive opening by vertical structuring element; (c) reconstructive opening by diagonal structuring element; (d) disjunctive opening; and (e) conjunctive opening.

Gray-scale area open is defined analogously to the binary case. The size-α **area open** of a gray-scale image f, denoted $f \circ (\alpha)_E$, is a τ-opening whose base consists of all E-connected sets containing α pixels. It is a connected operator. Moreover, since it satisfies the threshold decomposition property,

$$X_t(f \circ (\alpha)_E) = X_t(f) \circ (\alpha)_E, \tag{6.22}$$

it can be modeled as a stack filter in which at each level, only binary connected components containing at least α pixels are passed.

The area open is illustrated in Fig. 6.19, in which the top-hat transform constructed from the gray-scale area open is applied: (a) input image; (b) area open; (c) area-open top-hat transform; (d) top-view surface of input image; (e) top-view surface of area open; and (f) top-view surface of top-hat transform. Note how well the letters over the keys have been extracted.

6.7 Connected Alternating Sequential Filters

An important class of connected filters is composed of those generated from alternating openings and closings. The connected filter corresponding to the gray-scale ASF seen in Sec. 6.3 is the reconstructive alternating sequential filter of stage n, defined by

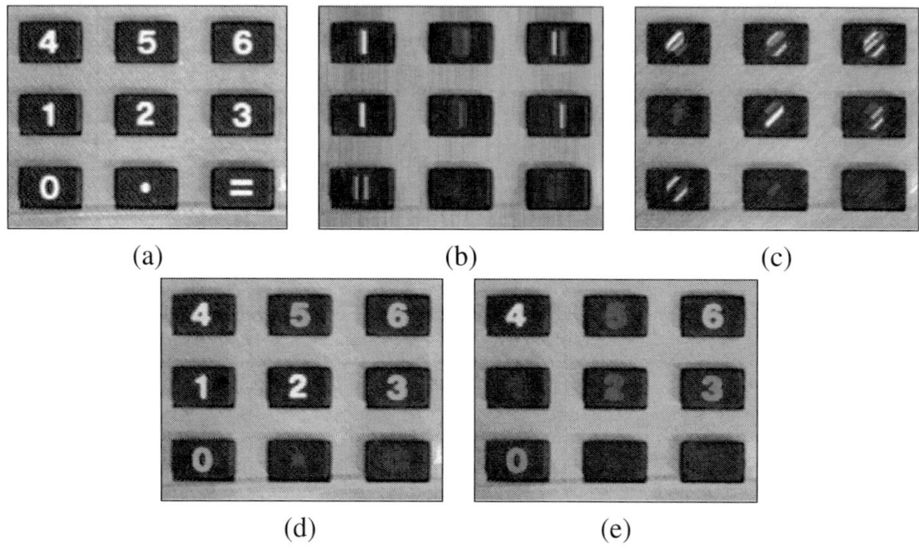

Figure 6.18 Disjunctive and conjunctive reconstructive opening: (a) input image, (b) opening by vertical structuring element, (c) opening by diagonal structuring element, (d) union of reconstructive openings by vertical and diagonal structuring elements, (e) intersection of reconstructive openings by vertical and diagonal structuring elements.

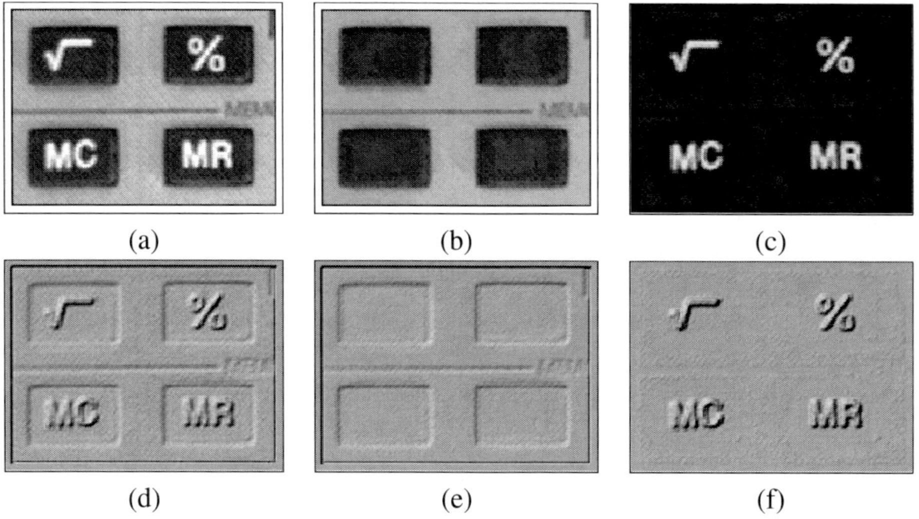

Figure 6.19 Gray-scale area open and top-hat: (a) input image, (b) area open, (c) area open top-hat, (d), (e) and (f) top-view surface of (a), (b) and (c), respectively.

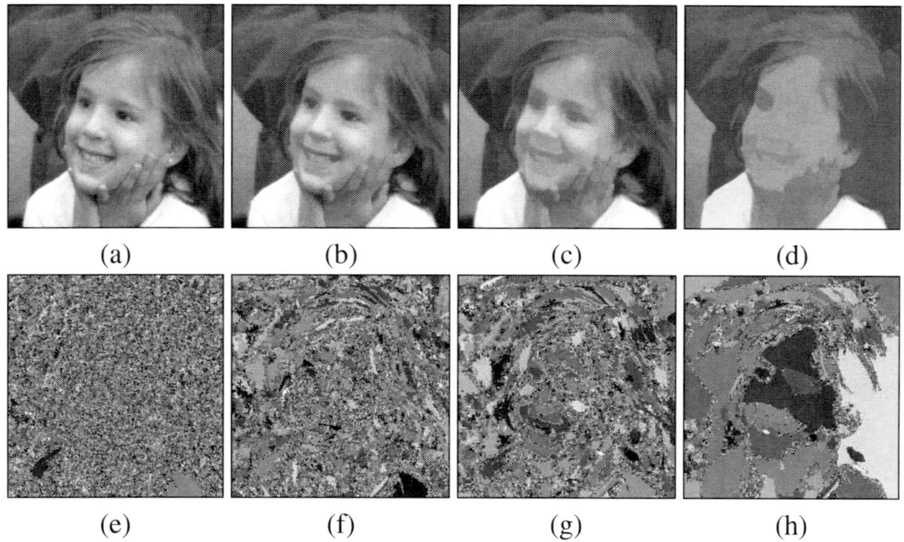

Figure 6.20 Connected pyramid with flat-zone image simplification by connected ASF filters of increasing stages: (a) input image, (b),(c),(d) reconstructive alternating sequential filter close-open of stages 2, 4, and 16, (e) flat-zones of original (35,335 regions), (f) stage 2 (21,647 regions), (g) stage 4 (18,490 regions), (h) stage 16 (9,460 regions).

$$ASF_{oc,B,D}^{n}(f) = (((((f \bullet_D B) \circ_D B) \bullet_D 2B) \circ_D 2B) \ldots \bullet_D nB) \circ_D nB,$$

$$(6.23)$$

or by

$$ASF_{co,B,D}^{n}(f) = (((((f \circ_D B) \bullet_D B) \circ_D 2B) \bullet_D 2B) \ldots \circ_D nB) \bullet_D nB.$$

$$(6.24)$$

The main property of the connected filter is that every flat zone of the input image is included in a flat zone of the output image. The flat zones of reconstructive alternating sequential filters of increasing sizes constitute a connected pyramid. A flat zone from a higher size filter contains flat zones of lower size filters. Figure 6.20 illustrates this property. The top row shows the original image and the image processed by reconstructive close-open ASFs of stages 2, 4 and 16. The bottom row shows the labeling of corresponding flat zones. The original image has 35,335 regions and the simplified images have 21,647 (stage 2), 18,490 (stage 4) and 9,460 (stage 16) regions. Notice how the shapes are well preserved along the scale space.

It is also possible to create an alternating sequential filter using a composition of area open followed by area close:

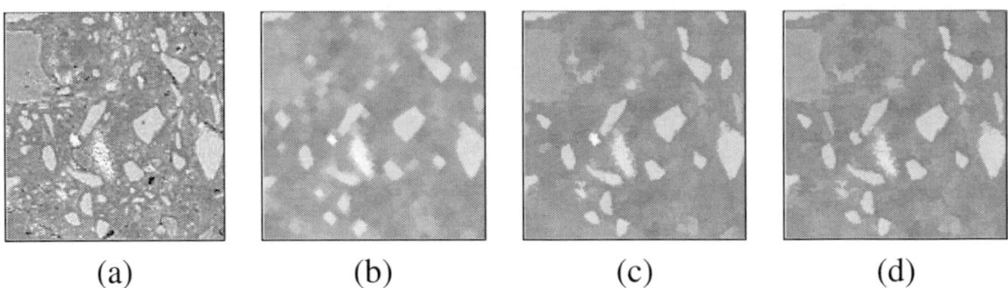

(a)   (b)   (c)   (d)

Figure 6.21 Connected flat alternating sequential filter: (a) input image, (b) alternating sequential filter close-open of stage 2, (c) reconstructive alternating sequential filter close-open of stage 2, (d) area open followed by area close.

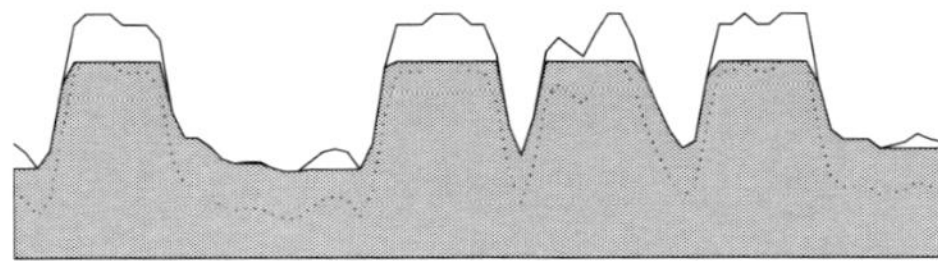

Figure 6.22 h-maxima: inf-reconstruction from the input signal subtracted by h. The solid line is the input signal, the dotted line is the signal subtracted by h and the signal over the shaded area is the h-maxima.

$$ASF^{area}_{co,a,D}(f) = (f \circ (a)_D) \bullet (a)_D. \tag{6.25}$$

Figure. 6.21 illustrates gray-scale image filtering using alternating sequential filters: (a) input image; (b) two-stage ASF based on close-open; (c) two-stage reconstructive ASF based on close-open; and (d) area open followed by area close.

6.8 Image Extrema

Erosion by a point structuring element of value h is equivalent to the pixelwise subtraction of h from the signal. The inf-reconstruction of f from f subtracted by h is called h-**maxima** and defined by

$$HMAX_{h,D}(f) = f \, \triangle_D \, (f - h). \tag{6.26}$$

It removes any dome with height less than h and decreases the height of the other domes by h. This effect is demonstrated in Fig. 6.22. Note that this operator is connected but not idempotent, as a further application will reduce the height of the domes. Neither does it exhibit the stack property. Nevertheless, it is one of the most popular operators for morphological gray scale processing. This operator is very powerful when used in composition with the regional maximum, soon to be introduced.

The dual operator of h-maxima is called h-**minima**:

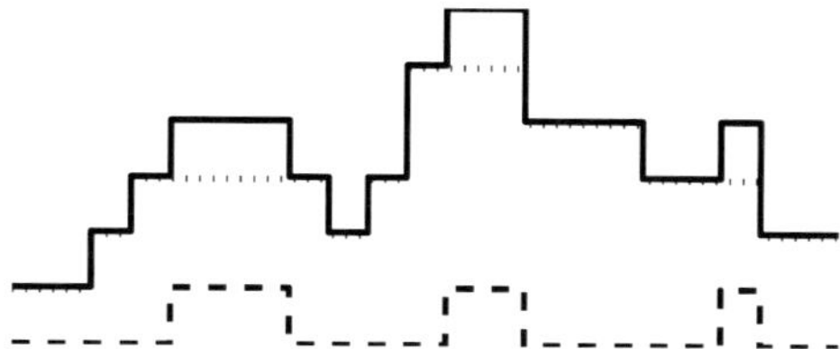

Figure 6.23 Regional maxima. The solid line is the input signal, the dotted line is the h-maxima with height 1. The subtraction of the h-maxima from the input signal marks the regional maxima flat zones.

$$HMIN_{h,D}(f) = f \nabla_D (f + h).$$ (6.27)

A **regional maximum** is a flat zone not adjacent to a flat zone with higher gray value. By duality, a **regional minimum** is a flat zone not adjacent to a flat zone with lower gray value. It is important not to confuse these concepts with **local maximum** and **local minimum**. Local maximum is a pixel property, not a regional property. A pixel is a local maximum if its value is greater than or equal to any pixel in its neighborhood.

There are many algorithms to compute regional maxima and minima. One of the simplest, yet efficient, methods is based on reconstruction. The regional maxima can be computed from the residue of the h-maxima of height 1:

$$RMAX_D(f) = f - HMAX_{1,D}(f).$$ (6.28)

By duality, the regional minima can be computed from the residue of the h-minima with height 1:

$$RMIN_D(f) = HMIN_{1,D}(f) - f.$$ (6.29)

Figure 6.23 illustrates the regional maximum computation using h-maxima.

6.9 Markers From Regional Maxima of Filtered Images

The next chapter will be devoted to morphological segmentation using the watershed transform. One of the crucial steps in the watershed transform is marker extraction. A marker must be placed in a representative sample of the region of the object to be extracted. In this section we will study some powerful methods for marker finding using the regional maxima (minima) of filtered images. One advantage of these methods is their independence of gray-scale thresholding values.

Typically, a real image presents a large number of regional maxima because of the inherent noise associated with the acquisition process. If the regional maximum operator is applied to a gradient image, then the situation is even worse. Filtering the domes of the image removes regional maxima. Dome filtering can be accomplished using opening, reconstructive opening, area open, or h-maxima. The choice of which filter to use is part of the design strategy.

Figure 6.24 shows the regional maxima of the input image following different filters: (a) input image; (b) regional maxima without filtering; (c) regional maxima following opening by a disk; (d) regional maxima following reconstructive opening by a disk; (e) regional maxima following area open; and (f) regional maxima following h-maximum. Note that the regional maxima are a binary image and as such the result can be seen as a segmentation strategy. This is an alternative to the common segmentation by thresholding.

The effect of the regional maxima reduction using connected filters can be illustrated by the connected components of the threshold decomposition of a gray-scale image. This is similar to building a cardboard mock-up of a topography of f. The cardboard is first cut into the shape of each isoline (lines of same height), then all pieces are stacked up to form the topography. Each cardboard piece is a connected component of a level set of f. Hence, we refer to these as **level components**. An antiextensive connected filter removes in full some level components from the top (never just from the middle). In contrast, a nonconnected filter, like opening by a disk, could diminish some level components by removing portions that the disk does not cover. A reconstructive opening removes all level components where the structuring element does not fit. The size-α area open removes all level components with area less than α. The **height** of a level component is the number of level components above it in the stacking plus 1. The h-maxima filter removes all level components with height below h. A level component is a regional maximum if it has height 1 [see Eq. (6.28)].

Figure 6.25(a) shows a signal f, with its 10 level components labeled in part (b). There are four regional maxima. Parts (c) and (d) show the height and **area** (number of pixels) of each level component, respectively.

If we define the **radius**, r, of a level component to be the maximum radius of a disk that fits inside of it, then reconstructive opening by a disk of radius r removes all level components with a radius less than r.

Area open is a size criterion, reconstructive opening is a shape criterion, and h-maxima is a contrast criterion. Other increasing criteria can be used to remove level components. A criterion that combines shape and contrast is volume. The **volume** of a level component is the sum of all the areas of the level components above it, including itself. This can be equivalently expressed by the integral of the umbra of the signal above the level component. The filter v-**maxima** removes all connected components with volume less than v. Figure 6.25(e) depicts the volume attribute of each level component of f.

To elucidate why the regional maxima are different depending on the type of filtering, we will consider the problem using a synthetic image of three Gaussian

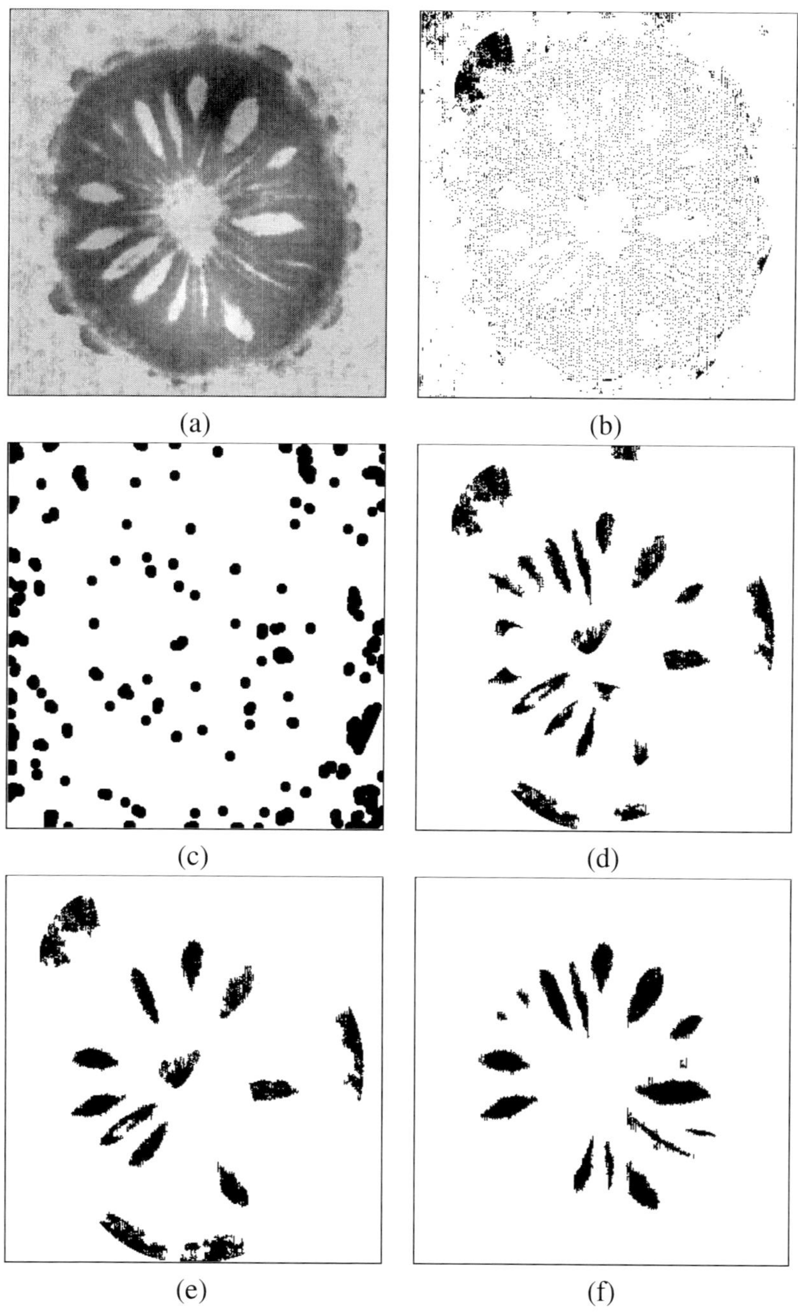

(a) (b)

(c) (d)

(e) (f)

Figure 6.24 Regional maxima using 8-connectivity: (a) input image, (b) regional maxima, (c) regional maxima after opening by a Euclidean disk of radius 3, (d) regional maxima after reconstructive opening by the same disk, (e) regional maxima after area open of 500 pixels, (f) regional maxima after h-maxima filtering with $h = 80$.

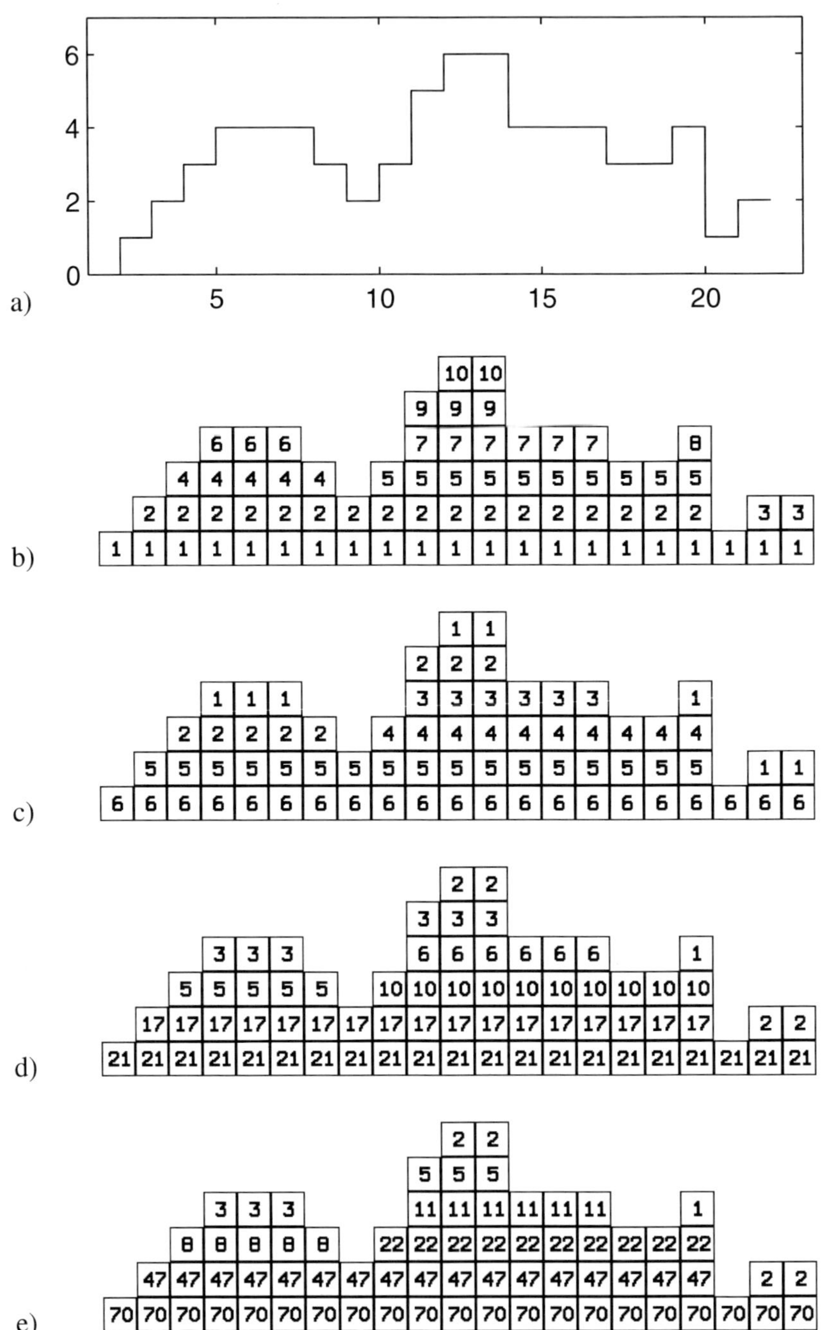

Figure 6.25 Features of connected components of threshold decompositions: (a) input signal, (b) level components, (c) height attribute, (d) area attribute, (e) volume attribute.

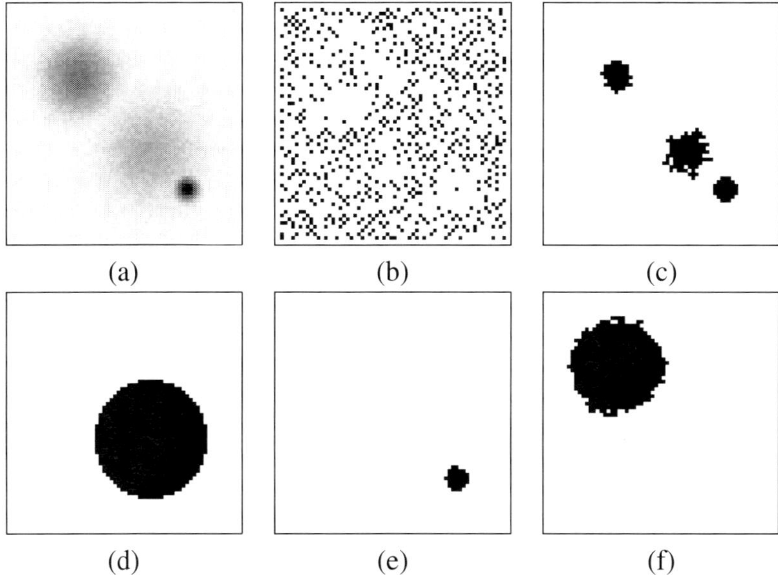

Figure 6.26 Regional maxima of synthetic noise image: (a) input image, (b) regional maxima, (c) regional maxima after reconstructive opening by small disk, (d) regional maxima after opening by a large disk, (e) regional maxima after h-maxima filtering, (f) regional maxima after v-maxima filtering.

peaks with some additive noise. Our interest is in design strategies to detect the domes. The matter is illustrated in Fig. 6.26: (a) synthetic image with added noise; (b) regional maxima appearing randomly over the image; (c) reconstructive opening by a small disk resulting in the regional maximum operator detecting just the three main domes; (c) opening by a larger disk resulting in selection of just the wider dome; (d) h-maxima resulting in just the tallest dome; and (f) the top-left dome resulting from v-maxima filtering.

6.10 Extinction Values

In the previous example, the exact values of the parameters for h-maxima, v-maxima and radius of the opening were chosen by trial and error. It is possible to find these values automatically by making a scale-space analysis using these operators. The idea is to apply the operators for all possible parameters and detect the least parameter value that makes a regional maximum vanish. One less than this value is called the **extinction value** λ of the regional maximum M. For a formal definition, if Ψ_λ is a connected operator with parameter λ that is decreasing in λ, meaning that $\mu \leq \lambda$ implies $\Psi_\lambda(f) \leq \Psi_\mu(f)$, then the extinction value of a regional maximum is the largest parameter of the connected filter that does not filter out the regional maximum, which means that after filtering, the regional maximum either remains as is or becomes a subset of a new regional maximum for all parameter values less than or equal to this maximum parameter. For the signal shown in Fig. 6.25, from left to right the extinction values of the four regional maxima are 2, 6, 1, and 1 for height; 5, 21, 1, and 2 for area; and 8, 70, 1, and 2 for volume. For instance, in the case of height, we see in Fig. 6.25(c) that the parameter 2 leaves the 3-pixel-peak domain within the new 5-pixel-peak domain, but the parameter 3 results in the regional maximum being at the next peak to the right so that the 3-pixel domain of the first peak is no longer a subset of the regional maximum. The second regional maximum is the most predominant extremum of all four using any attribute of height, area, or volume.

Computing the extinction values for the height, area, and volume of the image with the three Gaussian peaks with some additive noise shown in Fig. 6.26, the five highest extinction values in the image for each attribute are:

attribute	peak				
	top-left	central	brightest	noise1	noise2
height	117	40	**255**	12	11
area	445	**4,096**	75	75	75
volume	**137,108**	18,044	3,427	291	291

The largest extinction value for each attribute is marked in boldface. The tallest narrow peak is the most prominent peak relative to the height attribute, the middle peak is the one with the largest area, and the top-left one has the largest combined contrast and size criteria. Note that when using the area criterion, the narrow peak is confused with the noise.

Dynamics provide a tool for selecting significant extrema with respect to a contrast criterion. The **dynamic** of a maximum is the height we have to climb down from the maximum to reach another maximum of higher altitude. The concept of extinction values generalizes the dynamics concept. The height extinction value of an extremum is the dynamic. There are efficient algorithms to compute the extinction values using the component tree data structure or union-find algorithms.

The extinction technique is also useful for signal processing. Figure 6.27 illustrates regional maximum processing of a signal with four prominent peaks: (a) original signal with 10 regional maxima owing to the intrinsic signal noise; (b) the height extinction values (dynamics) of the regional maxima revealing the four most prominent peaks; (c) the regional maxima of the h-maxima signal, also revealing the four prominent peaks; and (d) the reconstruction from the four regional maximum with the highest dynamics. For 1D signals, using opening, area open, or reconstructive opening gives the same results if the structuring element is a single connected component.

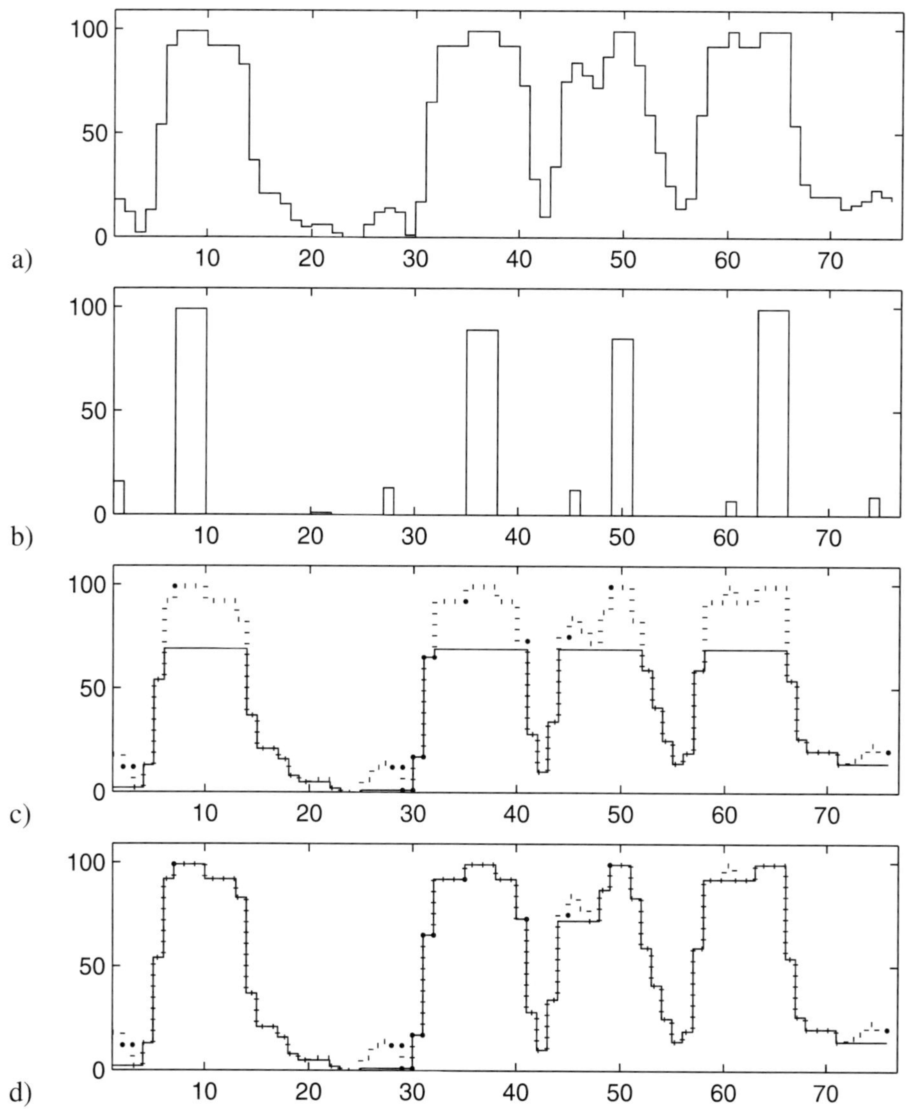

Figure 6.27 Regional maxima of real signal: (a) input signal, (b) dynamics of the regional maxima, (c) h-maxima of level 30 gives the four most relevant maxima, (d) reconstruction of the signal from the markers at the four most relevant regional maxima.

6.11 Demonstration

Here we present a complete demonstration of a real-world example using the **MT**. In this example, a satellite image of an airport is processed. The aim is to detect the runways, which are characterized by thin, long and straight features. An open top-hat is first used to enhance the runways. A rough thresholding followed by a thinning operator gives most of the thin structures of the image. They can be filtered by selecting only those features with sufficient length. This result is used as a marker to reconstruct the original gray-scale image, giving as output a gray-scale enhanced airport runway image, which can be finally thresholded.

The **MT** functions used in this demonstration are shown below.

name	description
`mmopenth:`	gray-scale open top-hat [Eq. (6.4)]
`mmareaopen:`	gray-scale area open [Eq. (6.22)]
`mminfrec:`	gray-scale inf-reconstruction [Eq. (6.13)]
`mmthin:`	sequential thinning (Fig. 4.6)
`mmaddm:`	addition
`mmunion:`	union
`mmthreshad:`	thresholding [Eq. (5.51)]
`mmgray:`	converts from binary to gray scale
`mmendpoints:`	hit-or-miss template for end point detection
`mmsebox:`	3×3 box structuring element
`mmsedisk(5):`	Euclidean disk structuring element of radius 5
`mmreadgray:`	read image file

The program is shown in Fig. 6.28 and all the images are shown in Figs. 6.29 and 6.30. The steps used to detect the airport runways are below. The program line numbers and figure parts are shown in brackets.

Image reading (line 1) The gray-scale aerial airport image is read (a).

Runways enhancing (lines 2–3) The top-hat is used to detect the runways as they are thin features (b). The disk of radius 5 (diameter 11) is chosen to detect features smaller than this size. For visualization, the top-hat image is brightened by 150 gray-levels (c).

Thresholding (line 4) A thresholding is applied to detect the features enhanced by the top-hat (d). This is a standard top-hat sequence.

Thinning and area open (lines 5–6) The thinning is used, as runways are long thin structures, to characterize the runways structure (e). The area open selects only very long features, with more than 1000 pixels (f). Note that

```
                        Program 6.1
01   a = mmreadgray('galeao.jpg')
02   b = mmopenth( a, mmsedisk(5))
03   c = mmaddm(b, 180) # just for visualization
04   d = mmthreshad( b,30)
05   e = mmthin(d)
06   f = mmareaopen(e, 1000, mmsebox())
07   g = mminfrec( mmgray(f), b)
08   h = mmthreshad( g, 20)
09   i = mmunion(a, mmgray(h))
```

Figure 6.28 Python code for the airport runway detection. The images from (a) to (i) are shown in Figs. 6.29 and 6.30.

an 8-neighborhood is used for the area open as the skeleton obtained from `mmthin` is 8-connected.

Reconstruction (line 7) The previous result is a sample of the runway pixels. It is used as a marker for gray-scale morphological reconstruction. The runways are enhanced in the reconstructed image (g).

Final thresholding (lines 8–9) A thresholding is applied to the reconstructed image, detecting the airport runways (h). For visualization purposes, the result is overlaid on the original (i).

6.12 Exercises

1. For signals, compare the morphological gradient of $f_1 = (1\ 2\ 3\ 4\ \mathbf{5}\ 6\ 7\ 8\ 9)$ and $f_2 = (1\ 2\ 3\ \mathbf{9}\ 5\ 6\ 7\ 8\ 9)$ using the structuring element $g = (0\ \mathbf{0}\ 0)$. What does this tell you about the effect of noise on the gradient?

2. Using the signals f_2 and g of the previous exercise, compute the open top-hat $f_2 \hat{\circ} g$.

3. Prove the top-hat complementary relation of Eq. (6.6).

4. Let

$$f = \begin{pmatrix} 0 & 0 & 0 & 0 & 0 \\ 6 & 5 & 6 & 0 & 1 \\ 7 & 0 & 7 & 0 & 2 \\ 4 & 0 & 0 & 0 & 3 \\ 5 & 3 & 2 & 8 & 9 \end{pmatrix}$$

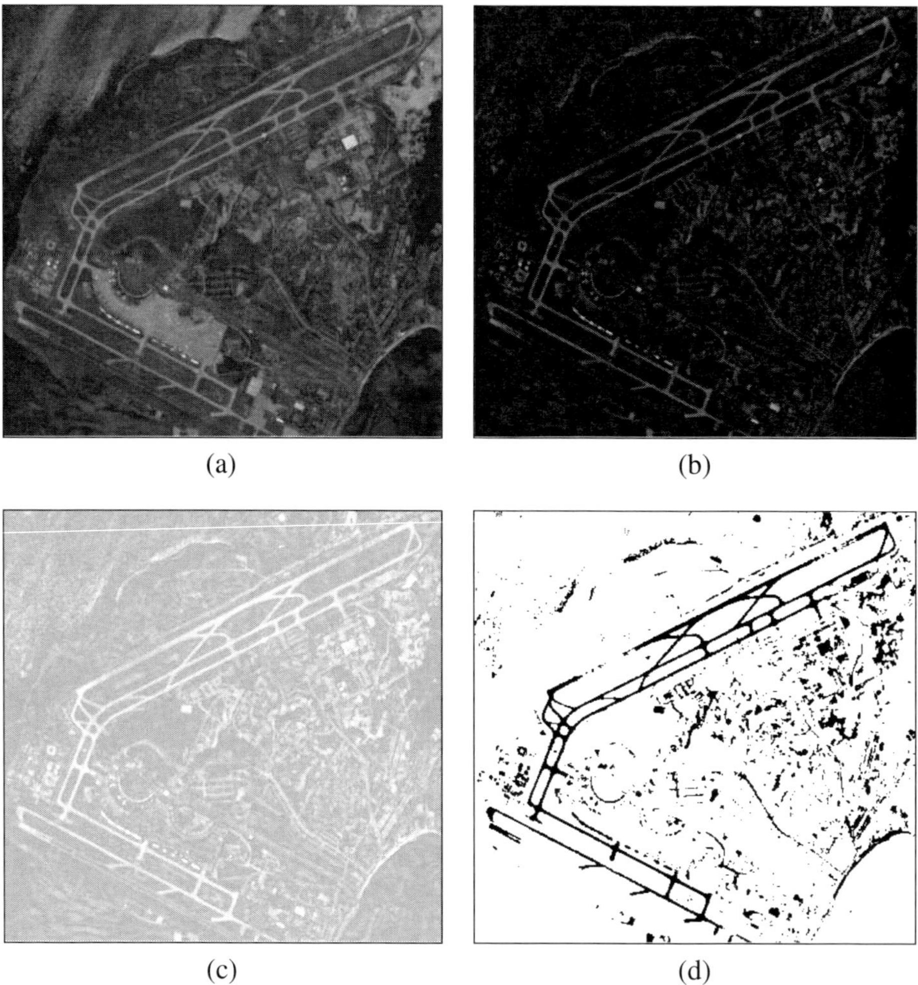

Figure 6.29 Airport runway detection. These images refer to the program of Fig. 6.28.

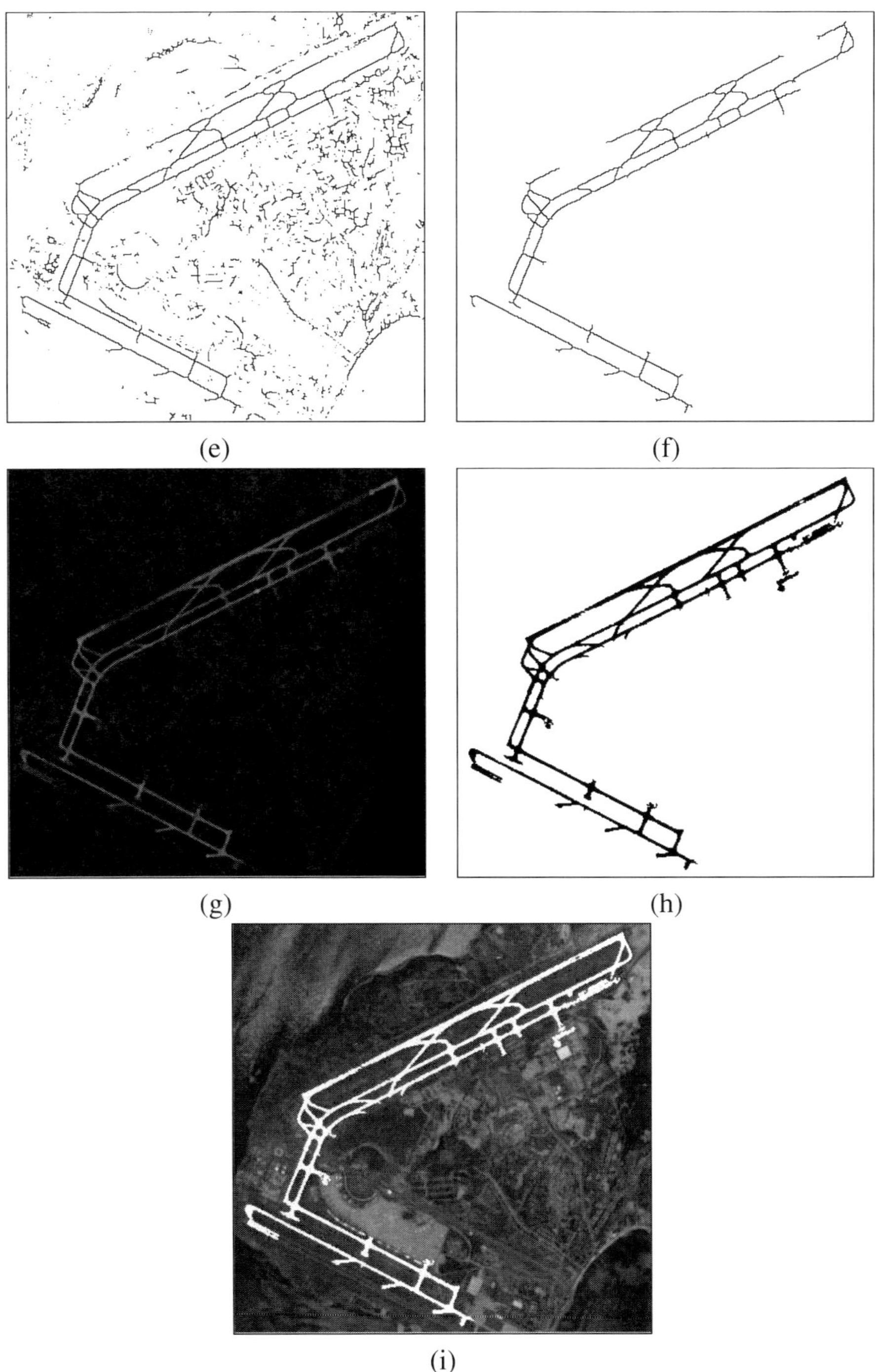

Figure 6.30 Airport runway detection. These images refer to the program of Fig. 6.28.

name	description
mmgradm:	morphological gradient [Eq. (6.1)]
mmopenth:	open top-hat [Eq. (6.4)]
mmcloseth:	close top-hat [Eq. (6.5)]
mmasf:	gray-scale ASF [Eqs. (6.9) and (6.10)]
mmcdil:	size-n geodesic dilation [Eq. (6.11)]
mminfrec:	inf-reconstruction [Eq. (6.13)]
mmcero:	size-n geodesic erosion [Eq. (6.15)]
mmsuprec:	sup-reconstruction [Eq. (6.17)]
mmclohole:	closing of holes (Fig. 6.14)
mmlabelflat:	flat zone labeling (Fig. 6.15)
mmareaopen:	area open [Fig. 6.19(b)]
mmasfrec:	connected ASF [Eqs. (6.23) and (6.24)]
mmhmax:	h-maxima [Eq. (6.26)]
mmhmin:	h-minima [Eq. (6.27)]
mmregmax:	regional maximum [Eq. (6.28)]
mmregmin:	regional minimum [Eq. (6.29)]
mmvmax:	v-maxima (Fig. 6.26)
mmregmin:	dynamics [Fig. 6.27(b)]

Figure 6.31 MT operators presented in this chapter.

and

$$
g = \begin{pmatrix} 0 & 0 & 0 & 0 & 0 \\ 0 & 0 & 0 & 0 & 0 \\ 0 & 0 & 9 & 0 & 0 \\ 0 & 0 & 0 & 0 & 0 \\ 0 & 0 & 0 & 0 & 0 \end{pmatrix}.
$$

Find $f \, \triangle_{E_4} \, g$, and $f \, \triangle_{E_8} \, g$.

5. Find $HMAX_{1,E_4}(f)$, and $HMAX_{1,E_8}(f)$, using the image f defined in the previous exercise.

6. Find $RMAX_{E_4}(f)$, and $RMAX_{E_8}(f)$, using the image f defined in Exercise 4. Verify Eq. 6.28.

7. Compute the extinction values for the height, area and volume of the image f defined in Exercise 4 in two situations: using structuring elements E_4, and E_8.

6.13   Laboratory Experiments

Figure 6.31 lists the **MT** functions discussed in this chapter.

1. Take the morphological gradient of a real-world gray-scale image and observe the effect of changing the size and shape of the structuring element

used for the internal and external gradients.

2. Propose and implement an alternative solution for the closing-of-holes illustration in Fig. 6.14 so that the three holes of the PCB at the bottom of the picture are also filled.

3. For the area open top-hat example illustrated at Fig. 6.19, compare the threshold ranges to detect the letters of the keyboard when applied to the original image and to the top-hat image.

4. Compute the number of flat-zones for the four images illustrated in Fig. 6.21.

References

1. J. T. Astola and P. Kuosmanen. Invariant signals of median and stack filters. In E. R. Dougherty and J. T. Astola, editors, *Nonlinear Filters for Image Processing*, pages 281–288. SPIE/IEEE Presses, Bellingham, WA, 1999.

2. J. T. Astola and Y. Neuvo. Statistical properties of discrete morphological filters. In E. R. Dougherty and J. T. Astola, editors, *Nonlinear Filters for Image Processing*, pages 93–120. SPIE/IEEE Presses, Bellingham, WA, 1999.

3. E.J. Breen and R. Jones. Attribute openings, thinnings, and granulometries. *Computer Vision and Image Understanding*, 64:377–389, 1996.

4. J. Crespo, H. Billhardt, J. Rodrigues-Pedrosa, and A. Sanandres. Methods and criteria for detecting significant regions in medical image analysis. In J. Crespo, V. Maojo, and F. Martin, editors, *Medical Data Analysis*. Springer-Verlag, 2001.

5. J. Crespo and V. Maojo. New results on the theory of morphological filters by reconstruction. *Pattern Recognition*, 31(4):419–429, 1998.

6. J. Crespo and R. W. Schafer. Locality and adjacency stability constraints for morphological connected operators. *Journal of Mathematical Imaging and Vision*, 7(1):85–102, 1997.

7. J. Crespo, J. Serra, and R.W. Schafer. Theoretical aspects of morphological filters by reconstruction. *Signal Processing*, 47(2):201–225, 1995.

8. E. R. Dougherty, Y. Chen, and A. Waks. Bayesian morphological peak estimation and its application to chromosome counting via fluorescence in situ hybridization. *Pattern Recognition*, 29(6):987–996, 1996.

9. J. Goutsias, H. J. A. M. Heijmans, and K. Sivakumar. Morphological operators for image sequences. *Computer Vision and Image Understanding*, 62:326–346, 1995.

10. A. Grigoryan, E. R. Kononen, L. Bubendorf, G. Hostetter, and O. Kallioniemi. Morphological spot counting from stacked images for automated analysis of gene copy numbers by fluorescence. *Biomedical Optics*, 7(1):109–122, 2002.

11. H. J. A. M. Heijmans. Easy recipes for morphological filters. In E. R. Dougherty and J. T. Astola, editors, *Nonlinear Filters for Image Processing*, pages 163–205. SPIE/IEEE Presses, Bellingham, WA, 1999.

12. H. J. A. M. Heijmans and L. Vincent. Graph morphology in image analysis. In E. R. Dougherty, editor, *Mathematical Morphology in Image Processing*, chapter 6, pages 171–203. Marcel Dekker, New York, 1993.

13. M. Khabou and P. D. Gader. Automatic target detection using entropy-optimized shared-weight neural networks. *IEEE Transactions on Neural Networks*, 11(1):186–194, 2000.

14. M. Khabou, P. D. Gader, and J. M. Keller. Ladar target detection using morphological shared-weight neural networks. *Machine Vision and Applications*, 11(6):300–305, 2000.

15. M. Khabou, P. D. Gader, and H. Shi. Entropy optimized morphological shared-weight neural networks. *Optical Engineering*, 38(2):263–273, 1999.

16. P. Maragos and F. Meyer. Nonlinear PDEs and numerical algorithms for modeling leveling and reconstruction filters. In M. Nielsen, P. Johansen, O.F. Olsen, and J. Weickert, editors, *Lecture Notes in Computer Science, Scale-Space Theories in Computer Vision*, 1682:363–374. Springer-Verlag, 1999.

17. F. Meyer and P. Maragos. Nonlinear scale-space representation with morphological levelings. *Journal of Visual Communication and Image Representation*, 11(3):245–265, 2000.

18. G. Piella and H. J. A. M. Heijmans. Adaptive lifting schemes with perfect reconstruction. *IEEE Transactions on Signal Processing*, 50(7):1620–1630, 2002.

19. J. F. Rivest, P. Soille, and S. Beucher. Morphological gradients. *Journal of Electronic Imaging*, 2(4):326–336, 1993.

20. C. Ronse. Order-configuration functions: mathematical characterizations and applications to digital and image processing. *Information Sciences*, 50(3):275–327, 1989.

21. C. Ronse. Removing and extracting features in images using mathematical morphology. *CWI Quarterly*, 11:439–457, 1998.

22. P. Salembier. Morphological multiscale segmentation for image coding. *Signal Processing*, 38(3):359–386, 1994.

23. P. Salembier, P. Brigger, J. R. Casas, and M. Pardas. Morphological operators for image and video compression. *IEEE Transactions on Image Processing*, 5(6):881–898, 1996.

24. P. Salembier and L. Garrido. Binary partition tree as an efficient representation for image processing, segmentation, and information retrieval. *IEEE Transactions on Image Processing*, 9(4):561–576, 2000.

25. P. Salembier and M. Kunt. Size-sensitive multiresolution decomposition of images with rank order based filters. *Signal Processing*, 27:205–241, 1992.

26. P. Salembier, A. Oliveras, and L. Garrido. Anti-extensive connected operators for image and sequence processing. *IEEE Transactions on Image Processing*, 7(4):555–570, 1998.

27. P. Salembier and J. Serra. Flat zones filtering, connected operators, and filters by reconstruction. *IEEE Transactions on Image Processing*, 4(8):1153–1160, 1995.

28. P. Salembier, L. Torres, F. Meyer, and C. Gu. Region-based video coding using mathematical morphology. *Proceedings of the IEEE*, 83(6):843–857, 1995.

29. J. Serra. Anamorphoses and function lattices (multivalued morphology). In E. R. Dougherty, editor, *Mathematical Morphology in Image Processing*, chapter 13, pages 483–523. Marcel Dekker, New York, 1993.

30. I. Shmulevich, K. Egiazarian, O. Yli-Harja, and J. T. Astola. Output distributions of stack filters using ordered binary decisions diagrams. *Journal of Signal Processing*, 4(2):195–200, 2000.

31. I. Shmulevich, V. Melnik, and K. Egiazarian. The use of sample selection probabilities for stack filter design. *IEEE Signal Processing Letters*, 7(7):189–192, 2000.

32. I. Shmulevich, J. L. Paredes, and G. R. Arce. Output distributions of stack filters based on mirrored threshold decomposition. *IEEE Transactions on Signal Processing*, 49(7):1454–1460, 2001.

33. I. Shmulevich, O. Yli-Harja, J. Astola, and A. Korshunov. On the robustness of the class of stack filters. *IEEE Transactions on Signal Processing*, 50(7):1640–1649, 2002.

34. I. Shmulevich, O. Yli-Harja, K. Egiazarian, and J. T. Astola. Output distributions of recursive stack filters. *IEEE Signal Processing Letters*, 6(7):175–178, 1999.

35. K. Sivakumar, M. J. Patel, N. Kehtarnavaz, Y. Balagurunathan, and E. R. Dougherty. A constant-time algorithm for erosions/dilations with applications to morphological texture feature computation. *Real-Time Imaging*, 6(3):223–239, 2000.

36. P. Soille. Spatial distributions from contour lines: an efficient methodology based on distance transformations. *Journal of Visual Communication and Image Representation*, 2(2):138–150, 1991.

37. P. Soille. Morphological partitioning of multispectral images. *Journal of Electronic Imaging*, 5(3):252–265, 1996.

38. P. Soille, E. J. Breen, and R. Jones. Recursive implementation of erosions and dilations along discrete lines at arbitrary angles. *IEEE Transactions on Pattern Analysis and Machine Intelligence*, 18(5):562–567, 1996.

39. S. Takriti and P. D. Gader. Local decompositions of gray-scale morphological templates. *Mathematical Imaging and Vision*, 2(1):39–50, 1992.

40. L. Vincent. Morphological algorithms. In E. R. Dougherty, editor, *Mathematical Morphology in Image Processing*, chapter 8, pages 255–288. Marcel Dekker, New York, 1993.

41. L. Vincent. Morphological grayscale reconstruction in image analysis: efficient algorithms and applications. *IEEE Transactions on Image Processing*, 2:176–201, 1993.

42. D. Wang and J. Ronsin. Bounded gray-level morphology and its applications to image representation. *IEEE Transactions on Image Processing*, 5:1067–1073, 1996.

43. Y. Won, P. D. Gader, and P. C. Coffield. Shared-weight neural networks based on mathematical morphology with applications to automatic target recognition. *IEEE Transactions on Neural Networks*, 8(5):1195–1204, 1997.

44. A. Yuille, L. Vincent, and D. Geiger. Statistical morphology and Bayesian reconstruction. *Journal of Mathematical Imaging and Vision*, 1(3):223–238, 1992.

Chapter 7

Morphological Segmentation—Watershed

The watershed transform is a key building block for morphological segmentation. In particular, a gray-scale segmentation methodology results from applying the watershed to the morphological gradient of an image to be segmented. The watershed methodology has become highly developed to deal with numerous real-world contingencies, and a number of implementation algorithms have been developed.

A classical approach for producing edge images is to apply a gradient and then threshold the resulting gradient image to produce a binary edge image. A salient difficulty with this approach is the selection of an appropriate threshold value. Even if we ignore the issue of producing false edges by choosing too low a threshold, problems remain. If the threshold is too low, then the edges can be very wide and require extensive thinning, which might leave them inaccurate; if the threshold is too high, then many edges will not be detected and those that are may be severely broken. The conundrum is illustrated by attempting to apply the thresholded-gradient procedure to segmenting the lean meat of an image of a raw beef steak. Figure 7.1 shows (a) the original image, (b) its morphological gradient, (c) the result of a low threshold, and (d) the result of a high threshold. Note how difficult it would be to construct a closed contour by choosing and tracking the thresholded edges.

The literature has devoted many complex methods to deal with this segmentation problem. The watershed approach is a simple yet powerful tool to solve this type of segmentation. Figure 7.2 illustrates the beef segmentation based on the watershed transform (with details to be subsequently explained). For this example, we use the watershed from markers. This watershed requires a set of markers. Each marker must be placed on a sample region of the object to be segmented. In this

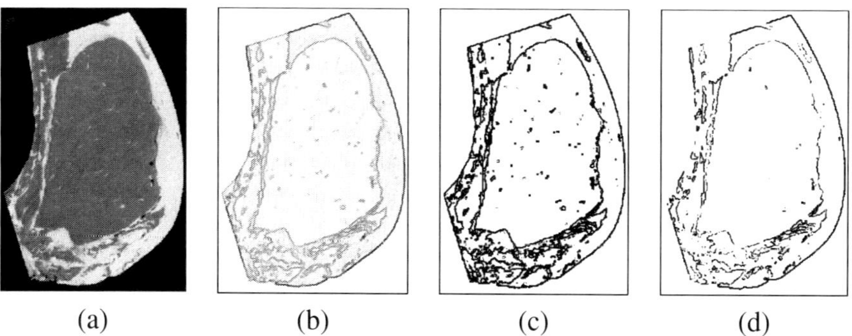

(a)   (b)   (c)   (d)

Figure 7.1 Segmentation by thresholded-gradient: (a) input image, (b) morphological gradient, (c) low threshold, (d) high threshold.

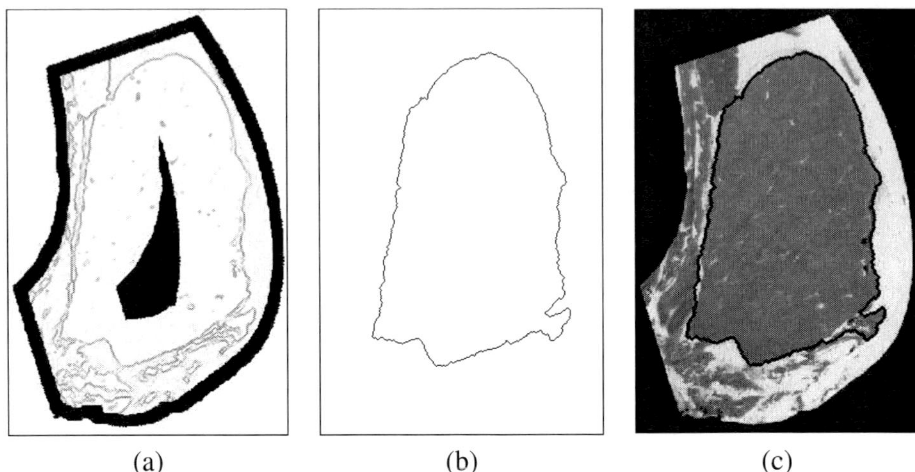

(a)   (b)   (c)

Figure 7.2 Segmentation by watershed: (a) inner and outer marker overlaid morphological gradient image, (b) watershed line, (c) illustration of the watershed line overlaid on input image.

case, we use two markers: one for the lean meat (inner marker) and the other for the fat (outer marker). These two markers are illustrated in Fig. 7.2(a), overlaid on the morphological gradient. Marker design in watershed-based segmentation is one of the most crucial steps for a successful solution. We have discussed in Chapter 6 several techniques to extract markers from images and we will use them in this chapter to illustrate different approaches for watershed-based segmentation.

As a motivation, we show in Fig. 7.2 a solution for the beef segmentation using the watershed from markers: (a) the gradient image overlaid by two markers, one inside and another outside of the object to segment, (b) the watershed line, and (c) the watershed line overlaid on the input image to illustrate the quality of the segmentation.

In fact, there are many watershed algorithms in the literature, and we will describe some commonly used ones here. Perhaps the most intuitive formulation of the **watershed transform** is the one based on a flooding simulation. Consider the input gray-scale image as a topographic surface. The problem is to produce the watershed lines on this surface. To do so, holes are punched in each regional minimum of the image. The topography is slowly flooded from below by letting water rise from each regional minimum at a uniform rate across the entire image. When the rising water coming from distinct minima is about to merge, a dam is built to prevent the merging. The flooding will eventually reach a stage when only the tops of dams are visible above the waterline, and these correspond to the **watershed lines**. The final regions arising from the various regional minima are called the **catchment basins**.

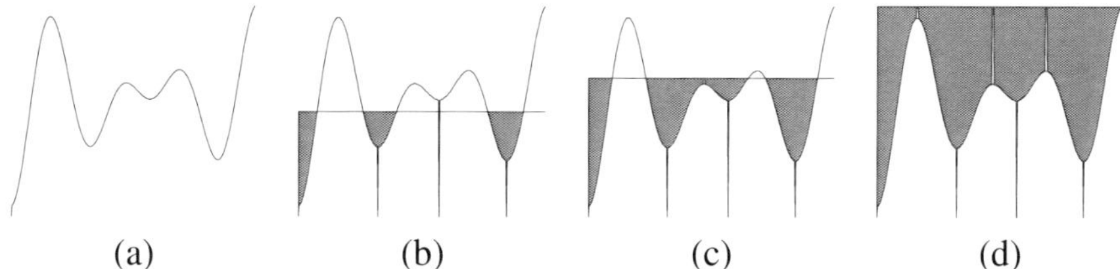

Figure 7.3 Flooding simulation of the watershed transform: (a) input signal, (b) punched holes at minima and initial flooding, (c) a dam is created when water from different minima is about to merge, (d) final flooding, three watershed lines and four catchment basins.

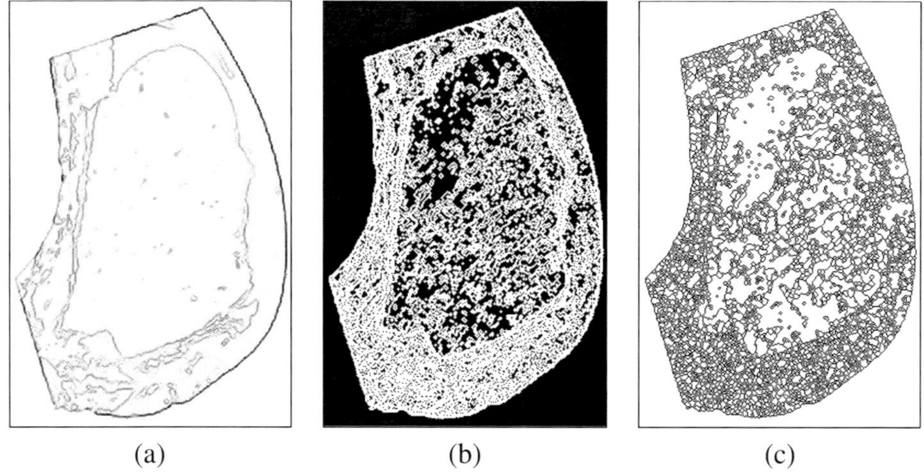

Figure 7.4 Watershed oversegmentation: (a) morphological gradient image, (b) regional minima, (c) watershed lines (oversegmentation).

Figure 7.3 illustrates this flooding process on a signal with four regional minima generating four catchment basins. The figure shows some steps of the process: (a) input image, (b) holes punched at minima and initial flooding, (c) dam created when waters from different minima are about to merge, and (d) final flooding yielding three watershed lines and four catchment basins.

For image segmentation, the watershed is usually applied on a gradient image. As real digitized images present many regional minima in their gradients, this typically results in a large number of catchment basins. This characteristic, called "watershed oversegmentation," is illustrated in Fig. 7.4: (a) the morphological gradient, (b) regional minima, and (c) catchment basins of the watershed transform originating from all regional minima. To avoid oversegmentation, one needs to apply a filter to the input image (usually a gradient) of the watershed transform to reduce the number of regional minima and thereby place flooding holes at desirable locations. This technique is described next.

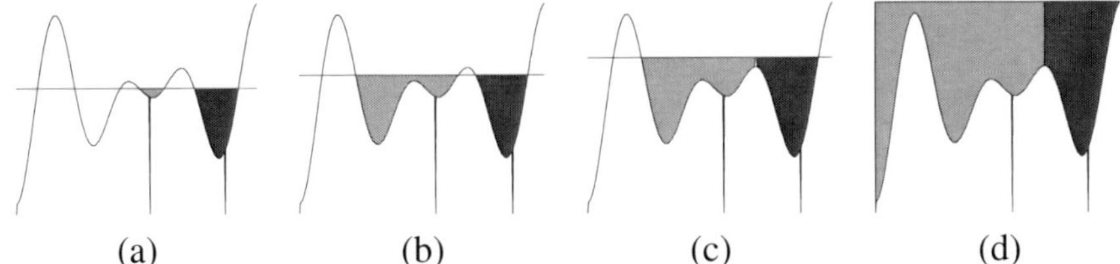

Figure 7.5 Flooding simulation of the watershed from markers: (a) punched holes at markers and initial flooding, (b) flooding a primitive catchment basin without marker, (c) a dam is created when water coming from different markers is about to merge, (d) final flooding, only one watershed line.

7.1 Watershed From Markers

The watershed from markers is a very effective way to reduce oversegmentation if one knows how to place the markers within the objects to be segmented. The watershed from markers can also be described as a flooding simulation process. In this case, holes are punched at the marker regions. Each marker is associated with a color. The topography is flooded from below by letting colored water rise from the hole associated with its color, this being done for all holes at a uniform rate across the entire image. If the water reaches a catchment basin with no marker in it, then the water floods the catchment basin without restriction. However, if the rising waters of distinct colors are about to merge, then a dam is built to prevent the merging. The colored regions are the catchment basins associated with the various markers. To differentiate these catchment basins from the ones obtained with the regular watershed transform, we call the latter **primitive catchment basins**.

Figure 7.5 illustrates the flooding of the watershed from markers in a signal. There are two markers placed into the two rightmost primitive catchment basins. Part (a) shows the two holes punched at the markers and some initial flooding. When the water rises, a primitive catchment basin without marker is flooded without creating a dam, as shown in part (b). In part (c), a dam is built to prevent the merging of waters coming from two markers. Finally, part (d) shows the final flooding with only one watershed line separating the two markers.

Returning to the beef segmentation illustration of Fig. 7.2, to use the watershed from markers we can place an inner marker in the center of the image, as we are sure the lean part of the beef is there, and we can place an outer marker near the edge of the whole beef. We repeat the same flooding process of the watershed from markers illustrated previously for the signal, but now in two dimensions using the beef example in Fig. 7.6: (a) the gradient image; (b) the gradient surface punched by different colors at the inner and outer markers; (c) initial flooding; (d) dam building to prevent the merging of different colors; (e) progressive flooding of many primitive catchment basins; and (f) the whole surface flooded and the final watershed line separating the catchment basins originating from the inner and outer

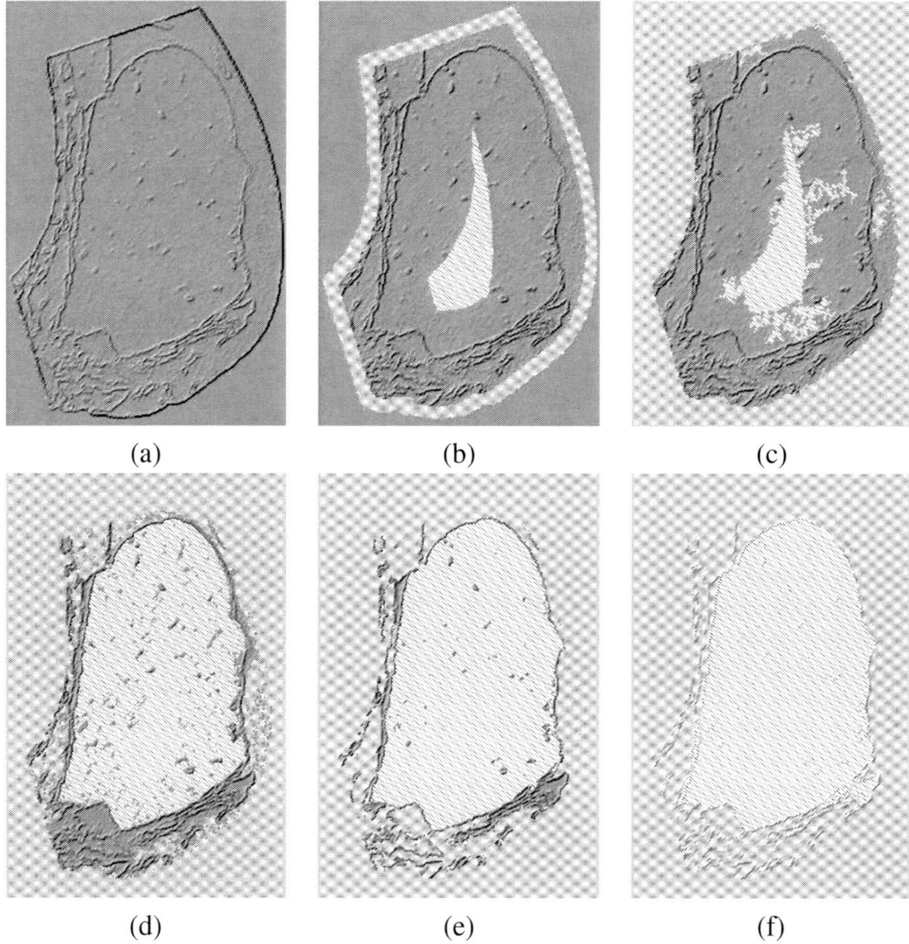

(a) (b) (c)

(d) (e) (f)

Figure 7.6 Watershed as a flooding process: (a) gradient image (shaded surface view), (b) holes punched at the inner and outer markers, (c) initial flooding, (d) when touching, builds a dam, (e) progressive flooding, (f) final flooding and inner and outer catchment basins.

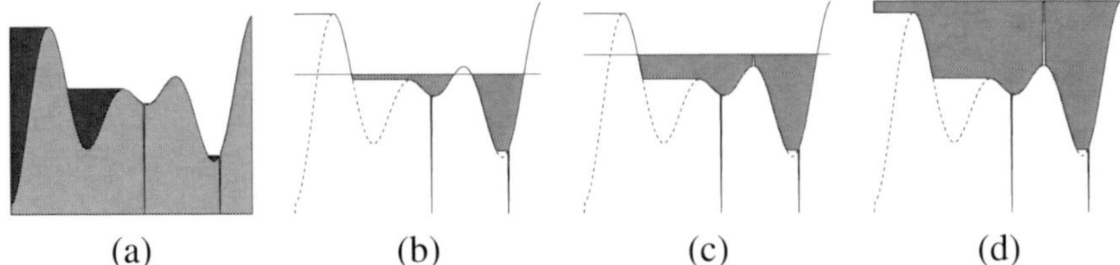

(a)   (b)   (c)   (d)

Figure 7.7 Watershed from markers as watershed on minima imposed by markers: (a) input image in light gray, sup-reconstruction in dark gray, (b) classical watershed on the reconstructed image, (c) a dam is created, (d) final flooding, one watershed line and two catchment basins.

markers.

The classical watershed transform can be constructed using the watershed from markers and vice-versa. If we place the markers for the watershed from markers at the regional minima of the input image, then we get the regular watershed transform. The three input parameters for the watershed from markers, $WSM_D(f, m)$, are: the input image f (usually a gradient), the marker image m, and the connectivity defined by the structuring element D. Then, the classical watershed of f, $WS_D(f)$, is given by

$$WS_D(f) = WSM_D(f, RMIN_D(f)). \tag{7.1}$$

To get the watershed from markers from the standard watershed transform is a bit more complicated. It can be achieved by reconstruction. The idea is to preprocess the input image guided by the markers in such a way that the filtered image has minima only at the markers. This transformation, called **minima imposition**, is given by

$$MI_D(f, m) = f \vee f \, \triangledown_D m^{\star}, \tag{7.2}$$

where $m^{\star}$ is 0 at the marker pixels (as defined by m) and ∞ for all other pixels.

The procedure is illustrated in the 1D example of Fig. 7.7: (a) the input signal and the sup-reconstructed signal, where the regional minima are guided by the markers; (b) initial flooding of the reconstructed image using the classical watershed based on regional minima, with the dotted line showing the original input signal for the purpose of illustration; (c) a dam created to avoid merging of the waters coming from two regional minima; and (d) the final flooding with two catchment basins, which is the same result obtained by using the watershed from markers, as illustrated in Fig. 7.5. Note in part (b) that the reconstructed signal has minima only at the markers.

While it is often the case that the watershed transform is applied to a gradient image, in some cases the input image itself is suitable for application of the watershed. This situation is illustrated in Fig. 7.8, which shows a watershed-based segmentation of a micrograph image of a cedar-tree cross section. The input image, shown in part (a), is already suitable for the watershed transform because the cells are separated by a thin line. If a direct watershed is computed, then the over-segmentation shown in part (b) is obtained. In the beef-segmentation example, the markers were easily placed in the inner and the outer parts of the beef. In the case of the cedar tree, the markers can be detected by regional minima simplification. As described in the previous chapter, the regional minima of a filtered image are important characteristics to detect. In this example, we filter the regional minima using three morphological approaches: the area close, the closing by a Euclidean disk, and the h-minima operator. All three filters are extensive and do not change the crest lines of the topographic surface, thereby preserving the places to detect the watershed lines. All three operators reduce the number of regional minima in the image. Following minima reduction, we can detect the minima of the filtered image and then use the watershed from markers where the detected minima form the markers; or we can apply the classical watershed directly on the filtered image. Parts (c), (d), and (e) of Fig. 7.8 show the watershed of the simplified image by area close, closing by a disk, and h-minima filtering, respectively. In this particular application, the watershed lines are rather similar among the three regional minima filtering approaches.

7.2 Watershed, Voronoi Diagram, and SKIZ

Making good use of the power of the watershed transform depends on creatively designing the input image and the markers. There are three main categories of input image: the gradient, the input image itself, and the negation of the distance transform. We now consider application of the watershed when computed on the negation of the distance transform.

Given a binary image, its **Voronoi diagram** is composed of lines that partition the plane into regions, each of which consists of the points closer to one particular grain (connected component) than to any of the other grains. The regions are called **influence zones**. The **Skeleton by Influence Zones** (or **SKIZ**) is composed of the boundaries of the various influence zones. The watershed transform is a suitable method to compute the Voronoi diagram and the SKIZ. The idea is to compute the classical watershed transform of the distance transform of the background of the objects. The catchment basins are the Voronoi regions and the watershed lines compose the SKIZ.

Figure 7.9 illustrates the computation of the Voronoi diagram of isolated points: (a) binary input image with points as grains, (b) distance function of the background to the points, and (c) watershed lines as the Voronoi diagram superimposed on the input image. This illustration gives a geometrically intuitive notion of the watershed transform. We can also observe the similarity of this image with many forms in biology. Many types of cells are grouped like Voronoi diagrams.

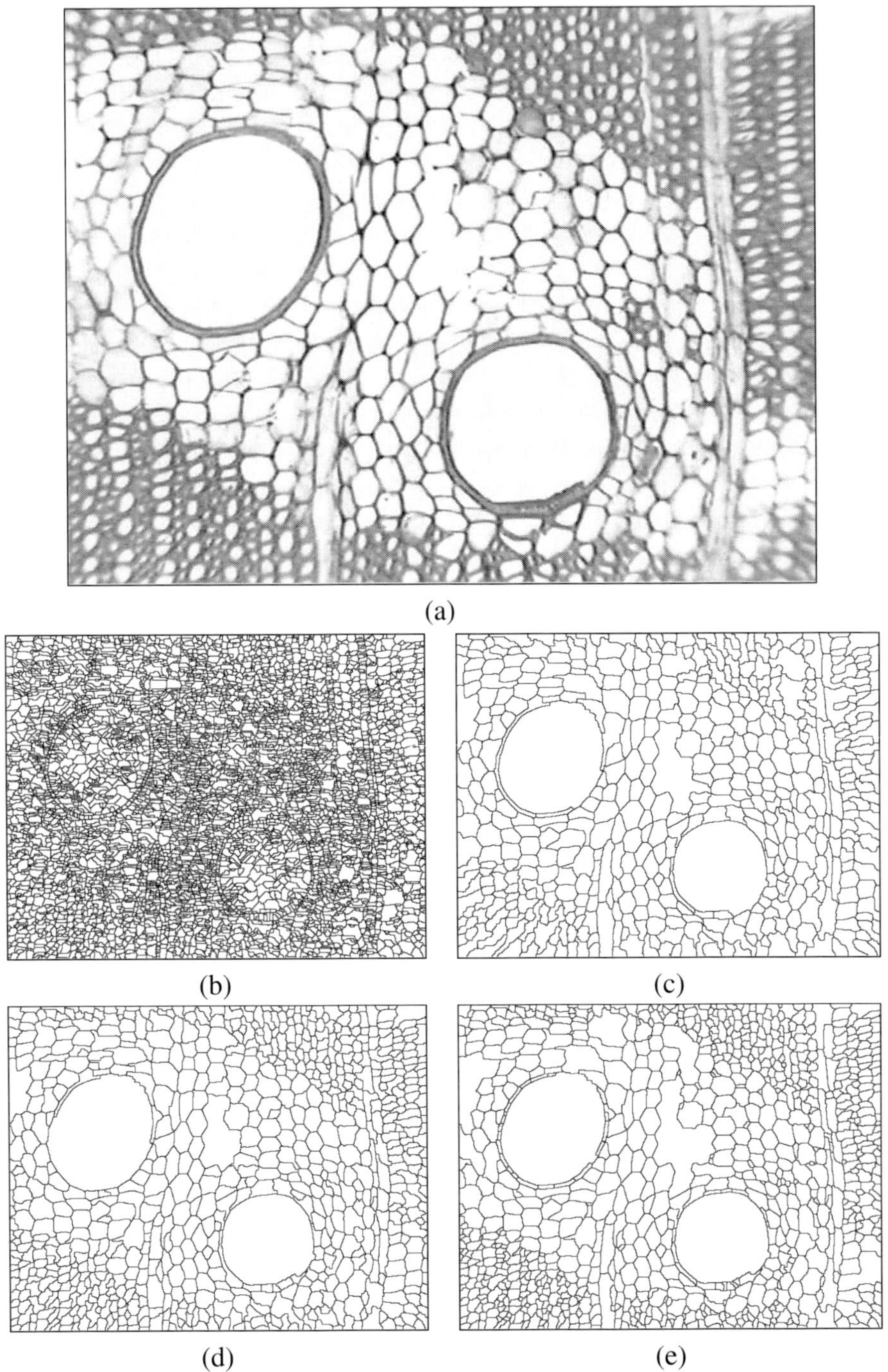

Figure 7.8 Watershed segmentation: (a) input gray-scale image of a micrograph of a cedar cross section, (b) watershed lines with typical over-segmentation, (c) minima reduction using area close, (d) minima reduction using close, (e) minima reduction using h-minima.

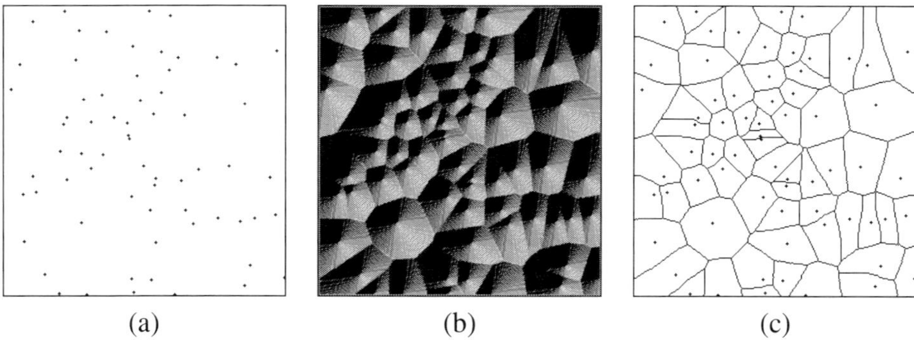

(a)   (b)   (c)

Figure 7.9 Voronoi diagram from watershed of the distance transform: (a) binary input image with points as grains, (b) distance function of the background to the grains, (c) watershed lines as the Voronoi diagram superimposed on the input image.

If the geodesic distance is used to define the Voronoi diagram instead of the Euclidean distance, then the boundaries of the geodesic influence zones form the **geodesic SKIZ** . An example of the geodesic SKIZ is given in Fig. 7.10, where we want to find the influence zones of the seeds in a fruit cross section: (a) input image overlaid by seeds, (b) geodesic distance of the points in the fruit to the seeds, and (c) watershed lines as the geodesic SKIZ overlaid on the input image.

7.3 Segmentation of Overlapped Convex Cells

One of the earliest purposes of the watershed transform was to address the problem of binary-image segmentation from the perspective of there being objects within the image that are touching or overlapping. For instance, in Fig. 7.11(a), there appears to be two objects overlapping to form a single connected component. Our goal is to segment the single component in a manner consistent with the integrity of each of the two objects. A key to this problem, and with many segmentation problems, is to find markers for each of the two objects.

The binary-image segmentation works in the following way and is illustrated in Fig. 7.11. The distance transform [part (b)] is computed and one marker is required for each cell. In this case of rounded cells, these markers can be extracted from the regional maxima of the distance function. Depending on the difficulty of this extraction, it may require filtering the distance function using an opening according to the methodology of marker extraction using regional maxima discussed in Sec. 6.9. The lines from the watershed transform of the negated distance function from the markers are used to cut the input binary image. The markers and the watershed lines are displayed in part (c) of Fig. 7.11 and the final segmentation is shown in part (d). The regional maxima of the distance transform compose the **ultimate erosion**, and the distance transform was one of the first methodologies used to extract

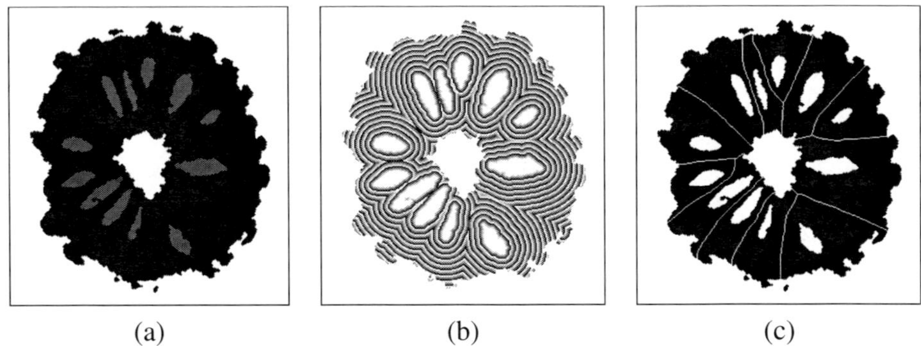

Figure 7.10 Geodesic SKIZ, the influence zones of the seeds in a fruit cross section: (a) input image overlaid by the seeds—this is the conditional image, (b) geodesic Euclidean distance of the points in the fruit to the seeds, (c) watershed lines as the geodesic SKIZ overlaid on the input image.

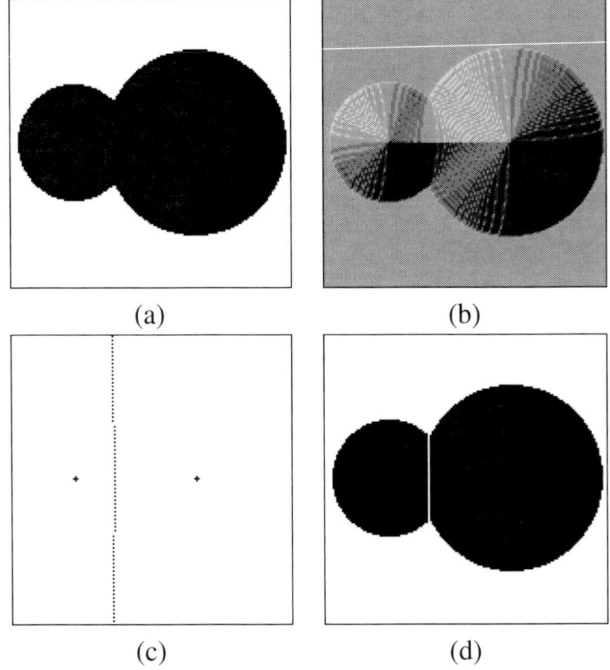

Figure 7.11 Segmentation of overlapped convex cells: (a) input image, (b) distance transform, (c) markers and watershed line, (d) watershed line used to separate the cells.

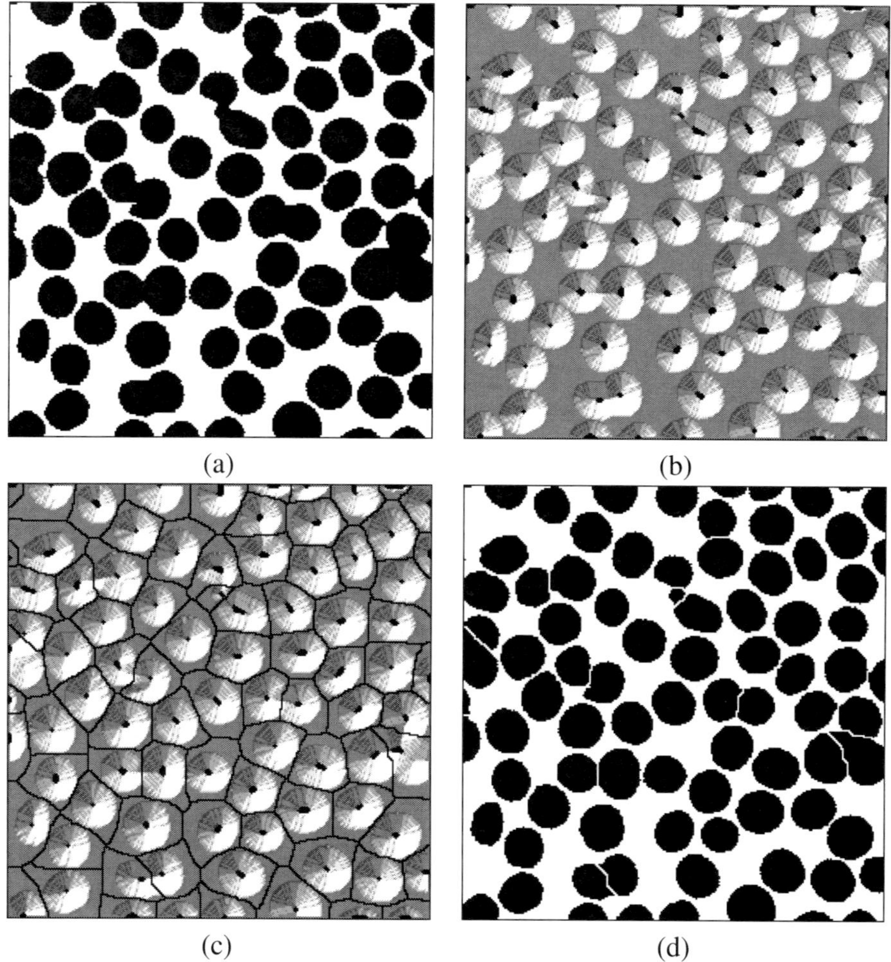

(a)

(b)

(c)

(d)

Figure 7.12 Segmentation of overlapped blood cells: (a) input image, (b) distance transform and the regional maxima as markers, (c) watershed lines of the negated distance function from markers, (d) watershed lines separating touching cells.

markers from touching convex cells.

The foregoing technique is applied to an image of blood cells in Fig. 7.12: (a) input image, (b) shaded view of the distance function overlaid by the regional maxima being used as markers, (c) the watershed lines overlaid on part (b), and (d) the input image segmented by the watershed lines.

7.4   Inner and Outer Markers

Typical watershed-based segmentation methods are used to segment cell-like objects from a gray-scale image. The general approach used to solve these problems is threefold: (i) preprocessing using a smoothing connected filter, (ii) object marker extraction (inner markers) and background marker (outer marker), and (iii) obtaining watershed lines of the morphological gradient from the markers. Usually the most crucial part is the extraction of object markers: if an object is not marked properly, then it will be missed in the final segmentation. In this section, we illustrate the methodology, in particular, marker extraction, using two examples. The first problem is to separate the keys of a keyboard calculator and the second is to extract cornea cells from a noisy image.

For detecting the keys in a gray-scale image of a calculator, the internal markers are the characters on the keys and the external markers are obtained by the SKIZ of the key characters. The full algorithm is illustrated in Fig. 7.13: (a) the input gray-scale image; (b) the opening top-hat to enhance the characters of the keyboard; (c) the threshold of the dilated top-hat to detect the characters, which are the inner markers for the watershed; (d) the Euclidean SKIZ of the inner markers to serve as the background marker; (e) the markers overlaid on the morphological gradient, which is the input to the watershed from markers; and (f) the watershed lines. Note that the dilation has been applied in part (c) to connect small objects in the same key, such as with the "AC" key. Also note that two watershed transforms are used in this segmentation, one for the SKIZ to compute the outer marker and the other to compute the final key segmentation.

The second example of watershed segmentation using inner and outer markers is slightly more complicated because the image is quite noisy and the inner and outer markers may touch in some places, but in essence the methodology is the same as the one used in the keyboard segmentation. The image in this example is a very poor quality microscopic image of a cornea tissue, shown in part (a) of Fig. 7.14. The cell markers are extracted by the regional maxima of the opening by a disk of the input image. This is the standard methodology described in Sec. 6.9. Note that a classical threshold is not used. The criterion used with regional maxima is mainly topological. We can model each cell as a small hill and we want to mark the top of each hill that has a base larger than the disk used in the opening. Parts (b) and (c) of the figure show the opened image and its regional maxima, respectively. The regional maxima constitute the inner markers. For the outer markers, instead of

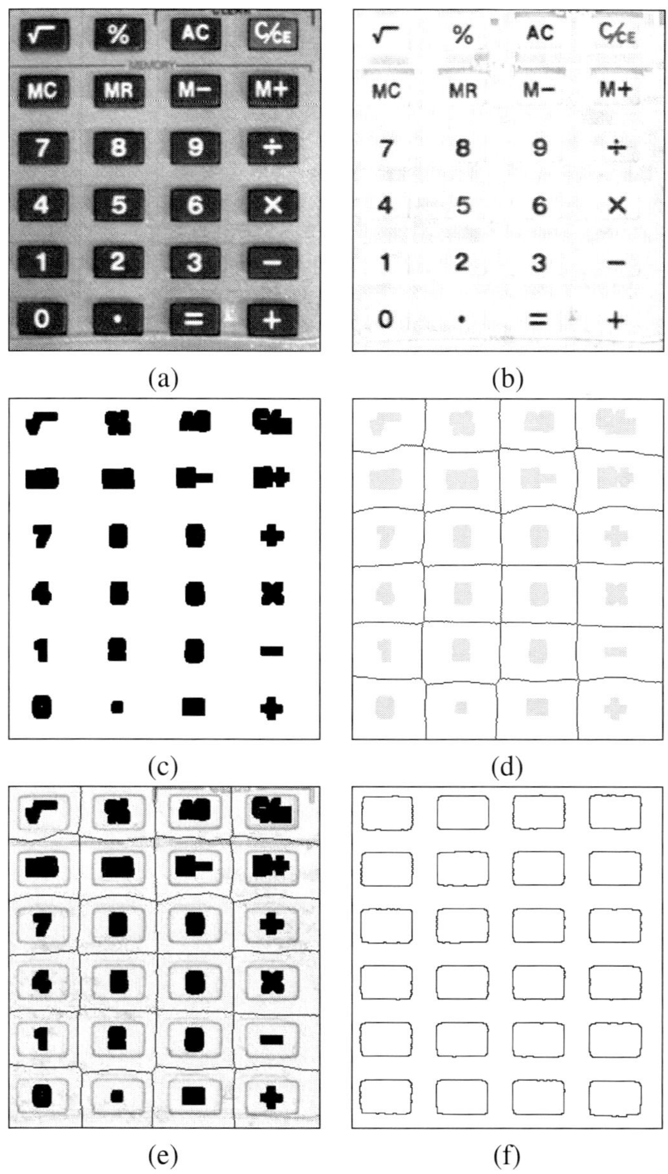

Figure 7.13 Segmentation of keys using watershed transform with inner and outer markers. (a) input image, (b) open top-hat, (c) threshold of dilated top-hat (inner markers), (d) Euclidean SKIZ (outer marker), (e) gradient overlaid by inner and outer markers, (f) watershed lines.

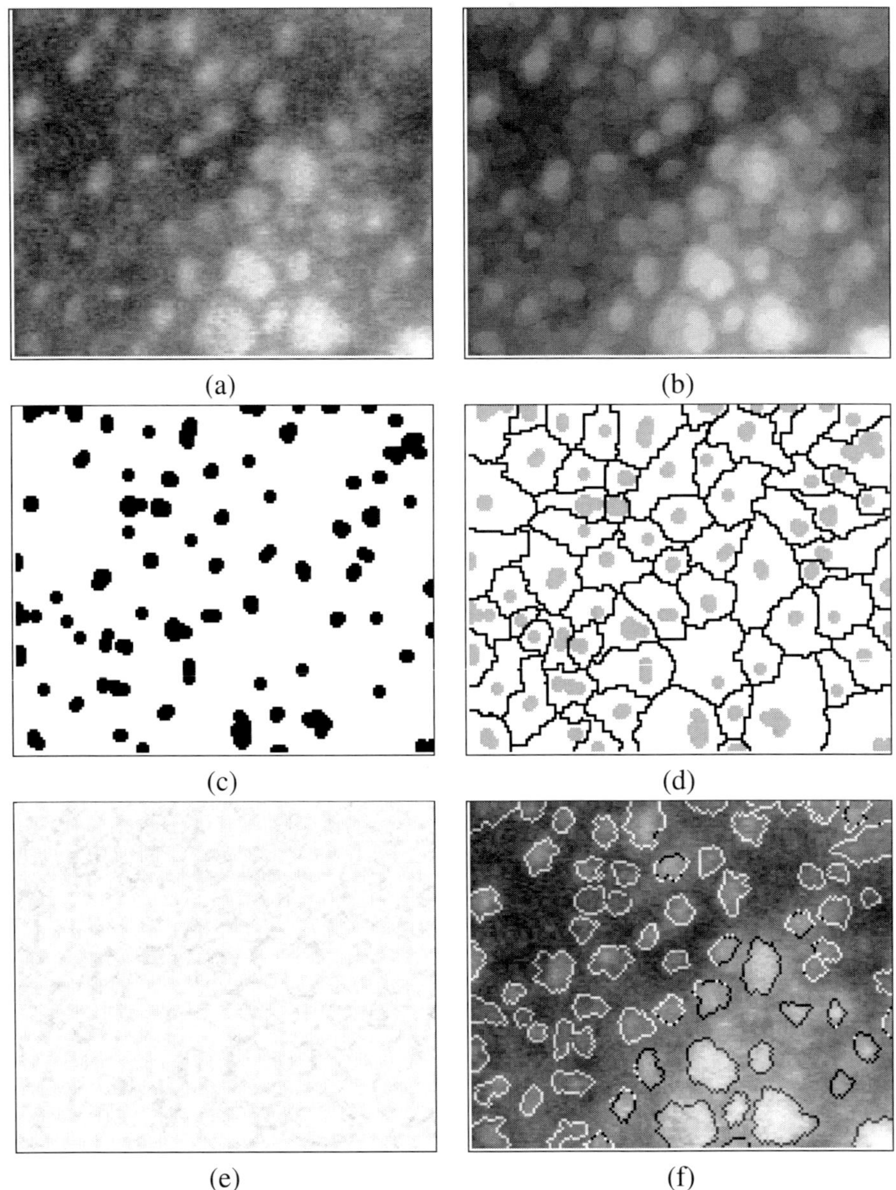

Figure 7.14 Segmentation of cornea cells from a noise image: (a) input image, (b) filtered by open, (c) regional maxima of the open (inner markers), (d) watershed lines of the negated input image from the inner markers (outer marker), (e) morphological gradient of input image, (f) final watershed lines overlaid on input image.

using the SKIZ as in the previous example, it is better to compute the watershed on the negation of the input image because we are interested in finding the influence zone of each cell. Negating the input image, the cells become basins, and taking the watershed transform from the inner markers yields the influence zones of the basins, as shown in part (d). These compose the background (outer) marker. One needs to be careful in combining both markers as they can touch each other at some points. We first label the inner markers with integers and then label the outer marker by 1 greater than the maximum inner-marker label. The final watershed lines are computed on the morphological gradient, shown in part (e). Although it is a very noisy gradient, the final watershed lines, which are overlaid on the input image and displayed in part (f), provide a satisfactory segmentation.

7.5 Hierarchical Watershed Transform

We present in this section a simple general hierarchical watershed transform based on the concept of dynamics of regional minima introduced in Sec. 6.10. Recalling that the input gray-scale image is a topographic surface, it is impossible to reach a point from a regional minimum to another point with a lower gray level without climbing. Suppose now that a progressive flooding occurs at a hole made at a regional minimum. As the water rises from this minimum, the flooding will expand to create a lake of a certain depth. It may happen that the lake invades neighboring basins as the water level increases. There will be a maximum lake before the flooding starts invading a basin with a lower regional minimum. The dynamic of a regional minimum is the depth of its maximum lake. The dynamic is the minimum height a point in the regional minimum has to climb to reach a lower regional minimum. Dynamics are illustrated in Fig. 7.15: (a) illustration of the maximum lake and the dynamic of one regional minimum for a signal; (b) the dynamics of all the regional minima added to the regional minima; (c) all dynamic values.

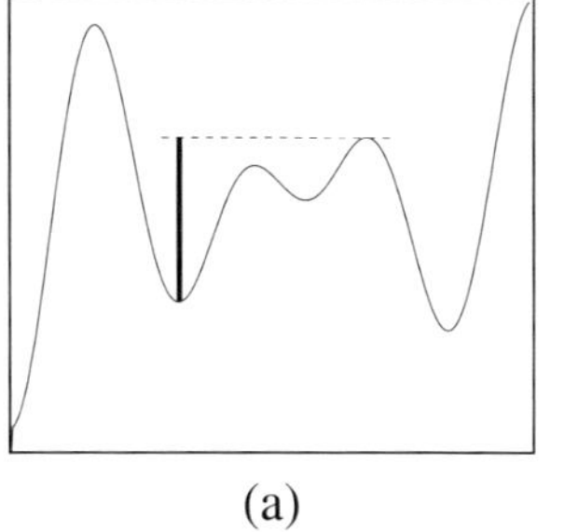 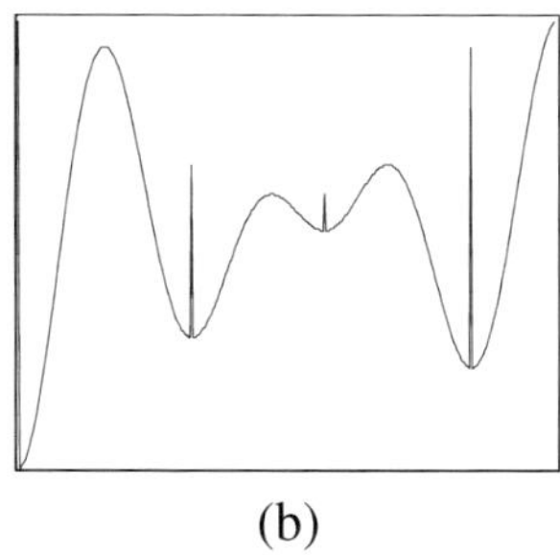 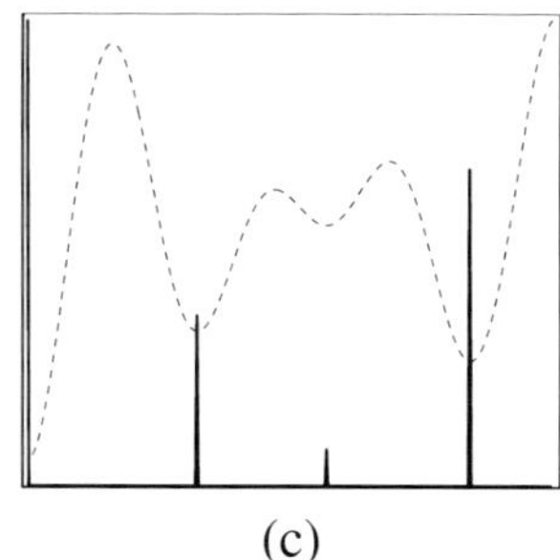

(a)   (b)   (c)

Figure 7.15 Dynamics: (a) dynamics is the height to raise a minima to reach a lower minima, (b) dynamics of each minima added to the minima, (c) dynamics of the signal with signal in dash for reference.

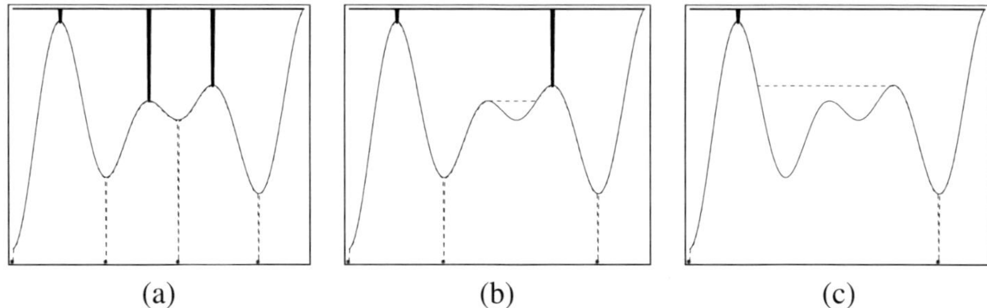

(a)   (b)   (c)

Figure 7.16 Hierarchical watershed based on dynamics: (a) finest partition: four minima, four catchment basins, (b) minima with lowest dynamic removed, three minima, three catchment basins, (c) two minima with lowest dynamics removed, two catchment basins.

A **hierarchical** or **multiscale watershed** (MSW) transform creates a set of nested partitions. The multiscale watershed presented here can be obtained by applying the watershed from markers to a decreasing set of markers. The **multiscale watershed at scale** s, $MSW(s)$, is the partition obtained when only minima with dynamics greater or equal to s are considered. The watershed at scale 1 (finest partitioning) is the classical watershed, made of the primitive catchment basins. As the scale increases, less markers are involved and the coarsest partition is the entire image obtained from a single marker at the regional minimum of largest dynamic. Figure 7.16 illustrates the multiscale watershed of the signal shown in Fig. 7.15, for which there are four regional minima with different dynamics, and consequently five intervals composing the finest partition: (a) finest partition; (b) next partition in the hierarchy, based on the three minima with the strongest dynamics; and (c) the next partition in the hierarchy, having the two deepest catchment basins in the signal. Note how in part (b) the minima imposition makes the catchment basin with the lowest dynamic disappear.

An example of the multiscale watershed applied to a real image is shown in Fig. 7.17: (a) input image; (b) the morphological gradient; (c) a mosaic image where the primitive catchment basins of the gradient are displayed with the gray-scale proportional to the dynamics of their regional minima; and (d) through (f) showing three levels in the hierarchy, with markers as the regional minima with dynamics above 3, 8, and 15, respectively. Note that the highest dynamic corresponds to the background, the second highest corresponds to the largest circular cell, and the third corresponds to the elongated cell above it. This observation is confirmed by the three most prominent objects shown in part (f).

Watershed-based segmentation can be used for image simplification if the image pixels values are replaced by the mean gray-scale values of their corresponding catchment basins. An example of image simplification of a real-life image is shown in Fig. 7.19. For this example, the partition used in the image simplification

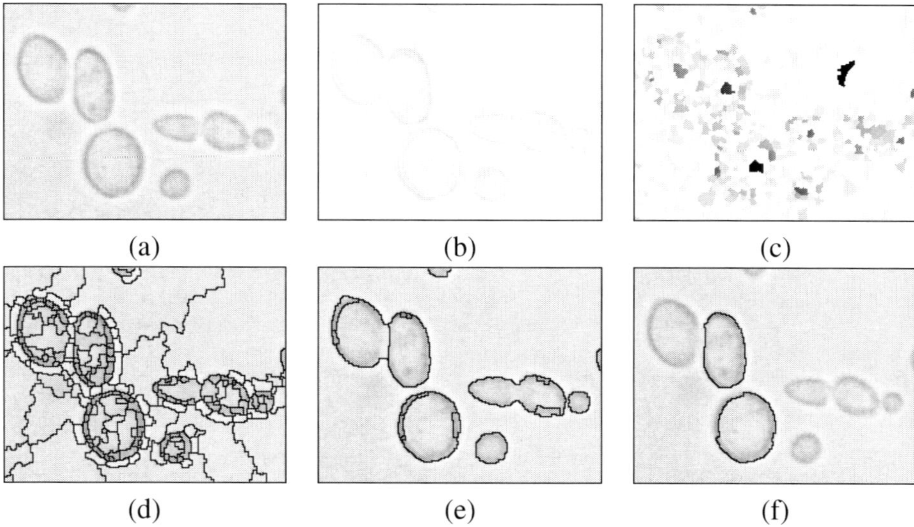

Figure 7.17 Multiscale watershed: (a) input image, (b) morphological gradient, (c) primitive catchment basins, with pixel gray-level proportional to its regional minima dynamics, (d–f) watershed from minima with dynamics above 3 (d), 8 (e), and 15 (f).

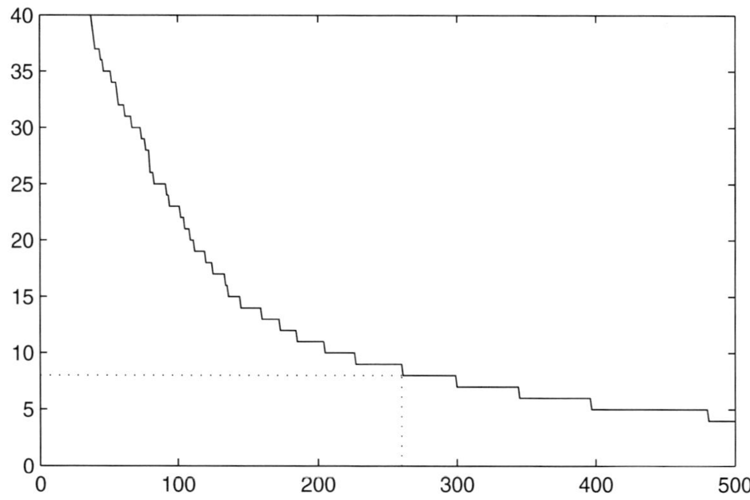

Figure 7.18 Dynamics versus number of most relevant regions for the example shown in Fig. 7.19.

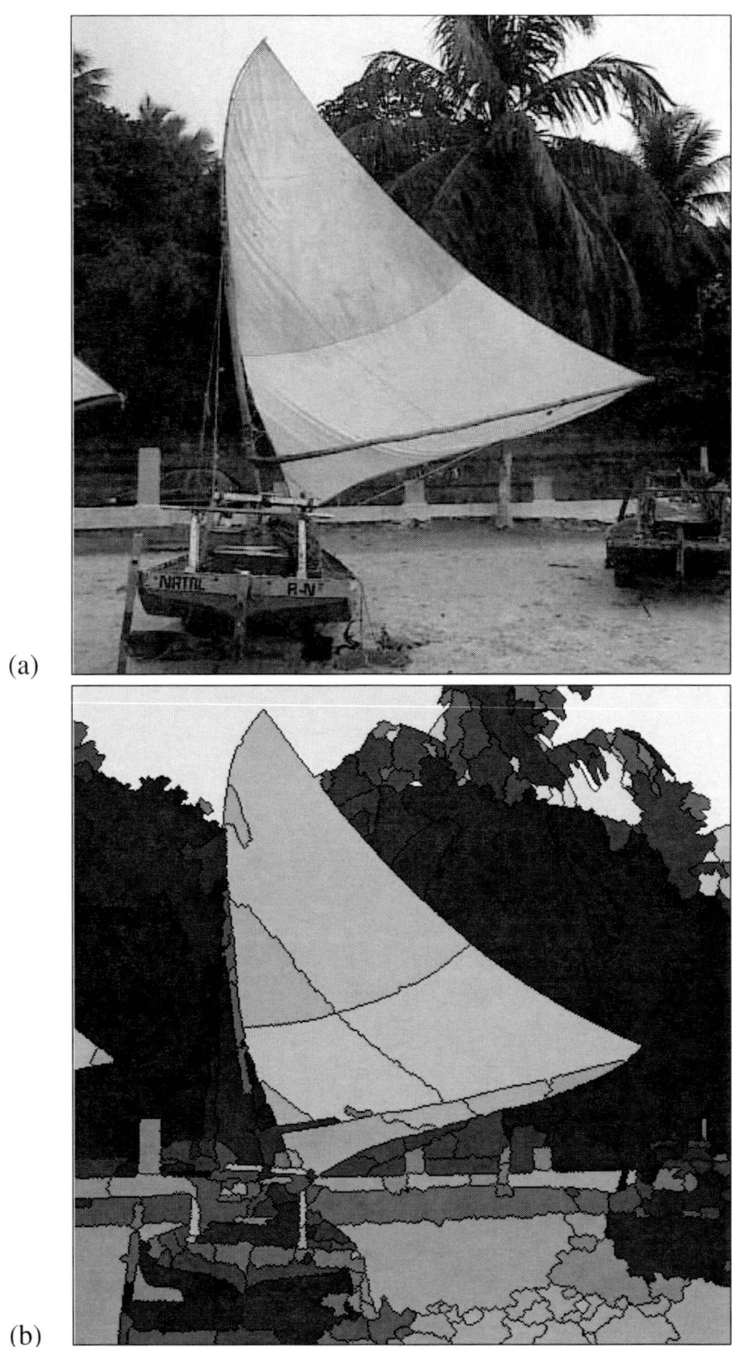

(a)

(b)

Figure 7.19 Image simplification using mean gray-values of the 260 most relevant catchment basins of the hierarchical watershed transform.

is chosen from one scale of the dynamic-based hierarchical watershed. The morphological gradient of the input image is first filtered by area close with 10 pixels. The dynamics of each regional minima of this filtered gradient are computed. It is possible to find the relationship between the contrast parameter of the h-minima filter and the number of output regions of the watershed applied to the result of this filter. For the example of Fig. 7.19, this relationship is shown in the plot of Fig. 7.18. This curve is built by sorting the dynamic values in descending order. The horizontal axis gives the numbers of regional minima with dynamics above a given value (vertical axis). Note that by using $h = 8$ for the h-minima, all minima with dynamics less than 8 are removed and just 260 minima with dynamics above 8 are preserved. The simplified image shown in Fig. 7.19 has been obtained by replacing the label of the catchment basin of the watershed by the average value of the original image pixels associated with that label.

7.6 Watershed Transform Algorithms

There are many watershed-transform algorithms in the literature, and this can be quite confusing for readers new to the subject. There are two types of algorithms that are most important: one is based on immersion simulation and the other is based on a minimum-cost path. It is difficult to have an algorithm definition that matches exactly a given implementation, or vice-versa. The algorithm presented here is based on the minimum-cost path for the watershed-from-markers transform, in which the definition and implementation are consistent. The classical watershed transform is obtained when the markers are the regional minima of the image.

The minimum path cost between two pixels p and q in the graph is given by the minimal cost of all the paths connecting p and q:

$$C^*(p,q) = min_i\{C(\pi_i(p \rightsquigarrow q))\}, \tag{7.3}$$

where $\pi_i(p \rightsquigarrow q)$ denotes a path from p to q. The cost of a simple connected path from $p_1 \rightsquigarrow p_n$ is given by a lexicographic cost $C(p)$, where the first component $C^1(p)$ is the maximum pixel value in the path and the second component $C^2(p)$ is the number of times the first component cost is the same before arriving at p_n:

$$\begin{aligned}
C(p_1, p_2, \ldots, p_n) &= [C^1(p_n), C^2(p_n)], \\
C^1(p_1) &= 0, \\
C^1(p_n) &= \max\{C^1(p_1), f(p_2), \ldots, f(p_n)\} \text{ for } n > 1, \\
C^2(p_n) &= \max\{j : C^1(p_n) = C^1(p_{n-j}), j = 0, 1, 2, \ldots, n-1\}.
\end{aligned} \tag{7.4}$$

Note that f is the input image and $C^1(p_n)$ is the maximum pixel value $f(p_i)$, p_i in the path $p_1 \rightsquigarrow p_n$.

The catchment basin CB_k associated to the marker L_k is given by the nodes p with less than or equal path cost from this marker than from any other marker,

$$CB_k = \{p : C^*(L_k, p) \leq C^*(L_j, p), j \neq k\}, \qquad (7.5)$$

where the image is modeled as a graph and each pixel is a node. $C^*(L, p)$ is the minimum-cost path from region L to pixel p, and this is the minimum-cost path from any pixel of region L to p,

$$C^*(L, p) = min\{C^*(l, p) : l \in L\}. \qquad (7.6)$$

An efficient yet simple implementation of this watershed definition is the following algorithm, $cb(f, L)$, where f is the input image, L is a labeled image where nonmarker pixels have the value 0, and the output of the algorithm is L, which shows the final catchment basin regions:

Function $L = cb(f, L)$
   f: input image
   L: labeled image (input and output)
1. *Initialization*
   for $(L(p) \neq 0)$ inHFQ$(p, 0)$
2. *Propagation*
   while HFQ **is not** empty
      $p \leftarrow$ outHFQ
      for each non-labeled q neighbor of p
         $L(q) \leftarrow L(p)$
         inHFQ$(q, f(q))$

The Hierarchical FIFO Queue (HFQ) has the following operations: inHFQ(p, v), insert pixel p with priority v; outHFQ, remove the pixel with the lowest priority with the FIFO policy for pixels at the same priority. This FIFO policy implements intrinsically the second component of the lexicographic cost of Eq. (7.4).

Two points are worth mentioning in the formulation of the catchment basins given in Eq. (7.5). First, there is no line definition for the watershed. Second, there are many possible optimal solutions because the criterion for a pixel belonging to a catchment basin is that its cost relative to the basin marker be less than or equal to its cost relative to any other marker. The watershed lines can be assigned to the pixels that have the same minimal cost to more than one marker. With this approach, the watershed lines can be thick. If one wants a one-pixel-thick line, a thinning of the thick watershed line can be obtained directly by an extension of the previous algorithm. In this new algorithm, shown below, ws is the binary image with lines, initialized at zero. Additionally, a flag indicates if a pixel is permanent or not. Initially, all pixels are nonpermanent; as they are removed from the queue they

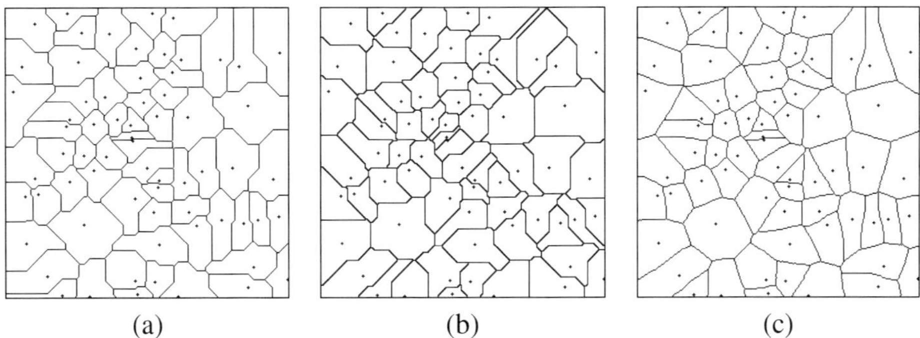

Figure 7.20 Watershed lines on a plateau. Input is a constant image, and markers are random pixels: (a) watershed using the hierarchical FIFO queue, 4-adjacency, (b) watershed using the hierarchical FIFO queue, 8-adjacency, (c) watershed on the negation of the Euclidean distance to the markers (Euclidean SKIZ).

become permanent in accordance with the Dijkstra algorithm terminology. A watershed line is placed on a permanent neighbor of a recently made permanent pixel if the pixels belong to different catchment basins and neither of them is already a watershed pixel. The propagation step of the modified algorithm is

2. *Propagation*
    while HFQ **is not** empty
        $p \leftarrow$ outHFQ; make p permanent
        for each unlabeled q neighbor of p
            if q **is not** permanent
                $L(q) \leftarrow L(p)$
                inHFQ$(q, f(q))$
            else
                if $L(q) \neq L(p)$ **and** $ws(q) = 0$ **and** $ws(p) = 0$
                    $ws(q) \leftarrow 1$

We end this section by illustrating the behavior of the watershed algorithm using the hierarchical FIFO queue. Figure 7.20 shows the watershed lines on a constant image with the markers being random points, like the one used in the illustration of the SKIZ (Fig. 7.9). Parts (a) and (b) of Fig. 7.20 show the watershed lines using the algorithm implemented with a hierarchical FIFO ordered queue with 4- and 8-adjacency, respectively. These are equivalent to the city-block and chessboard SKIZ, respectively. Part (c) shows the Euclidean SKIZ for comparison.

```
name                description
mmhistogram:        histogram
mmclose:            closing
mmareaopen:         area open [Eq. (6.22)]
mmareaclose:        area close
mmhmin:             h-minim [Eq. (6.13)]
mmgradm:            morphological gradient
mmwatershed:        watershed
mmthreshad:         thresholding [Eq. (5.51)]
mmsubm:             subtraction
mmunion:            union
mmintersec:         intersection
mmneg:              negation
mmgray:             converts from binary to gray scale
mmsebox(2):         5 x 5 square structuring element
mmreadgray:         read image file
```

Figure 7.21 MT functions used the program shown in Fig. 7.22.

7.7   Demonstrations

Concrete analysis

We present a complete demonstration of a real-world example using the **MT**. This example shows an image analysis technique to detect the anhydrous and aggregate phases from a polished concrete section observed by a scanning electron microscope (SEM) image. The anhydrous phase appears as white grains, while the aggregate appears in the image as homogeneous medium-gray grains. Two aspects are particularly interesting in this example; use of the watershed on a histogram to find a thresholding value, which illustrates the use of the watershed for 1D data, and the use of the oversegmentation feature of the watershed to select the homogeneous regions in the image. The main steps for this analysis are (i) anhydrous detection by automatic threshold analysis, (ii) homogeneous grain detection using a watershed technique, and (iii) detecting aggregates as homogeneous grains not from the anhydrous phase.

The **MT** functions used in this demonstration are shown in Fig. 7.21, the program is shown in Fig. 7.22, and all the images are shown in Figs. 7.23 and 7.24. The steps used to detect the phases of the concrete micrograph are below. The program line numbers and figure parts are shown in brackets.

Image reading (line 1) The SEM image of a polished concrete section is read (a). The anhydrous phase is composed of the white pores, while the aggregate phase is composed of the medium-gray homogeneous pores.

Histogram (line 2) The histogram has a small peak in the white region related to the anhydrous phase. The histogram is negated (b) so the watershed transform can be used in the next step to detect the valley between the two predominant populations of pixels.

Automatic threshold from histogram (lines 3–8) The threshold value is extracted using the watershed transform. The aim is to detect the middle valley of the histogram. As the histogram is negated, we need to extract the middle peak of the signal. For this, the histogram is filtered so that only the two predominant peaks of the original histogram originate the regional minima. This is accomplished by filtering using a closing of length 5 (c) (note that we can use the 2D structuring element for 1D filtering as we use the bounded close), and an h-minima of 10 (d). Note that the closing by a structuring element of length 5 removes regional minima of width 5 units in the gray scale and the h-minima removes any regional minima with depth less than 10 counts in the histogram. The watershed line is a single pixel (e). Its coordinate gives the threshold value. For visualization the watershed line is unioned with the negated histogram (f). The threshold value is used to binarize the image (g).

Anhydrous grains (line 9) The area open removes grains smaller than 20 pixels as they are considered as noise (h).

Homogeneous region detection (lines 10–12) The watershed is applied on the filtered gradient as a classical watershed methodology step. The gradient of the input image is computed in (i). The filter applied to the gradient has been chosen to be a contrast h-minima removing any regional minima of depth less than 10 units in the gray scale (j). This parameter has been selected empirically and chosen such that the aggregates appear as a single region (k).

Aggregate detection (lines 13–14) The negation of the watershed lines gives the watershed regions. First, the anhydrous grains are removed as they are also homogeneous regions (l). After that the image is filtered using an area alternating filter which removes any objects with area smaller than 300 pixels and fills any hole with area smaller than 50 pixels (m). These are the aggregate regions.

Final display (line 15) The anhydrous grains are displayed as white (255) and the aggregates as black (0) in the final image (n).

Silver-halide T-grain crystals

Consider the electron micrograph of silver-halide T-grain crystals in emulsion shown in Fig. 7.26(a). Automated crystal analysis involves the segmentation of the grains for measurement. The segmentation of this image looks simple at first,

```
                              Program 7.1
01   a = mmreadgray('csample.jpg');
02   b = mmneg( mmhistogram(a));
03   c = mmclose( b, mmsebox(2));
04   d = mmhmin( c, 10);
05   e = mmwatershed( d);
06   t = nonzero( e); # coord.  of nonzero values
07   f = mmunion(b,mmgray(e,'uint16')); # for visualization
08   g = mmthreshad( a, t);
09   h = mmareaopen( g, 20); # anhydrous grains
10   i = mmgradm( a);
11   j = mmhmin( i, 10);
12   k = mmwatershed(j);
13   l = mmsubm(mmneg(k), h);
14   m = mmareaclose(mmareaopen( l,300), 50);
15   n = mmunion(mmintersec(a,mmneg(mmgray(m))),mmgray(h));
```

Figure 7.22 Python code for the concrete analysis. The images from (a) to (n) are shown in Figs. 7.23 and 7.24.

but a careful analysis reveals that the image has several factors that make the segmentation difficult. The image has a strong illumination gradient, the gray-scale values for the crystal grains are at the level of the background, the image has strong white "shadows" noise, it has a wide range of grain sizes, and there are overlapping and touching grains. Despite all this, the standard watershed-based segmentation can be used to obtain very good results. The two key points here are the background correction and the enhancement of the dark contours using a close top-hat with a disk diameter larger than the thickness of these contours.

The **MT** program for this demonstration is shown in Fig. 7.25, and all the images are shown in Fig. 7.26. The steps used to detect the crystals are below. The program line numbers and figure parts are shown in brackets.

Image reading (line 1) The silver-halide-T-grain-crystals-in-emulsion gray-scale micrograph image is read (a). The crystals are surrounded by dark contour lines that are to be used for the watershed-based segmentation. The preprocessing stage is to enhance only these lines.

Illumination correction (lines 2–3) The illumination gradient is estimated by an alternating sequential filtering with a large structuring element and a large stage (b). The input image is divided by this gradient estimation normalized by its minimum value (c). This division is used to guarantee that the dark contours have the same depths, so that in the dark regions of the image, the depths of the dark contours are increased by this procedure. This is necessary to have the same amount of segmentation when applying the h-minima filtering in step 5.

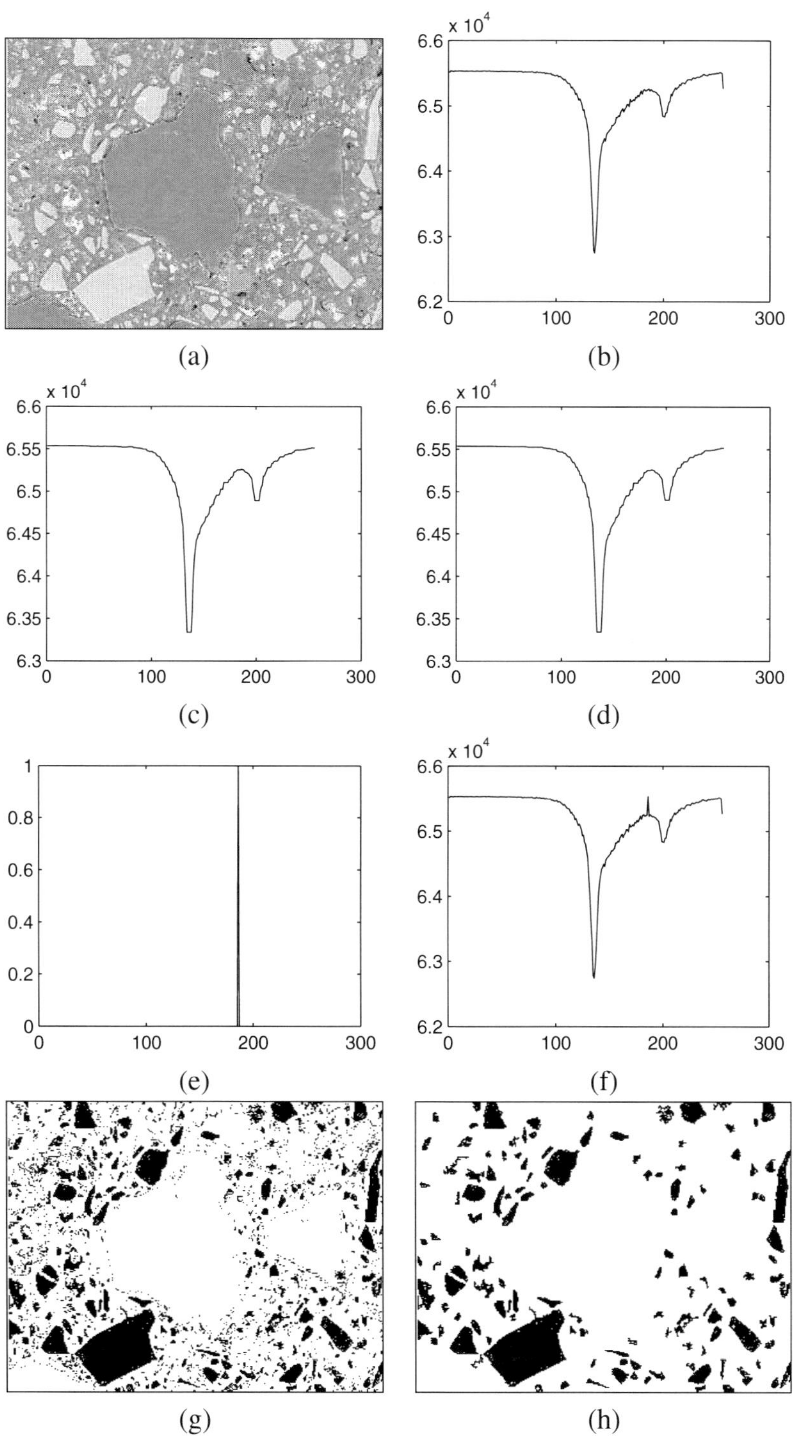

Figure 7.23 Concrete analysis. These images refer to the program of Fig. 7.22.

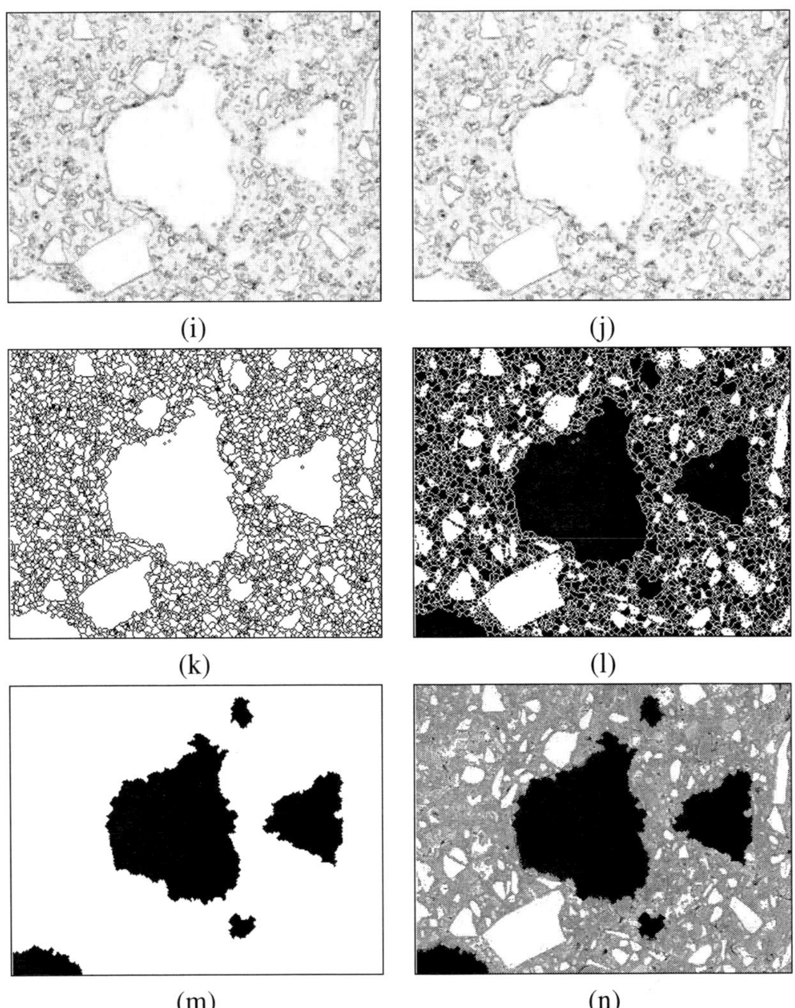

Figure 7.24 Concrete analysis. These images refer to the program of Fig. 7.22.

```
                        Program 7.2
01   a = mmreadgray('crystals.tif');
02   b = mmasf(a,'oc',mmsedisk(2,'2D','octagon'),10);
03   c = (mmstats(b,'min') * a)/b;
04   d = mmcloseth( c,mmsedisk(4,'2D','octagon'));
05   e = mmhmin(mmareaclose(d,7),8);
06   f = mmwatershed(e);
07   g = mminfrec(mmframe(f),mmneg(f));
08   h = mmintersec(f,mmneg(mmclose(g)));
09   i = mmunion(mmgray(h),a);
```

Figure 7.25 Python code for the T-grain crystals analysis. The images from (a) to (i) are shown in Figs. 7.26.

Contour enhancing (line 4) A classical close top-hat detects the dark contours. The size of the structuring element must be larger than the thickness of the dark contours (d). Note that the white areas of the image do not affect the detection of the dark contours.

Watershed segmentation (lines 5–6) The purpose of the watershed transform is to detect the dividing lines of dark regions. In this case, it is not necessary to compute the gradient as the contour lines are already enhanced. We apply the classical minima simplification by cascading an area close and an h-minima (e). The choice of parameters for the filters used in this simplification is very crucial and it has been found by trial and error. Application of the watershed on the simplified contour image gives the watershed lines (f).

Edge off (lines 7–8) A procedure similar to removing grains connected to the image frame described in Fig. 3.15 is used (g). After that, the intersection of the watershed lines and the grains not connected to the image frame gives the final contour of grains not touching the image frame (h).

Display (line 9) For display purposes, the contours are overlaid on the original image (i).

7.8 Exercises

1. Apply the watershed from regional minima to the signal (16 15 10 3 4 9 **9** 15 11 8 4 7 12 18).

2. Apply the watershed to the signal of the previous exercise but using as the two single markers the third point and last point from the left, which have values 10 and 18, respectively.

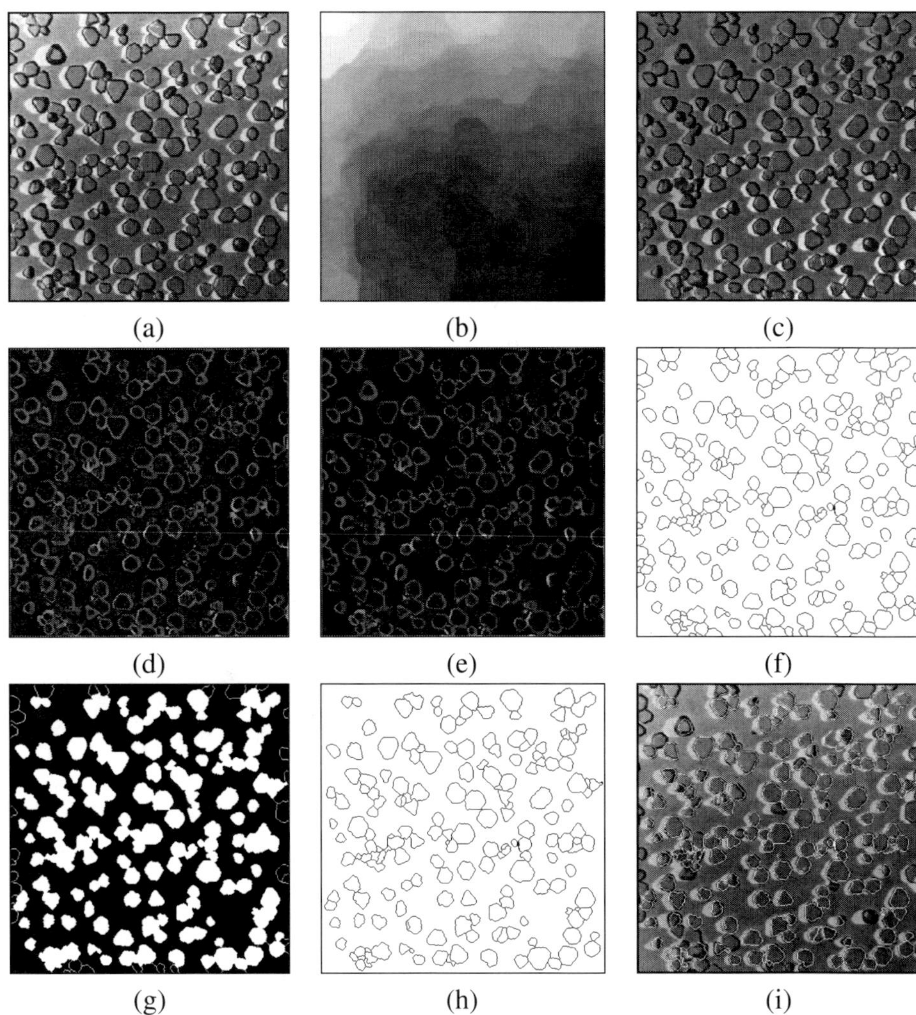

Figure 7.26 Silver-halide T-grain crystals analysis. These images refer to the program of Fig. 7.25.

name	description
mmcwatershed:	watershed from markers (algorithm $cb(f, L)$)
mmwatershed:	watershed [Eq. (7.1)]
mminpos:	minimum imposition [Eq. (7.2)]
mmskiz:	SKIZ (Fig. 7.9)

Figure 7.27 MT operators presented in this chapter.

3. Extinction values can be computed for regional maxima of the distance transform. In this case, what are the geometrical meanings of the dynamics and the area extinction values?

7.9 Laboratory Experiments

Figure 7.27 lists the **MT** functions discussed in this chapter.

1. Compute the dynamics and the area extinction values of the image along the number of regions in the image, like the plot of Fig. 7.18, but using the image in the example shown in Fig. 6.24.

2. Modify the automatic threshold from histogram detection in the concrete analysis demonstration such that instead of filtering using a closing, use the extinction value concept to detect the two most relevant valleys in the negated histogram.

References

1. S. Beucher and F. Meyer. The morphological approach to segmentation: the watershed transformation. In E. R. Dougherty, editor, *Mathematical Morphology in Image Processing*, chapter 12, pages 433–481. Marcel Dekker, New York, 1993.

2. A. Bieniel, A. Moga. An efficient watershed algorithm based on connected components. *Pattern Recognition*, 33:907–916, 2000.

3. R. A. Lotufo and A. X. Falcão. The ordered queue and the optimality of the watershed approaches. In J. Goutsias, L. Vincent, and D. Bloomberg, editors, *Mathematical Morphology and its Application to Image and Signal Processing*, volume 12 of *Computational Imaging and Vision*, pages 341–350. Kluwer Academic Publishers, Dordrecht, 2000.

4. F. Meyer. Topographic distance and watershed lines. *Signal Processing*, 38:113–125, 1994.

5. F. Meyer and S. Beucher. Morphological segmentation. *Journal of Visual Communication and Image Representation*, 1(1):21–46, 1990.

6. L. Najman and M. Schmitt. Geodesic saliency of watershed contours and hierarchical segmentation. *IEEE Transactions on Pattern Analysis and Machine Intelligence*, 18(12):1163–1173, 1996.

7. J.B.T.M. Roerdink and A. Meijster. The watershed transform: Definitions, algorithms and parallelization strategies. *Fundamenta Informaticae*, 41(1–2):187–228, 2000.

8. R. E. Sequeira, F. J. Preteux. Discrete Voronoi diagrams and the SKIZ operator: A dynamic algorithm. *IEEE Transactions on Pattern Analysis and Machine Intelligence*, 19(10):1165–1170, 1997.

9. L. Vincent and P. Soille. Watersheds in digital spaces: an efficient algorithm based on immersion simulations. *IEEE Transactions on Pattern Analysis and Machine Intelligence*, 13:583–598, 1991.

10. D. M. Wang. A multiscale gradient algorithm for image segmentation using watersheds. *Pattern Recognition*, 30:2043–2052, 1997.

Chapter 8

Granulometries

The granulometric method discovered by Georges Matheron provides a morphological method for characterizing granular images by means of how they are sieved through sieves of various size and shape. Besides application to grains (particles), the method is effective for texture and shape analysis.

Imagine a sieve through which one might pan gold. If an image is considered as a collection of grains, then whether an individual grain will pass through the sieve depends on its size and shape relative to the mesh of the sieve. By increasing the mesh size while keeping the basic mesh shape, more and more of the image will pass through, the eventual result being that no more grains remain. Of course, this sieving model does not fully describe even a granular image, for in a real image the grains will likely overlap; nevertheless, it does serve as a means to approach the removal of nonconforming image structure, and can be further developed to obtain image signatures based on the rate of sieving. It is this sieving model that one should keep in mind throughout the chapter.

8.1 Granulometries Generated by a Single Opening

We begin by considering an elementary but basic type of granulometry for Euclidean images. Euclidean granulometric theory does not directly apply to digital processing, and application to digital images requires care.

It can be shown that if B is convex and $r > s > 0$, then rB is sB-open; that is, $rB \circ sB = rB$. Consequently, $A \circ rB$ is a subset of $A \circ sB$. If we think of the image A falling through the holes sB and rB, more will fall through the hole rB, thereby yielding a more diminished filtered image. Indeed, since rB is sB-open, filtering by both, in either order, simply yields $A \circ rB$.

If we consider $t > 0$ as a variable, the class of operators Ψ_t defined by $\Psi_t(A) = A \circ tB$ is called a **granulometry**, and the primitive B is said to be the **generator** of the granulometry. If $\Omega(t)$ is the area removed by opening by tB, then

$$\Omega(t) = v[A] - v[A \circ tB], \tag{8.1}$$

where v denotes area (or volume). $\Omega(0) = 0$ and $\Omega(t)$ is an increasing function of t. Under the assumption that A is not of infinite extent (which is certainly reasonable for image processing), $\Omega(t) = v[A]$ for sufficiently large t. $\Omega(t)$ is called a **size distribution**.

A normalized size distribution is defined by

$$\Phi(t) = \frac{\Omega(t)}{\Omega(\infty)} = \frac{\Omega(t)}{v[A]}. \tag{8.2}$$

$\Phi(t)$ increases from 0 to 1 and can be shown to be a probability distribution function. Thus, its derivative $d\Phi(t)$ is a probability density. Both Φ and $d\Phi$ are known as the **pattern spectrum** of the image relative to the granulometry (or, relative to the generator). The moments of $d\Phi$ are known as **granulometric moments**. As will be discussed in a subsequent section, these are employed as image signatures.

For an illustration of the manner in which the pattern spectrum can be used to define a shape factor, consider any convex set B. If we apply the granulometry generated by B to B itself, then $B \circ tB = B$ for $t \leq 1$ and $B \circ tB$ is empty for $t > 1$. Hence, $\Phi(t) = 0$ for $t \leq 1$ and $\Phi(t) = 1$ for $t > 1$. The pattern spectrum $d\Phi(t)$ consists of a single unit impulse at $t = 1$. Thus, its mean is 1 and its variance and skewness are both 0. Were we to apply the granulometry generated by B to a different single-component shape A, not simply a scalar multiple of B, then the pattern spectrum would not consist of a single impulse. The degree to which the pattern spectrum differs from a unit impulse can be taken as a measure of the degree to which the shape A differs from the shape B. Parameters that measure this difference include the variance and the entropy of the pattern spectrum.

Figures 8.1 (a) and (b) contain a 2×2 square and a 2×2 square with a 1×1 square adjoined, respectively. The unnormalized size distributions for the shapes relative to a 1×1 square are depicted in Figs. 8.1 (c) and (d), respectively, and the pattern spectra are depicted in Figs. 8.1 (e) and (f), respectively. Since the 2×2 square is simply a scalar multiple of the underlying generator, its pattern spectrum is a single spike, whereas the pattern spectrum of the shape of Fig. 8.1(b) is not. Note that the pattern spectrum does not fully characterize a shape: the position of the adjoined smaller square has no effect on the pattern spectrum, so long as it does not overlap the larger square; indeed, it need not even be connected to the larger square.

To practically apply the granulometric method we first have to adapt it to digital images. This cannot be done directly owing to two difficulties regarding the Cartesian grid: first, the lack of an appropriate notion of convexity; second, the inability to apply scalar multiplication by arbitrary real numbers. In the second instance, even if we restrict ourselves to scalar multiplication by integers, there are still difficulties, since digital images without holes may have holes after scalar multiplication.

8.2   Discrete Size Distributions

The method of granulometric generation we now discuss is applicable to both Euclidean images and discrete images; however, its main purpose is digital implementation.

Consider a sequence $\{E_k\}$, $k = 1, 2, \ldots$, of structuring elements of increasing size, where E_0 consists of a single pixel and E_{k+1} is E_k-open for all k. The latter assumption insures that $S \circ E_{k+1}$ is a subimage of $S \circ E_k$. Consequently, opening

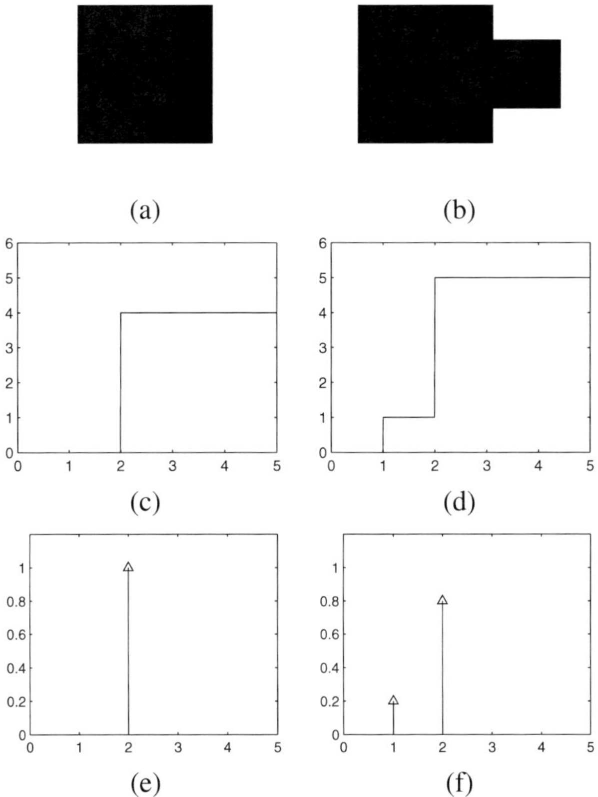

Figure 8.1 Granulometry: (a) 2×2 square image, (b) 2×2 square with a 1×1 square adjoined, (c) unnormalized size distribution of (a), (d) unnormalized size distribution of (b), (e) pattern spectrum of (a), (f) pattern spectrum of (b).

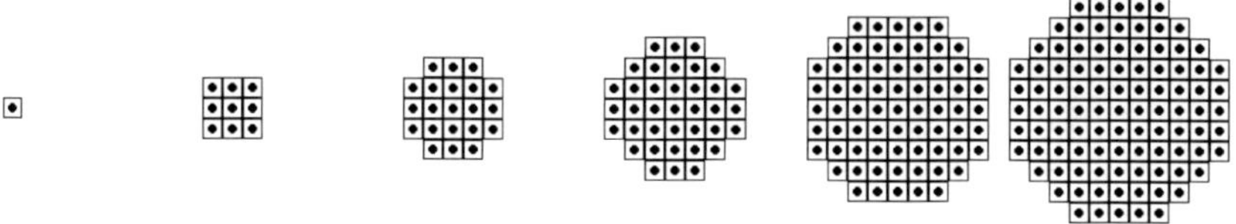

Figure 8.2 Family of octagonal disks with increasing radius of 0, 1, 2, 3, 4, and 5 pixels.

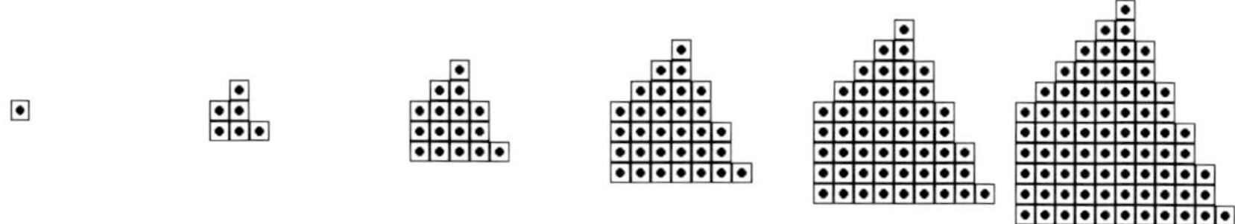

Figure 8.3 Family of disks with increasing radius of 0, 1, 2, 3, 4, and 5 pixels generated from kE.

in turn by the structuring elements yields a decreasing sequence of images:

$$S \circ E_1 \supset S \circ E_2 \supset S \circ E_3 \supset \dots \tag{8.3}$$

For each k, let $\Omega(k)$ be the number of pixels removed by opening. $\Omega(k)$ is the number of pixels in $S - S \circ E_k$. Then $\Omega(k)$ is an increasing function of k, $\Omega(0) = 0$, and $\Omega(k)$ gives the original pixel count in S for sufficiently large k. Applying the normalization of Eq. (8.2) with k in place of t yields a normalized size distribution $\Phi(k)$. It is a discrete probability distribution function that possesses a discrete derivative

$$d\Phi(k) = \Phi(k+1) - \Phi(k), \tag{8.4}$$

which is a discrete density (probability mass function) and defines the **discrete pattern spectrum**.

A key practical issue concerns the construction of structuring elements for which E_{k+1} is E_k-open for all k. An example of such a sequence is the "octogonal" sequence shown in Fig. 8.2. For a systematic approach to construction, recall that A is B-open if and only if there exists a set D such that $A = D \oplus B$. Thus, a sequence of the desired type results from choosing a primitive E, letting E_0 be a single pixel, and then defining $E_1 = E, E_2 = E \oplus E, E_3 = E \oplus E \oplus E, \dots$. Figure 8.3 illustrates this kind of construction.

As an illustration consider Fig. 8.4(a), in which digital "balls" of four sizes are randomly dispersed about the image. The generating sequence $\{E_k\}$, from which the four balls generating the image are drawn, consists of digital balls of increasing size (the first being a single pixel). As ever larger balls are employed for the opening structuring elements, the grains (balls) in the image are sieved from the

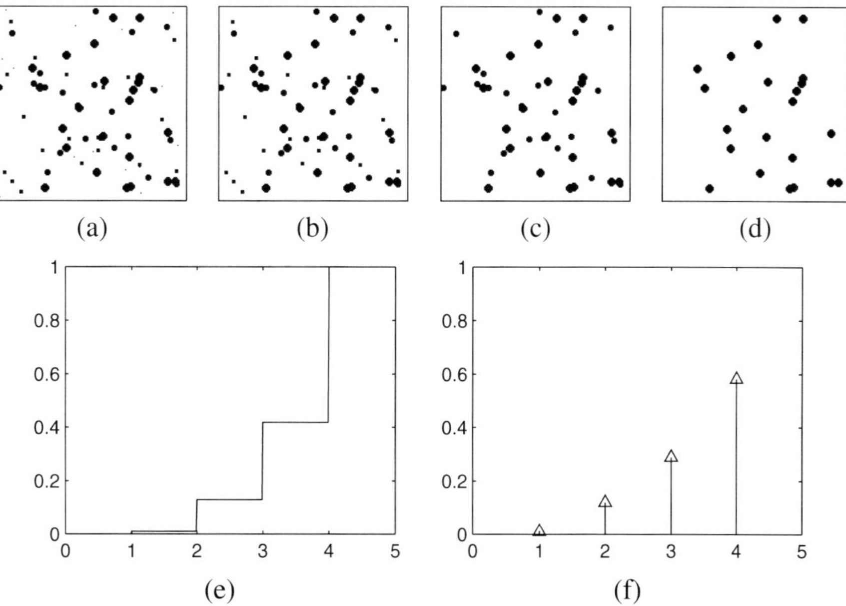

Figure 8.4 Granulometry with four octagonal disks: (a) random image with digital balls of four sizes, (b) opening by E_1, (c) opening by E_2, (d) opening by E_3, (e) pattern spectrum $\Phi(k)$, (f) derivative pattern spectrum $d\Phi(k)$.

image. As the structuring element sequence passes each of the four balls that generate the image, translations of the specific structuring element are sieved from the image, as illustrated in parts (b), (c), and (d) of Fig. 8.4. Although it is possible (and likely in real images) for overlapping grains to create larger, irregular compound grains that are not so regularly sieved by the granulometry, such is not the case in our simulated illustration (but it will be in a real-image example to be discussed shortly). The pattern spectrum, $\Phi(k)$, and its discrete derivative, $d\Phi(k)$, are depicted in Fig. 8.4 parts (e) and (f), respectively. Notice in this simulated example how $d\Phi(k)$ consists of four impulses. These correspond to the four ball sizes and their heights correspond to the relative image areas sieved at the four stages of the granulometry in which they were eliminated.

We now show a granulometry of a real image. The image is a micrograph of a cross section of a bamboo tree. Figure 8.5(a) shows the binary image and parts (b) and (c) show its derivative pattern spectrum and the derivative pattern spectrum of the background, both using a family of octagonal disks which approximate the Euclidean digital balls but with the property required by the granulometry. We can see that the image has a cluster of blobs of small size and several blobs of larger size. The analysis of the background is often important to characterize the shape distributions. We note that we also have a number of background blobs of small size spread in the small blobs of the foreground image and several background blobs of larger size.

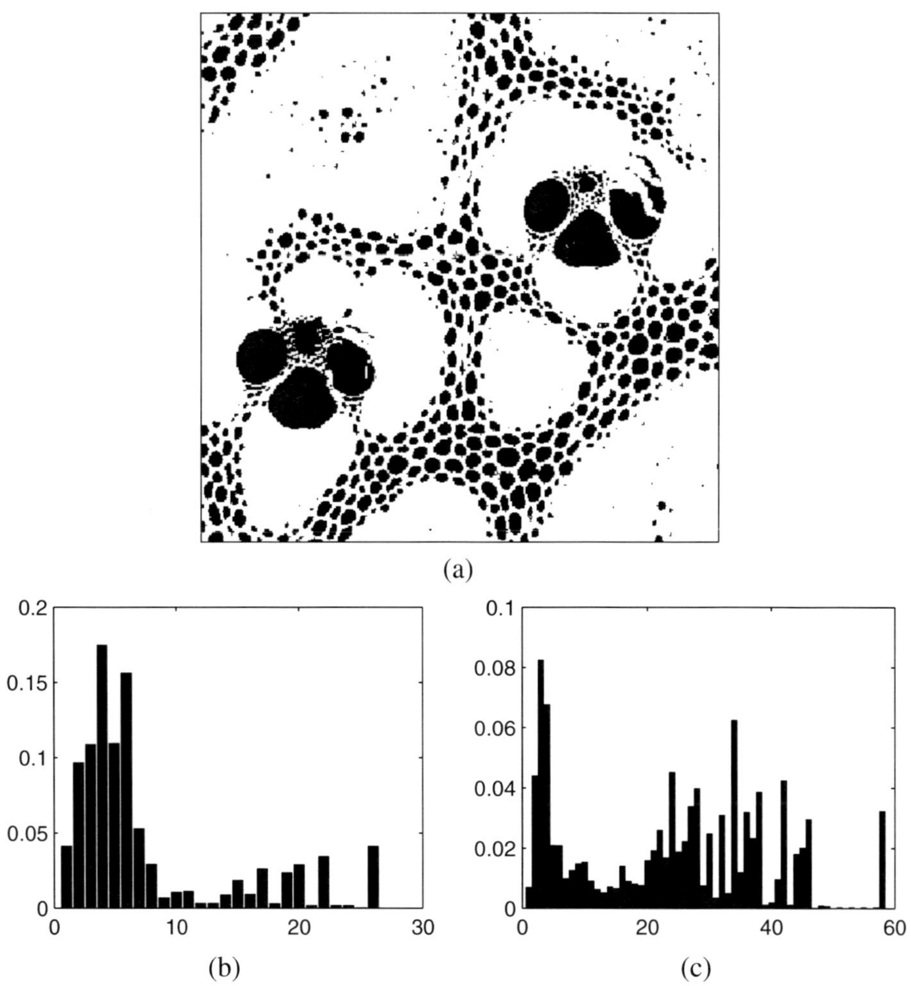

(a)

(b)   (c)

Figure 8.5 Granulometry of a bamboo image: (a) binary image of a micrograph of a bamboo transversal cross section, (b) derivative pattern spectrum, (c) derivative pattern spectrum of the negated image.

8.3 The Open and Discrete-Size Transforms

There are two transforms in use that are closely related to granulometries. If $\mathcal{E} = \{E_k\}$ is a sequence such that E_{k+1} is E_k-open for all k, then the granulometry of image S forms a decreasing sequence of images [Eq. (8.3)]. We can stack this sequence creating the **open transform**:

$$O_{\mathcal{E}}(S)(x) = \max\{k : x \in S \circ E_{k-1}\}. \tag{8.5}$$

The open transform maps each point into one greater than the index of the largest structuring element for which the opened image contains the point.

The definition applies at once to reconstructive openings, in which case all points in the same connected component possess the same transform value. Hence, for reconstructive openings the open transform labels the components according to the largest structuring element that fits within the component.

A useful property of the open transform is that its normalized histogram gives the derivative pattern spectrum as it computes the number of pixels in the opening by E_{k-1} but not in the opening by E_k. In the histogram computation, we need only to compute the occurrence of pixels different than zero as they are the ones that belong to the object. The pattern spectrum mean is the mean of the nonzero pixels of the open transform. The open transform gives a concise visualization of its correspondent granulometry.

A cross section at level k of the open transform gives

$$\begin{aligned} \Xi_{\mathcal{E}}(S)(k) &= \{x : O_{\mathcal{E}}(S)(x) = k\}, \tag{8.6} \\ &= S \circ E_{k-1} - S \circ E_k, \end{aligned}$$

which defines the **discrete size transform** DST. The collection $\{\Xi_{\mathcal{E}}(S)(k)\}$ is known as the **discrete granulometric spectrum** of S relative to the structuring-element sequence. Taking pixel counts on both sides of the preceding equation shows that the pattern spectrum is given by

$$\begin{aligned} d\Phi(k) &= \Phi(k+1) - \Phi(k), \tag{8.7} \\ &= \frac{\Omega(k+1) - \Omega(k)}{v[S]}, \tag{8.8} \\ &= \frac{v[\Xi_{\mathcal{E}}(S)(k+1)]}{v[S]}, \tag{8.9} \end{aligned}$$

which is the normalized histogram of level $k + 1$ of the open transform.

Figure 8.6 illustrates these concepts, with the structuring-element family being of the form $\{kE\}$, where E is the octagonal disk of radius 5: (a) the input drop-shaped image; (b) the open transform, which provides a visualization of the

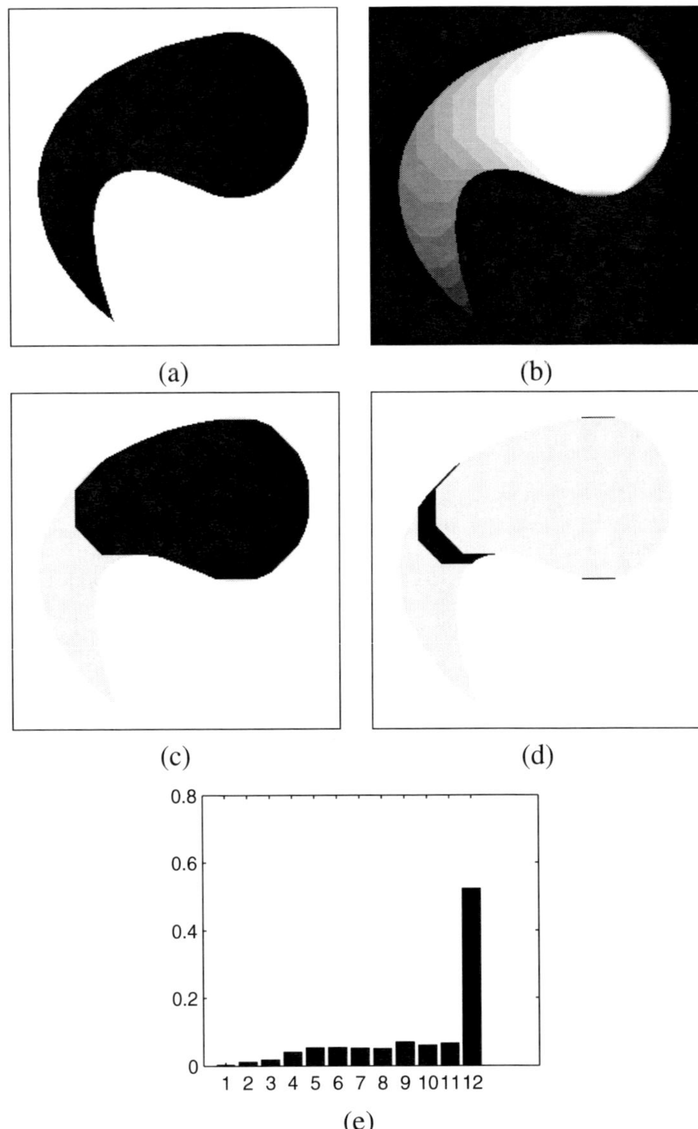

Figure 8.6 Open transform using the octagonal disk of radius 5 as the basic disk for the family $\mathcal{E} = \{kD(5)\}$. (a) input binary image S, (b) open transform $O_{\mathcal{E}}(S)$, (c) open with radius 7 as thresholding of $O_{\mathcal{E}}(S) \geq 8$, (d) cross section 7 of the open transform $\Xi_{\mathcal{E}}(S)(7)$, (e) pattern spectrum $d\Phi(k)$ as the histogram of the open transform.

granulometry; (c) the opening as a thresholding of the open transform; (d) the cross section at level 7; and (e) the pattern spectrum as the normalized histogram of nonzero pixels of part (b).

For a real-world example that illustrates how granulometries can be employed to measure changes in particle-distribution processes, consider Figs. 8.7(a), (b), and (c), which show thresholded toner-particle distributions resulting from an electrophotographic process. In part (a), the toner particles are fairly uniformly spread across the image, whereas in parts (b) and (c) the particles suffer from increasing degrees of agglomeration, a typical problem with electrophotographic processes. Granulometries have been applied to the images using a digital-ball generating sequence, and the resulting pattern spectra are shown in parts (d), (e), and (f) of Fig. 8.7. Notice how the agglomeration has resulted in a shift of the pattern spectra to the right, especially with regard to skewing. We might expect this to result in significant changes in the mean, variance, and skewness of the pattern spectra. In fact, hypothesis tests can be based on these granulometric moments to determine whether, owing to agglomeration, the electrophotographic process is out of control. Finally, the open transforms for the three toner-particle images are shown in parts (g), (h), and (i) of Fig. 8.7.

8.4 Granulometries on Random Binary Images

In the foregoing toner example each actual image represents only a single selection from the population of images being generated by the electrophotographic process. In effect, what confronts us when analyzing the electrophotographic process is a random-image process. Each actual image is only a realization of the process, and it is the overall image process that is of concern. Because images in the process are binary, the image process is a **random binary image**.

Owing to the randomness of the image process, the size distribution $\Omega(t)$ and pattern spectrum $\Phi(t)$ are actually random functions (stochastic processes): each realization of the image process yields its own particular size distribution and pattern spectrum, which are realizations of the size-distribution and pattern-spectrum processes, respectively. We can take the mean of these size distributions as a descriptor of the random image itself, not merely of its realizations. This is the **mean size distribution** (**MSD**) of the image process. It is denoted by $M(t)$ and it is not normalized by the image area. The derivative of the MSD, called the **granulometric size density** (**GSD**) and denoted $H(t)$, plays a key role in the statistical design of opening-based filters.

Each pattern-spectrum realization has its own particular (granulometric) moments. Thus, the granulometric moments (mean, variance, skewness, etc.) are themselves random variables. And now for the interesting part: since these granulometric moments are random variables, they possess their own statistical distributions, and these in turn possess their own moments. Thus, we arrive at the moments

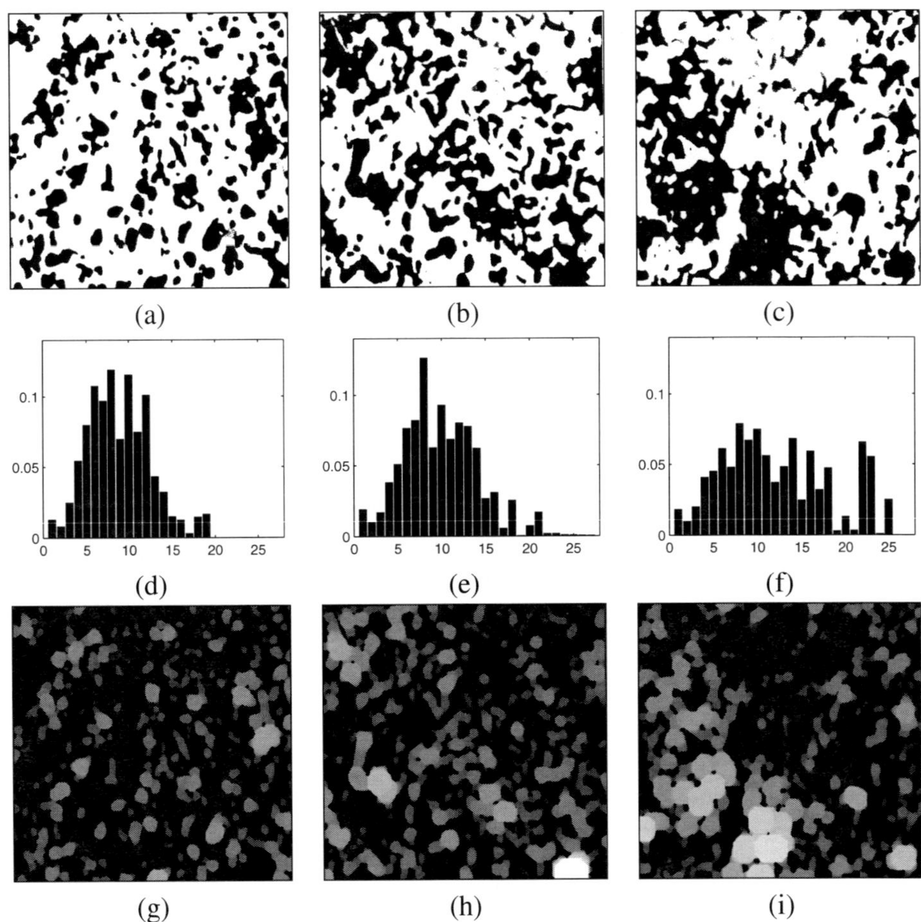

Figure 8.7 Granulometry of toner particles: (a) toner particles with little agglomeration, (b) toner particles with modest agglomeration, (c) toner particles with excessive agglomeration, (d), (e) and (f) corresponding pattern spectra, (g), (h) and (i) corresponding open transform.

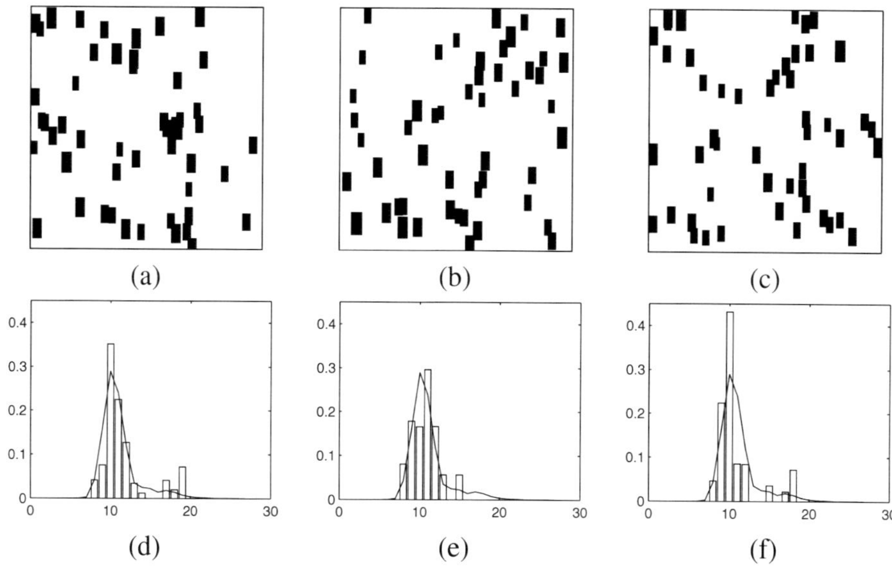

Figure 8.8 Granulometry of random images: (a)–(c) three realizations, (d)–(f) corresponding pattern spectra overlaid by pattern spectrum means computed on 100 random images.

of the granulometric moments. For instance, if we let PSM denote the mean of the pattern spectrum, then PSM is a random variable and it has a mean μ_{PSM} and a standard deviation σ_{PSM}.

To illustrate the preceding concepts, consider a random binary image formed in the following manner: r is a random variable possessing a gamma distribution with mean 20 and variance 5, and 50 rectangles, each with width r and height $2r$, are uniformly randomly tossed upon the image frame. Three realizations of the process are shown in Fig. 8.8 parts (a) through (c), where it should be recognized that r is actually rounded off to an integer value so as to count pixels. Each time the process is run, a different realization occurs. If we consider a digital granulometry with structuring element E_k being a $2k$ by k rectangle, then each image realization yields its own size distribution and pattern spectrum. The latter are shown for the realizations of Fig. 8.8 parts (a) through (c) in parts (d) through (f), respectively, as bar graphs. These graphs are overlaid by the means of 100 pattern spectra relative to 100 random images.

Each realization of the pattern spectrum possesses a mean and variance. The question arises: What is the mean μ_{PSM} of the pattern-spectrum means and the mean μ_{PSV} of the pattern-spectrum variances? In the present case we have estimated both of these means by generating 100 random images, computing 100 pattern spectra, and finding the 100 means and variances of the pattern spectra. We found the average of the 100 means to be 11.04 and the average of the 100 variances to be 5.36. Taking the usual statistical perspective, these serve as estimates

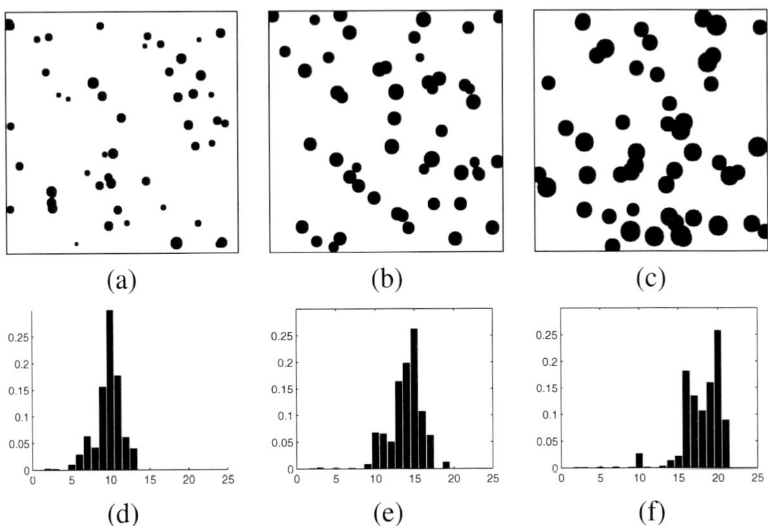

(a)   (b)   (c)

(d)   (e)   (f)

Figure 8.9 Granulometry of random disks: (a) disks with mean radius of 10, (b) disks with mean radius of 15, (c) disks with mean radius of 20, (d) through (f) corresponding pattern spectra.

of the actual means of the pattern-spectra means and variances. Although we will not go into the matter here, given certain assumptions regarding the random-image process, it is possible to obtain asymptotic (and in some cases exact) expressions for the desired means.

8.5   Granulometric Classification

Granulometric classification is accomplished by running a granulometry on an image, computing the granulometric moments, and using these moments as features as inputs to a classifier. For a straightforward illustration of the methodology, we consider three random images consisting of randomly placed balls of random radii. The average number of balls is the same for each image process, but the radii of the processes are normally distributed with means of 10, 15, and 20, respectively, each having a variance of 5. Realizations of each random image are shown in Figs. 8.9 (a) through (c), and their pattern spectra are shown in parts (d) through (f), respectively. It is possible to satisfactorily discriminate between the random images by using the pattern-spectrum mean of a granulometry generated by a unit ball. By randomly generating samples of images and thereby obtaining samples of the PSMs for the three processes, we obtain the three distributions of the PSMs. Classification is accomplished by classifying an observed image realization according to which interval contains its empirically computed PSM.

Figure 8.10 illustrates shape classification using granulometries. Part (a) shows the input image with three numbers: 3, 4, and 5. Applying the granulometry by a

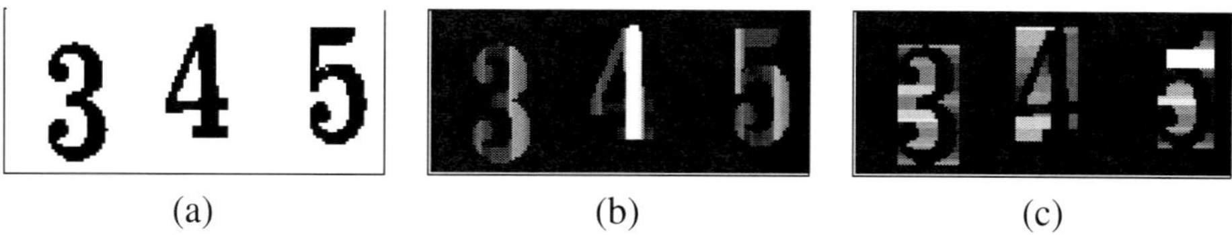

(a)   (b)   (c)

Figure 8.10 Granulometry shape classification: (a) input image of numbers, (b) vertical line differentiates number four from the others, (c) horizontal line of the residue of the shape bounding box differentiates the number five from the others.

family of vertical lines, the mean PSMs of the shapes are 5.7, 11.4, and 6.1 for the numbers 3, 4, and 5, respectively, as the number 4 is the only one that has a long vertical line in its shape decomposition [see Fig. 8.10(b)]. It is a bit harder to differentiate the numbers 3 and 5, as their shape decompositions are very alike. One option is to use the granulometry of the background in the bounding box of each shape. In this bounding box, the granulometry by a family of horizontal lines gives the mean PSMs as 3.6, 3.6, and 4.9, respectively, for the numbers 3, 4, and 5. The open transform of this last granulometry is shown in Fig 8.10(c).

Extending granulometric classification to pixel classification can provide texture-based image segmentation. Pixel classification based on texture requires that a pixel be classified according to the nature of the surrounding image region. Specifically, a parametric texture feature at a pixel is actually a measure based on the image values in some window about the pixel itself. If we assume a texture image to be a random process composed of small texture primitives, then it would seem that a granulometric approach to classification would prove fruitful, so long as the granulometries are based on the texture primitives.

In the standard global approach to granulometries, the entire image is successively opened and at each stage an image pixel count (area) is taken. To measure image texture local to a given pixel, rather than take the pixel count across the whole image we can take the count in a window about the pixel, the result being a **local granulometric size distribution** at each pixel. Normalization yields a **local pattern spectrum** at each pixel, and each of these possesses moments. The window should be kept as small as possible to make the classifier sensitive; however, it must be kept large enough to avoid misclassification owing to variability. If a subregion of the image possesses homogeneous texture, then it is likely that the moments remain somewhat stable across the subregion. Hence, differing subregions characterized by different textures can be differentiated based on the local-pattern-spectrum (granulometric) moments. If we confine ourselves to the pattern-spectrum mean, then for each pixel x we have a mean PSM_x and pixel regions are segmented based on differing PSM_x values. As discussed in the previous section, an image is treated as a random process, so that each PSM_x is a random variable, and our actual segmentation depends on the PSM_x distributions being

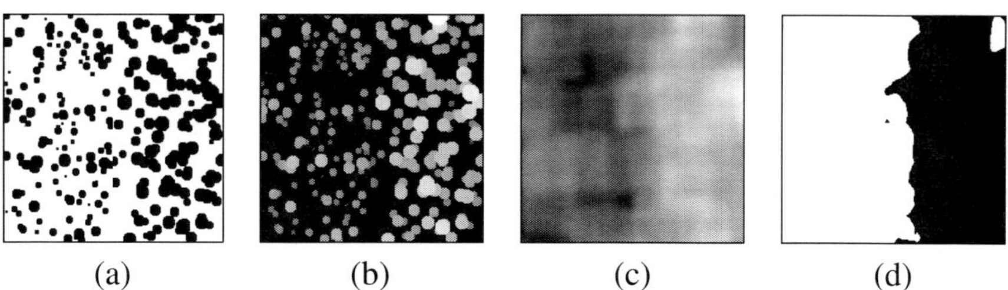

(a)   (b)   (c)   (d)

Figure 8.11 Local granulometry: (a) input image (256×256), left part with disks with mean radii of 5, right part with mean radii of 7, (b) open transform, (c) local mean of the pattern spectra in a windows of 60×60, (d) thresholding local mean at 6.5.

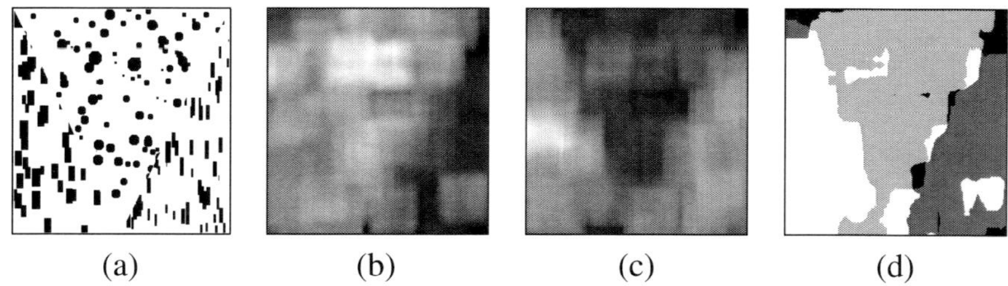

(a)   (b)   (c)   (d)

Figure 8.12 Local granulometry: (a) input image with three different synthetic textures, (b) local means of the pattern spectra using disks, (c) local means of the pattern spectra using vertical lines, (d) final segmentation.

identically distributed across a given texture subregion. In an application such as the present one, statistical analysis of pattern-spectra moments becomes crucial. Here we confine ourselves to some illustrations of the methodology.

In Fig. 8.11(a), each side of the image consists of randomly dispersed balls possessing random radii, the difference being that on the left side the mean radii is 5 and on the right side the mean radii is 7. Part (b) shows the open transform generated by the family of octagonal balls. The local pattern-spectra means in a window of 60×60 are shown in part (c). These can be efficiently computed by a moving-window average of the open transform. The segmentation, shown in part Fig. 8.11(d), is based on the local pattern-spectra means above 6.5.

For more complicated situations, classification usually requires more than just the local pattern-spectra means; the variances and other higher-order moments might be necessary. The choice of structuring element sequence is also important; in fact, one might require a number of local size distributions generated by various structuring element sequences. An example of such a situation occurs in Fig. 8.12(a). A ball sequence can separate the thick rectangles and balls from the thin rectangles, and a vertical-linear sequence one pixel wide can separate the rectangles from the balls. Using the local pattern spectra resulting from each of these sequences together yields the segmented image of Fig. 8.12(d).

Table 8.1 Classification accuracies, in percentage, for independent data.

Input	Classified as							
	d102	d103	d20	d62	d65	d68	d75	d84
d102	99.80	0.00	0.00	0.00	0.00	0.00	0.20	0.00
d103	0.00	93.94	0.00	0.00	0.65	0.00	0.00	5.41
d20	0.00	0.00	100.00	0.00	0.00	0.00	0.00	0.00
d62	0.00	0.70	0.00	99.17	0.00	0.14	0.00	0.00
d65	0.00	0.00	0.00	0.00	93.76	0.00	6.24	0.00
d68	0.01	0.00	0.00	0.00	0.00	99.99	0.00	0.00
d75	0.00	0.00	0.00	0.00	0.00	0.00	100.00	0.00
d84	0.00	0.00	0.00	0.00	0.47	0.00	0.00	99.53

We now apply local granulometric classification to a realistic setting using the binarized texture images of Fig. 8.13. Five structuring elements are employed to construct granulometries: vertical line, horizontal line, 45-deg line, -45-deg line, and circular disk. Three pattern-spectrum moments are used: mean, variance, and skewness. Granulometries are applied to both the foreground (black) and background (white), so that each structuring element produces two granulometries and six features, there being 30 features in all. We apply Gaussian maximum-likelihood classification in which the mean vector and covariance matrix for each feature vector is used (not pooled). The left side of each image is used for training. Thus, classification on dependent and independent data is achieved by applying the classifier to the left and right sides of the images, respectively. Overall classification accuracies on dependent and independent data are 100% and 98.3%, respectively. Table 8.1 shows the individual results for classification on independent data.

8.6 General Granulometries

A family of openings by a multiplicatively parameterized opening is the most commonly employed kind of granulometry, but there are many others. In this section we explore some other types.

A collection of image operations $\{\Psi_t\}, t > 0$, is called an **algebraic granulometry** if (1) Ψ_t is antiextensive for all t, (2) Ψ_t is increasing for all t, and (3)

$$\Psi_t \Psi_s = \Psi_s \Psi_t = \Psi_{max\{t,s\}}. \tag{8.10}$$

If we view t as a sieving parameter for sieves of increasing mesh size, we see the genesis of the three properties: (1) the image remaining after any sieving operation is a subset of the original; (2) if one image is a subset of another, then the sieved images maintain the same subset relation; (3) if Ψ_t and Ψ_s are two sieves in the process, the order of sieving does not matter, the remaining image being the same as if one were only to sieve through the largest of the mesh sizes.

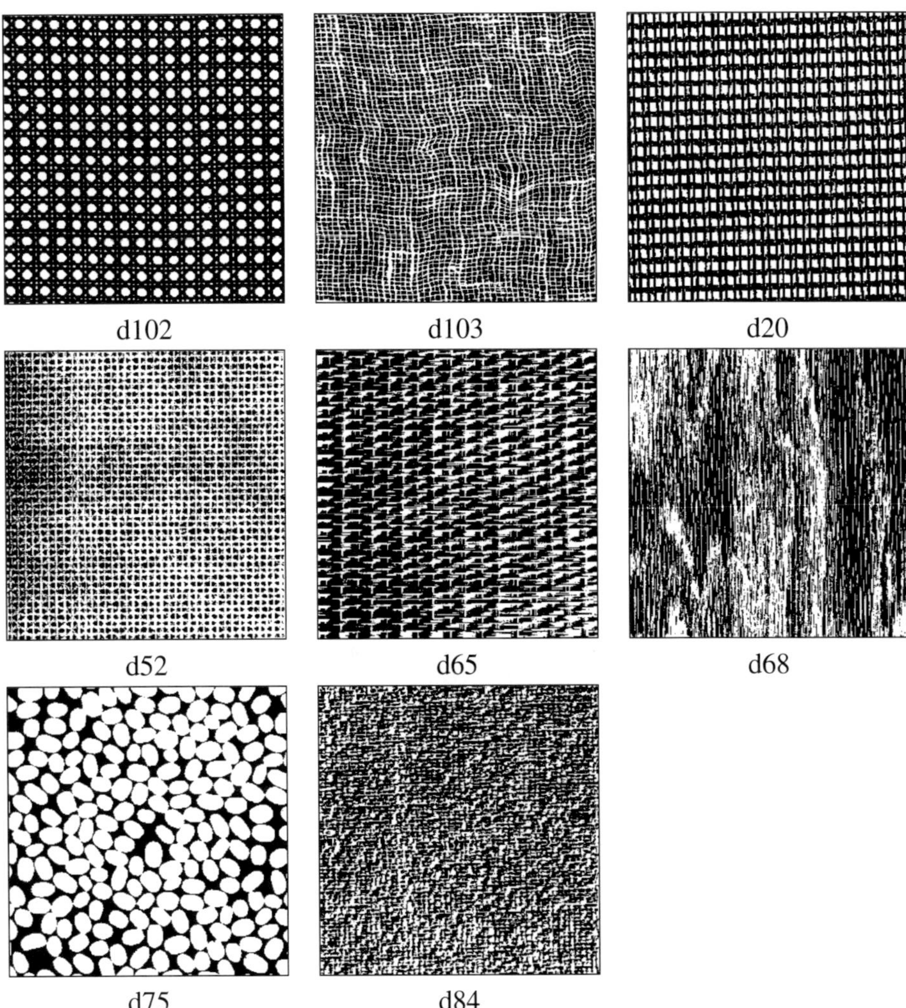

d102　　　　　　　　d103　　　　　　　　d20

d52　　　　　　　　d65　　　　　　　　d68

d75　　　　　　　　d84

Figure 8.13 Binary texture images.

Two further properties can be deduced from the three basic postulates: if $r \geq s$, then $\Psi_r(A)$ is a subset of $\Psi_s(A)$; and if $r \geq s$, then the invariant class of Ψ_r is a subclass of the invariant class of Ψ_s. The latter subclass relation between the operators is called **invariance ordering**. In line with the sieving model, there will be less residue remaining after sieving with the larger sieve, and those images that are invariant under the larger sieve must also be invariant under the smaller sieve.

The elementary opening granulometries $\{A \circ tB\}$, B convex, satisfy the three granulometric postulates. They also satisfy two other fundamental properties. They are (4) translation invariant, and (5) they satisfy the **Euclidean property**, namely,

$$\Psi_t(A) = t \times \Psi_1(t^{-1}A), \tag{8.11}$$

for any $t > 0$ and any binary Euclidean image A. Property (5) is the most interesting: it says that there is a **unit sieve**, Ψ_1, and that any other sieve in the process can be evaluated by first scaling the image by the reciprocal of the parameter, filtering by the unit sieve, and then rescaling. If one thinks of sieving particles through a mesh, the property appears quite intuitive. If an algebraic granulometry $\{\Psi_t\}$ satisfies property 4, then it is simply called a **granulometry**. If it satisfies properties 4 and 5, then it is called a **Euclidean granulometry**. A basic proposition regarding granulometries is that a parameterized family $\{\Psi_t\}$ is a granulometry if and only if Ψ_t is a τ-opening for all $t > 0$ and the operators are invariance ordered.

Every Euclidean granulometry can be expressed in terms of openings, an elementary opening generated by a convex primitive being the simplest. While we leave a full discussion to more complete texts, we make note that the most important example of a Euclidean granulometry is a union of openings, each by a parameterized convex primitive:

$$\Psi_t(A) = \bigcup_{k=1}^{n} A \circ tB_k. \tag{8.12}$$

$\mathcal{G} = \{B_1, B_2, \ldots, B_n\}$ is called the **generator** of the granulometry. Such a union of parameterized openings satisfies properties 1 through 5. So, too, do other expressions composed of openings, but they can be more complicated. The key to the relative simplicity of Eq. (8.12) is that the generator is composed of convex shape primitives. If we compare the preceding equation with Eq. (2.18), we see that each operator in the family $\{\Psi_t\}$ is a τ-opening with base $\mathcal{B}_t = \{tB_1, tB_2, \ldots, tB_n\}$. All of the definitions regarding size distributions apply directly, with the basic size distribution $\Omega(t)$ being the area removed by Ψ_t. Classification and segmentation can be performed using more general Euclidean granulometries.

Rather than using the same parameter for each set in the generator, one can allow each opening to have its own parameter. Letting $\mathbf{t} = \{t_1, t_2, \ldots, t_n\}$, a **multivariate granulometry** is formed by the union

$$\Psi_{\mathbf{t}}(A) = \bigcup_{k=1}^{n} A \circ t_k B_k. \tag{8.13}$$

Allowing more parameters can increase sensitivity to image differences and thereby provide better classification. The size distribution definitions need to be adapted to a vector parameter, and this involves some subtleties. We will not pursue the matter here.

8.7　Logical Granulometries

Granulometries can be constructed using reconstructive instead of ordinary openings. We can proceed simply by making the openings in Eq. (8.12) reconstructive to obtain the **disjunctive granulometry**

$$\Psi_t(A) = \bigcup_{k=1}^{n} A \circ_E tB_k, \tag{8.14}$$

where E gives the connectivity for the reconstructive opening and is usually the 3×3 square structuring element centered at the origin. Fixing t yields a disjunctive opening [Eq. (3.15)]. The family $\{\Psi_t\}$ formed by the union of disjunctive openings is a true granulometry because each Ψ_t is a τ-opening and the operators are invariance ordered. Pattern spectra, and related applications, can be applied using a disjunctive granulometry. Since grains (connected components) are either left intact or removed in full, the pattern spectrum $\Phi(t)$ is a step function with jumps created by grain deletions.

An illustration of the reconstructive pattern spectrum using octagonal-ball structuring elements is shown in Fig. 8.14: (a) input binary image, (b) open transform, (c) reconstructive open transform, (d) maximal balls, (e) pattern spectrum, and (f) reconstructive pattern spectrum. The reconstructive open transform can be computed from the open transform by simply taking the maximum value of the open transform among the points in a grain.

Disjunctive granulometries can also be constructed using area openings, since these too act on the level of connected components. Figure 8.15 shows the area open transform and the pattern spectrum for the image of Fig. 8.14. Because most areas are unique among the grains, displaying the pattern spectrum for increments of a single pixel is not informative. Hence, we have shown the histogram using steps of 100.

If we were to replace the union in Eq. (8.12) by an intersection, the result would not be a granulometry because Ψ_t would not be idempotent and therefore not be a τ-opening, and the family of operators would not be invariance ordered. However, if we replace opening by reconstructive opening, then the intersection is a

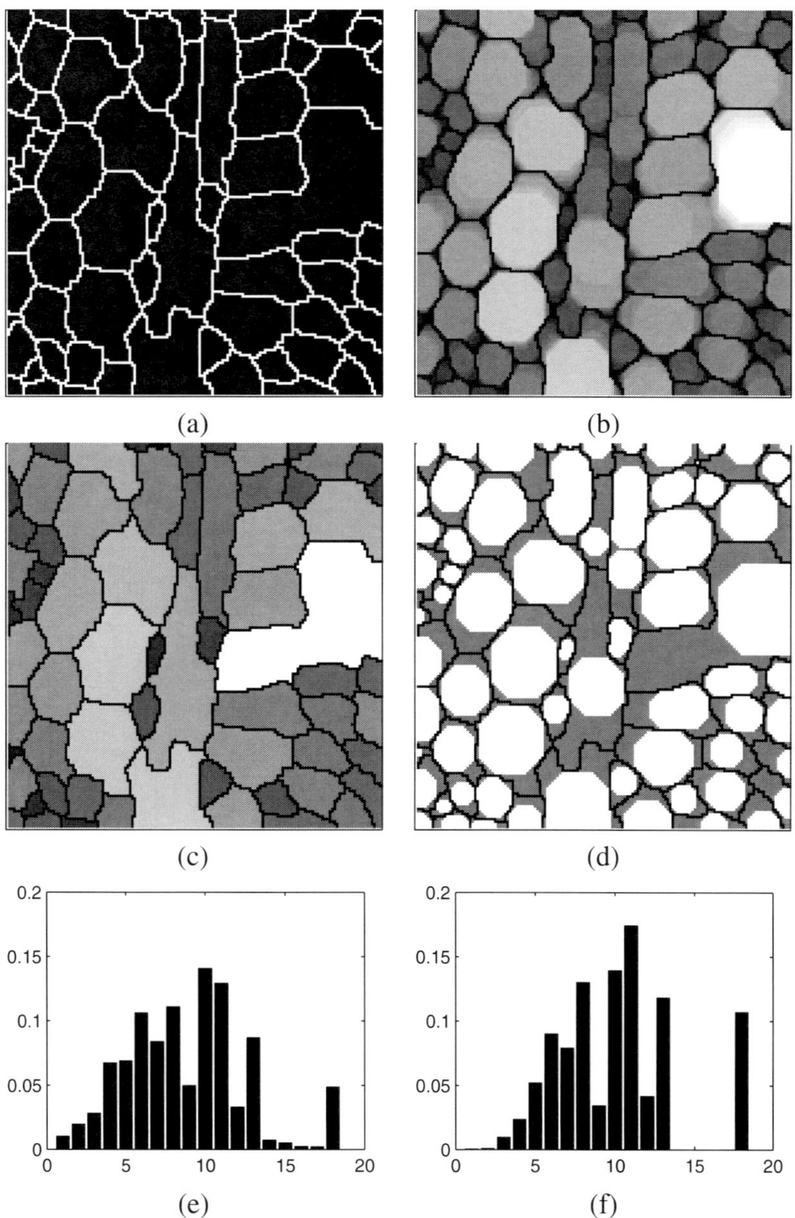

Figure 8.14 Reconstructive granulometry: (a) input binary image S, (b) open transform $O_{\mathcal{E}}(S)$, (c) reconstructive open transform, (d) maximal balls, (e) pattern spectrum, (f) reconstructive pattern spectrum.

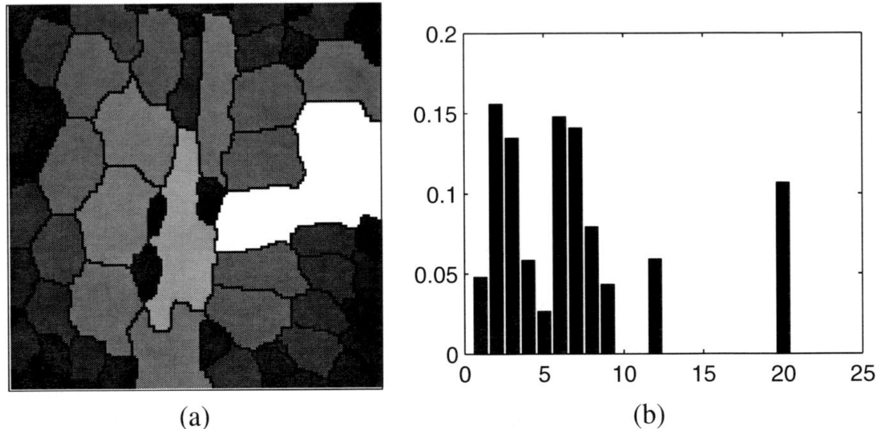

Figure 8.15 Disjunctive granulometry: (a) area open transform, (b) pattern spectrum (area step = 100).

granulometry. A **conjunctive granulometry** is defined by

$$\Psi_t(A) = \bigcap_{k=1}^{n} A \circ_E tB_k.$$ (8.15)

For fixed t, Ψ_t is a conjunctive opening.

Conjunction and disjunction can be combined to form a **logical granulometry**:

$$\Psi_t(A) = \bigcup_{k=1}^{n} \bigcap_{j=1}^{m_k} A \circ_E tB_{k,j}.$$ (8.16)

For fixed t, Ψ_t is a logical opening. The family satisfies the conditions of a granulometry.

The **granulometric size** of a grain (connected component) G relative to the logical granulometry $\{\Psi_t\}$ is the largest value of t for which $\Psi_t(G) = G$. If t_0 is the granulometric size of G, then $\Psi_t(G) = G$ for any $t \leq t_0$ and $\Psi_t(G) = \emptyset$ for any $t > t_0$. If the granulometry is generated by a single set B then there exists a translation of tB contained within G if and only if $t \leq t_0$. In this case, the granulometric size of G gives one greater than the value of the reconstructive open transform for each point in G.

Optimization of granulometric filters involves finding the value of the parameter t that provides the best filtering of the images under consideration, the goal being to pass the desired signal and not pass the noise. Different methods, both optimal and adaptive, have been developed to find good values of the parameter. Rather than treat optimization of granulometric filters in their own right, in the next section we consider the more general issue of finding optimal granulometric bandpass filters.

8.8 Discrete Granulometric Bandpass Filters

A disjunctive granulometry is a family of disjunctive openings. Selecting a value of t means selecting a particular disjunctive opening that removes grains with granulometric size less than t. This kind of sizing filter can be extended to pass and remove grains in a more complicated way than simply passing larger grains.

If $\{\Psi_t\}$ is a disjunctive granulometry and $T = \{t_k\}$ is a sequence of points such that $0 = t_0 < t_1 < t_2 < \ldots$ and $t_k \to \infty$ as $k \to \infty$, then every set S is partitioned according to

$$S_k = \Psi_{t_k}(S) - \Psi_{t_{k+1}}(S), \tag{8.17}$$

for $k = 0, 1, 2, \ldots$. Specifically,

$$S = \bigcup_{k=0}^{\infty} S_k, \tag{8.18}$$

and $S_k \cap S_{k'}$ for $k \neq k'$. The family $\{S_k\}$ is called the **discrete granulometric spectrum** of S relative to $\{\Psi_t\}$ and T, and S_k is the kth **spectral band**. Note that what we have here is a slightly more general form of the open transform. The discrete pattern and granulometric spectra are related by $\Phi'_S(k) = v[S_k]/v[S]$. For digital processing, we assume $T \subset \{0, 1, 2, \ldots\}$.

A **discrete granulometric bandpass filter (GBF)** Λ_Π is defined by choosing a **pass set** of orders $\Pi \subset T$ and defining

$$\Lambda_\Pi(S) = \bigcup_{k \in \Pi} S_k. \tag{8.19}$$

Π^c is called the **fail set**. Λ_Π is defined by selecting spectral bands to pass (whereas a linear bandpass filter selects frequency bands).

An optimal discrete GBF is one that minimizes an error criterion in some random image model. Here we consider error to be the mean (expected) area of the symmetric difference between the output of the GBF and the ideal image,

$$e[\Lambda_\Pi] = \overline{v}[\Lambda_\Pi(S \cup N) \triangle S], \tag{8.20}$$

where $\overline{v}$ denotes the mean (average) area and $\triangle$ denotes the symmetric difference $[A \triangle B = (A - B) \cup (B - A)]$. If we assume that the image is a disjoint union of grains, some of them forming the signal S and others forming the noise N, so that the observed signal is $S \cup N$, then it is not hard to see that the error of the GBF is the sum of the mean areas of the noise grains in the pass set and the mean areas of the signal grains in the fail set:

$$e[\Lambda_\Pi] = \sum_{t_k \in \Pi} \overline{v}[N_{t_k}] + \sum_{t_k \in \Pi^c} \overline{v}[S_{t_k}]. \tag{8.21}$$

Error is minimized by the following decision rule: t_k is in the pass set if and only if

$$\overline{v}[N_{t_k}] \leq \overline{v}[S_{t_k}]. \tag{8.22}$$

The optimal pass set determines the optimal filter, Λ_Ψ, relative to the spectrum associated with $\{\Psi_t\}$ and T.

Equation (8.22) can be reformulated in terms of the size distribution of S. Using the relationship

$$v[S_{t_k}] = \Omega_S(t_k) - \Omega_S(t_{k+1}), \tag{8.23}$$

and taking the means, t_k is in the pass set for the optimal GBF if and only if

$$M_N(t_{k+1}) - M_N(t_k) \leq M_S(t_{k+1}) - M_S(t_k), \tag{8.24}$$

where M_S and M_N are the mean size distributions for the signal and noise, respectively. Dividing both sides by $t_{k+1} - t_k$ yields

$$\frac{M_N(t_{k+1}) - M_N(t_k)}{t_{k+1} - t_k} \leq \frac{M_S(t_{k+1}) - M_S(t_k)}{t_{k+1} - t_k}. \tag{8.25}$$

Thus, the decision rule for t_k being in the optimal pass set can be expressed in terms of the difference quotients of the mean size distributions.

Were the mean size distributions of the signal and noise differentiable, then letting $t_{k+1} - t_k \to 0$ would yield a rule involving the derivatives of the mean size distributions; namely, Eq. (8.25) would reduce to a rule that says "t_k is an element in the optimal pass set if and only if $H_N(t_k) \leq H_S(t_k)$, where H_S and H_N are the granulometric size densities (GSDs) of the signal and noise, respectively." To proceed in this way one must first define the granulometric size density for a continuous parameter t, which would indeed be natural considering the fact that granulometries are defined in terms of a continuous parameter. A full mathematical development involves extending the concept of granulometric spectra to continuous t and also considering issues of differentiability. Let us simply note that all of this can be done, and for real-world models the mean size distribution, which is now a function of t, is differentiable, its derivative being the granulometric size density. In sum, for continuous granulometric bandpass filters, t is in the optimal pass set if and only if

$$H_N(t) \leq H_S(t). \tag{8.26}$$

Because the decision criterion involves a derivative, the natural setting for studying GBFs is with a continuous parameter.

In a typical situation, the decision inequality will be satisfied over some intervals and not over others. For instance, it may be that $H_N(t) \leq H_S(t)$ for $a \leq t \leq b$.

Then the interval $[a, b]$ is called the **passband** of the optimal filter, and the optimal filter is given by

$$\Lambda_\Psi(S \cup N) = \Psi_a(S \cup N) - \Psi_b(S \cup N). \tag{8.27}$$

Ψ_a passes grains whose granulometric size exceeds a and Ψ_b deletes grains whose granulometric size exceeds b.

An illustration of this kind of logical GBF is shown in Fig 8.16. Part (a) shows an image of disjoint balls and randomly rotated rectangles, where the balls are the signal and the rectangles are the noise. Part (b) shows the reconstructive open transform of the image and part (c) shows the granulometric size densities of the signal and noise using the granulometry generated by the octagonal disk family. The passband is obtained from these densities. Finally, part (d) shows the input image filtered by the logical GBF and part (e) shows the error as the symmetric difference of the filtered and the signal image.

For real-world images there may be insufficient data to estimate the GSDs. Adaptive methods have been designed for this situation. The **MT** program for a real-world image demonstration is shown in Fig. 8.17, using the segmented silver-halide T-grain crystals image of Fig. 7.26. The intermediate images of this demonstration are shown in Fig. 8.18. The steps used to filter the crystals are below. The program line numbers and figure parts are shown in brackets.

Image reading (line 1) The silver-halide-T-grain-crystals-in-emulsion binary image is read (a). This image is segmented using the watershed technique described in the previous chapter (Program 7.2).

Open transform (line 2) Open transform of the reconstructive open by octagonal disk family (b).

Pattern spectrum (line 3) The pattern spectrum of the reconstructive open by the octagonal disk family is shown in (c).

GBF filtering (line 4) Granulometric bandpass filter choosing t in interval $[7, 10]$ (d). This interval has been found by an adaptive design method.

Display (line 5) For display purposes, the contours of the original image are overlaid on the filtered image (e).

We close this section by noting that, while we have applied GBFs in the context of logical granulometries, the theory does not have such a restriction and everything mentioned here applies for ordinary granulometries, the difference being that grains are not passed in full, but instead are diminished by the manner in which the structuring elements fit within them.

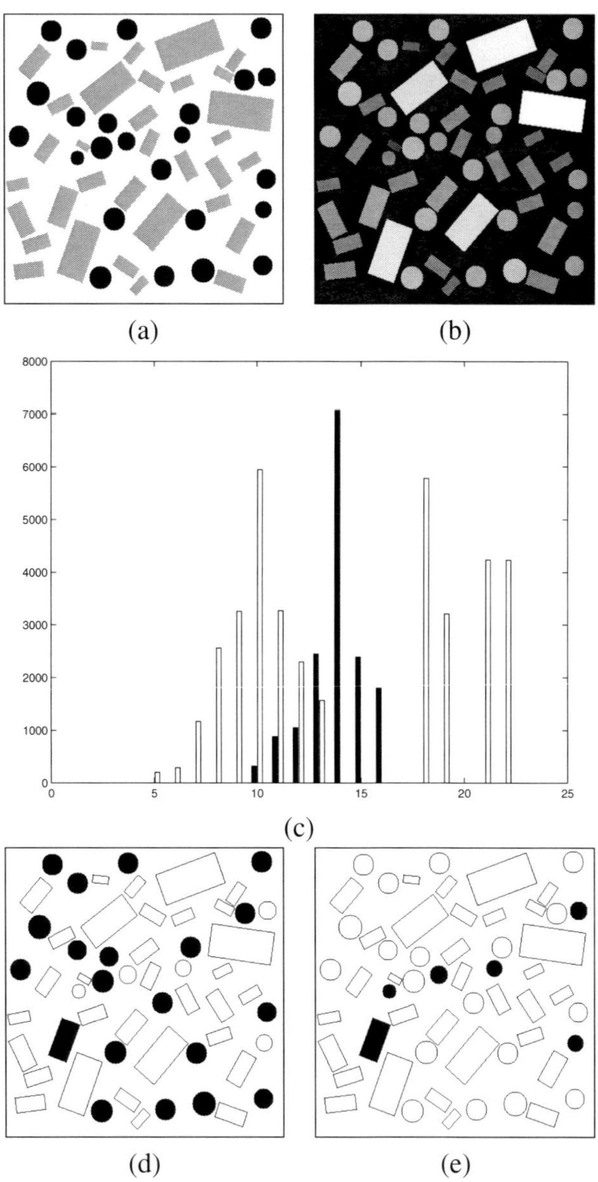

Figure 8.16 Granulometric bandpass filter: (a) input image, signal in black and noise in gray-scale; (b) open transform of the reconstructive open by octagonal disk family; (c) estimation of H_S (filled bars) and H_N; (d) optimum filter choosing t in interval $[13, 16]$, where the borders of the removed grains are shown to help visualize the filtering; (e) minimum error from symmetric difference (3822 pixels).

```
                      Program 8.1
01   a = mmreadgray('crystals_bin.tif')
02   b = mmopentransf(a,'octagon-rec')
03   c = mmpatspec(a,'octagon-rec')
04   d = mmcloseth( c,mmsedisk(4,'2D','octagon'))
05   e = mmunion(d,mmgradm(a,mmsecross(0)))
```

Figure 8.17 Python code for the T-grain crystals GBF filtering. The images from (a) to (e) are shown in Figs. 8.18.

8.9 Gray-Scale Granulometries

A good deal of the granulometric theory for binary images extends to gray-scale images. While much of the extension is straightforward, there are some subtleties. A key issue is the replacement of scalar multiplication of a set by an appropriate scalar multiplication of a function. For a positive real number t and a function f, we define the function $t * f$ by

$$(t * f)(x) = tf\left(\frac{x}{t}\right). \tag{8.28}$$

In terms of umbrae and surfaces

$$U[t * f] = tU[f], \tag{8.29}$$

$$S[tA] = t * S[A]. \tag{8.30}$$

A key property of $t * f$ is its relation to translation:

$$t * (f_x + y) = (t * f)_{tx} + ty. \tag{8.31}$$

Geometrically, $t * f$ is constructed from f by taking each point (u, v) on the graph of f, and considering (u, v) as a 3D vector (2D for signals). If we scalar multiply to obtain (tu, tv), then $(t * f)(tu) = tv$. For openings, there is the following fundamental relation between scaling the signal and scaling the structuring element:

$$\left[\left(\frac{1}{t}\right) * f\right] \circ g = f \circ (t * g). \tag{8.32}$$

Now consider a set $\mathcal{G} = \{g_1, g_2, ..., g_n\}$ of functions that are concave down. Because they are concave down, their umbrae are convex. A parameterized family defined by a supremum of the form

$$\Psi_t(f) = \bigvee_{k=1}^{n} f \circ (t * g_k), \tag{8.33}$$

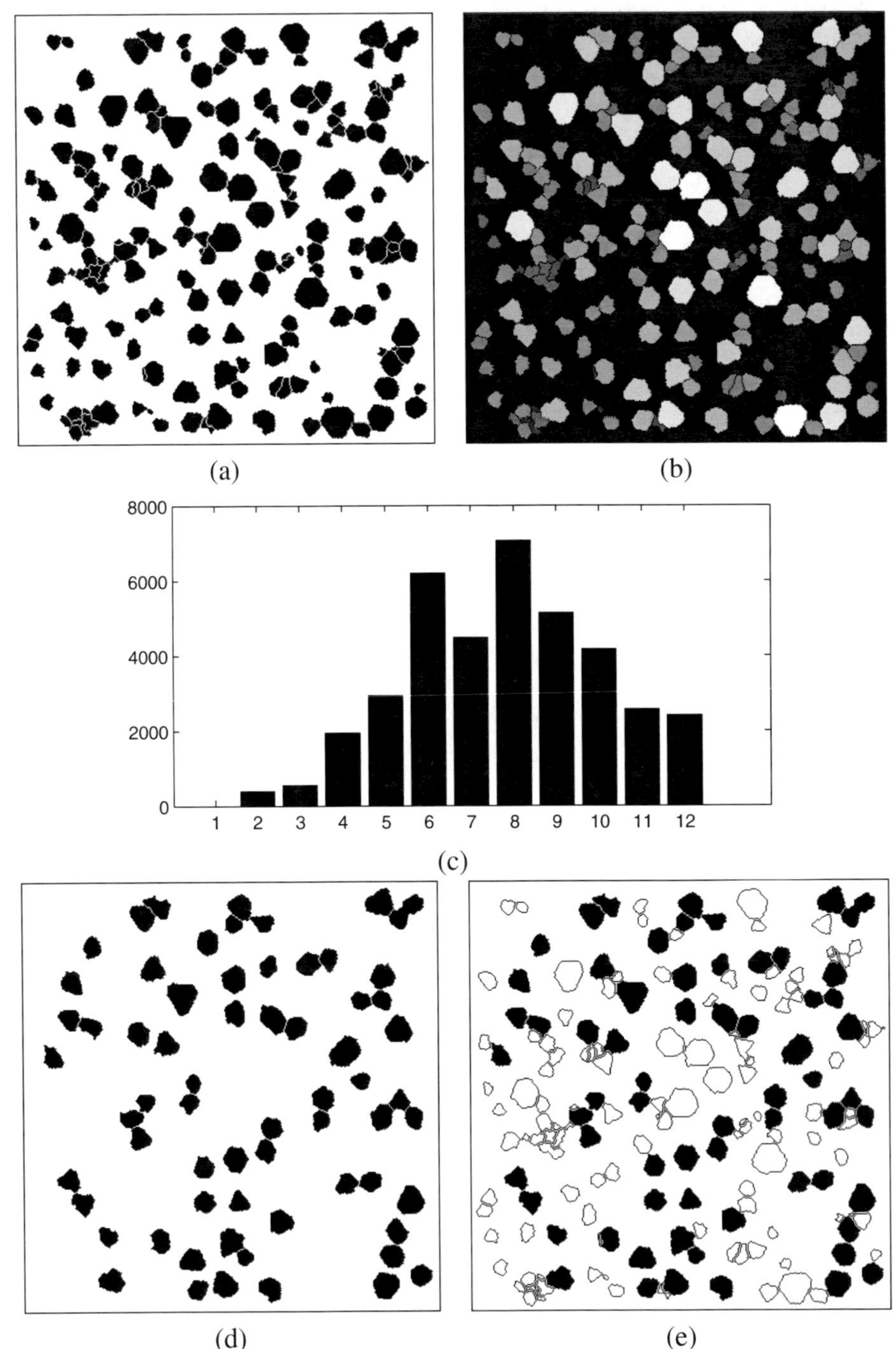

Figure 8.18 Silver-halide T-grain crystals GBF filtering. These images refer to the program of Fig. 8.17.

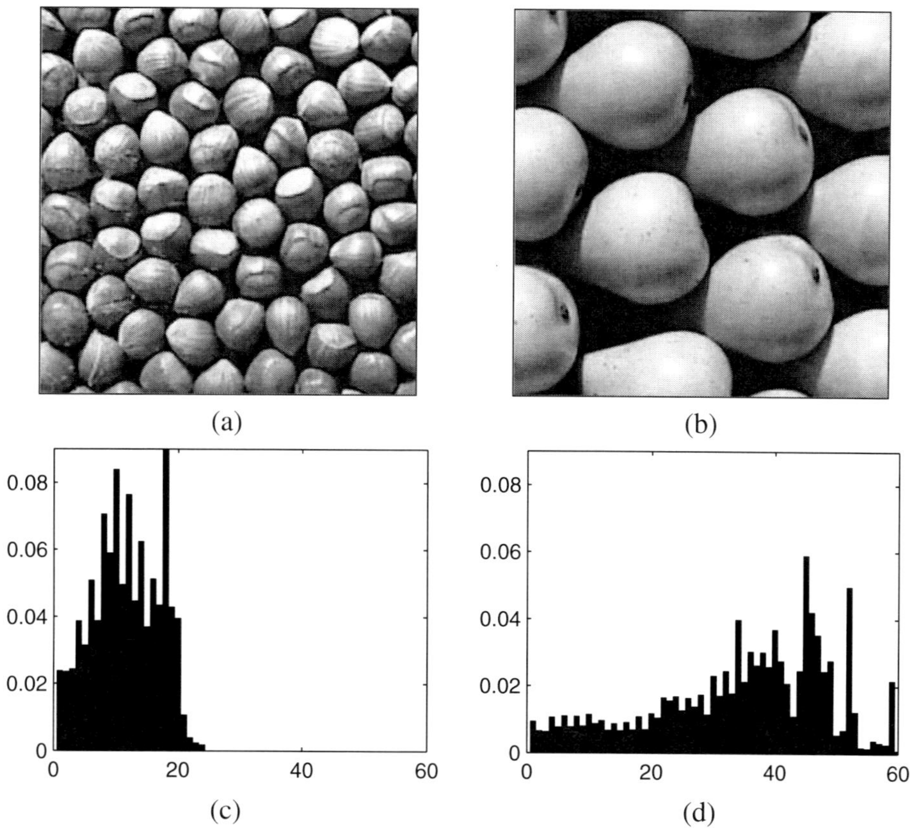

Figure 8.19 Gray-scale granulometry with nonflat octagonal balls: (a) image of nuts, (b) image of pears, (c) pattern spectrum of the nuts image, (d) pattern spectrum of the pears image.

defines a gray-scale granulometry. There are more general forms of gray-scale granulometries, but we will not discuss them here. The simplest gray-scale granulometry is a single parameterized opening $\Psi_t(f) = f \circ (t*g)$. For fixed t, the supremum defines a gray-scale τ-opening, which is a supremum of ordinary openings. A τ-opening is translation invariant, increasing, antiextensive, and idempotent.

Size distributions are defined in an analogous manner to binary granulometric size distributions by using the volume between the graph of the image and a fixed plane parallel to the domain.

Figure 8.19 shows two gray-scale images and their pattern spectra corresponding to a family of nonflat octagonal structuring elements (Fig. 8.20): (a) image of nuts; (b) image of pears; (c) pattern spectrum for nut image; and (d) pattern spectrum for pear image. Notice how the larger spherical structure in the pear image has resulted in a pattern spectrum distributed more to the right than that of the nut image. These two images can easily be discriminated via their pattern-spectrum means.

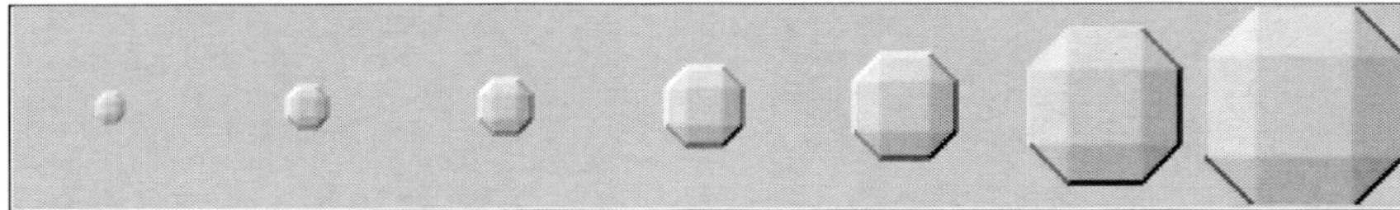

Figure 8.20 Family of nonflat octagonal balls with increasing radius of 4, 6, 8, 12, 16, 24, and 32 pixels.

A local gray-scale granulometry is illustrated in Fig. 8.21 using a signal: (a) input signal; (b) open transform of the signal's umbra transform using a linear structuring-element sequence to illustrate the effect of flat structuring elements; (c) open transform of the signal's umbra using a diamond-shaped structuring-element sequence to illustrate the effect of nonflat structuring elements; (d) pattern spectrum corresponding to linear structuring elements; (e) pattern spectrum corresponding to diamond-shaped structuring elements; (f) local-pattern-spectrum means for a 10-pixel window for the flat structuring elements; and (g) local-pattern-spectrum means for a 10-pixel window for the nonflat structuring elements.

An image example of a local gray-scale granulometry is shown in Fig. 8.22: (a) input image; (b) local pattern-spectrum means corresponding to a gray-scale granulometry generated by flat octagonal structuring elements using a 10×10 window; and (c) like part (b) but with nonflat octagonal balls.

In the **MT**, the functions `mmopentransf`, and `mmpatspec` implement the open transform and the discrete pattern spectrum, respectively. Figure 8.23 shows the list of these operators discussed in this chapter.

8.10 Exercises

1. Suppose a binary image consists of nonoverlapping disks of radii 3, 6, 6, 7, 7, 7. Find the pattern spectrum relative to opening by a disk of unit radius, and find the pattern-spectrum mean.

2. What may happen to the discrete pattern spectrum if the disk E_{k+1} is not E_k-open for all k?

3. Design a family of quasi-Euclidean discrete disks that satisfy the property E_{k+1} being E_k-open for all k?

4. How can the maximal balls [e.g. those displayed in Fig. 8.14(d)] be determined from the open transform?

5. A multivariate granulometry [Eq. (8.13)] is a true granulometry. Show that it is invariance ordered.

6. As noted in the text, if the union of Eq. (8.13) were changed to an intersection, the operators would not be idempotent (see Exercise 11), nor would the

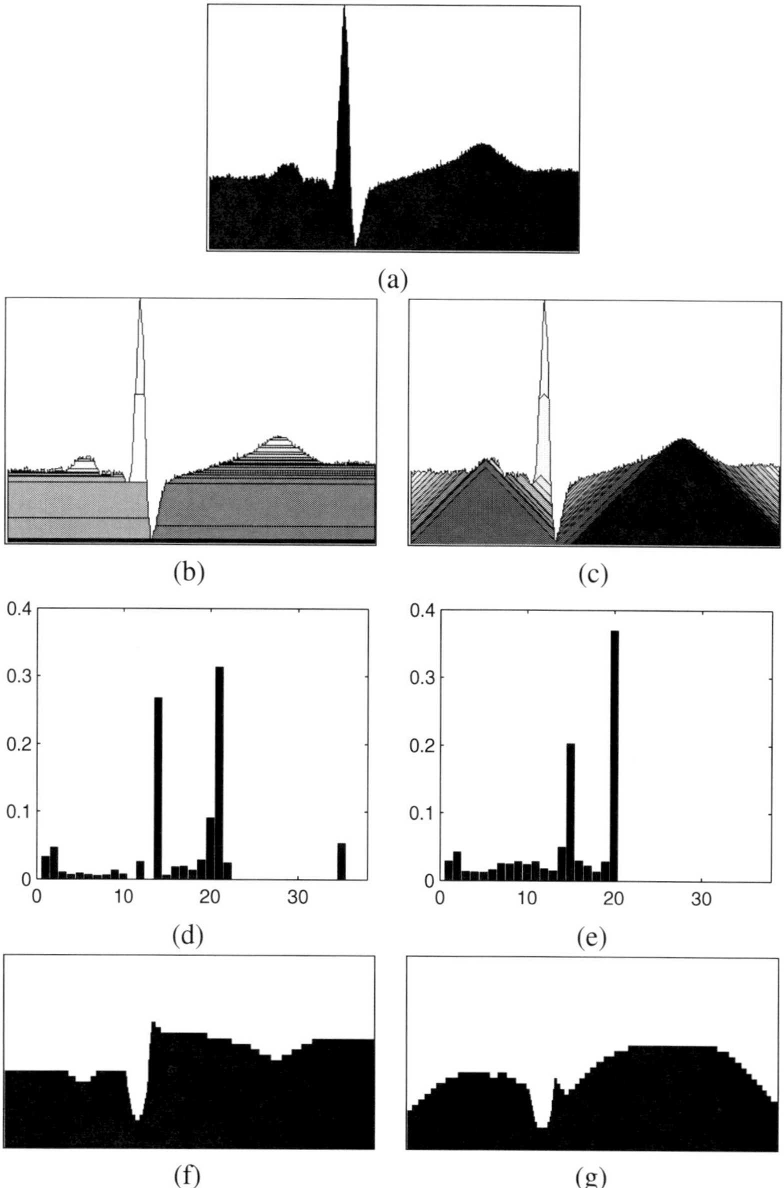

Figure 8.21 Signal local gray-scale granulometry illustration: (a) input signal, (b) open transform of its umbra using linear structuring element to simulate flat structuring element, (c) open transform of its umbra using diamond structuring element to simulate nonflat city-block ball structuring element, (d) pattern spectrum of (b), (e) pattern spectrum of (c), (f) mean radius of the local granulometry of (b) around a window of 10 pixels, (g) similar to (f) but applying a nonflat structuring element (c).

(a)

(b)   (c)

Figure 8.22 Local gray-scale granulometry: (a) input image, (b) first moment (mean radius) of local gray-scale granulometry with flat octagonal disks computed with 10×10 window, (c) similarly as (b), but using nonflat octagonal balls.

name	description
mmopentransf:	open transform [Eq. (8.5)]
mmpatspec:	pattern spectrum [Eq. (8.4)]

Figure 8.23 MT operators presented in this chapter.

family be invariance ordered. Give an example to show that it need not be invariance ordered.

7. Show that disjunctive and conjunctive granulometries are invariance ordered.

8.11 Laboratory Experiments

1. Compute the open transform and the pattern spectrum of the image in Fig. 8.6 using: a) a horizontal structuring element, and b) a vertical structuring element.

2. The discrete Euclidean disks $D(k)$ cannot be used to generate a granulometry. Write a small program to identify the radii k where $D(k)$ is not $D(k-1)$-open.

References

1. G. Ayala, M. E. Diaz, and L. Martinez-Costa. Granulometric moments and corneal endothelium status. *Pattern Recognition*, 34(6):1219–1227, 2001.

2. S. Baeg, S. Batman, E. R. Dougherty, V. Kamat, N. D. Kehtarnavaz, S. Kim, A. Popov, K. Sivakumar, and R. Shah. Unsupervised morphological granulometric texture segmentation of digital mammograms. *Journal of Electronic Imaging*, 8(1):65–75, 1999.

3. Y. Balagurunathan and E. R. Dougherty. Optimal linear granulometric estimation for random sets. *Pattern Recognition*, 35:1315–1325, 2002.

4. Y. Balagurunathan and E. R. Dougherty. Granulometric parametric estimation for the random boolean model using optimal linear filters and optimal structuring elements. *Pattern Recognition Letters*, 24:283–293, 2003.

5. Y. Balagurunathan, E. R. Dougherty, S. F. Isovi-Bilinski, and N. Vdovi. Morphological granulometric analysis of sediment images. *Image Analysis and Stereology*, 20:87–99, 2001.

6. S. Batman and E. R. Dougherty. Size distributions for multivariate morphological granulometries. *Optical Engineering*, 36(5):1518–1529, 1997.

7. S. Batman and E. R. Dougherty. Morphological granulometric estimation of random patterns in the context of parameterized random sets. *Pattern Recognition*, 34(6):1207–1217, 2001.

8. S. Batman, E.R. Dougherty, and F. Sand. Heterogeneous morphological granulometries. *Pattern Recognition*, 33(6):1047–1057, 2000.

9. Y. Chen and E. R. Dougherty. Gray-scale morphological granulometric texture classification. *Optical Engineering*, 33(8):2713–2722, 1994.

10. Y. Chen and E. R. Dougherty. Optimal and adaptive reconstructive granulometric bandpass filters. *Signal Processing*, 61(1):65–81, 1997.

11. Y. Chen and E. R. Dougherty. Markovian analysis of adaptive reconstructive multiparameter τ-openings. *Mathematical Imaging and Vision*, 10(3):253–267, 1999.

12. Y. Chen, E. R. Dougherty, S. Totterman, and J. Hornak. Classification of trabecular structure in magnetic resonance images based on morphological granulometries. *Magnetic Resonance Medicine*, 29(3):358–370, 1993.

13. E. R. Dougherty. Euclidean gray-scale granulometries: representation and umbra inducement. *Journal of Mathematical Imaging and Vision*, 1:7–21, 1992.

14. E. R. Dougherty. Optimal binary morphological bandpass filters induced by granulometric spectral representation. *Mathematical Imaging and Vision*, 7(2):175–192, 1997.

15. E. R. Dougherty. Optimal conjunctive granulometric bandpass filters. *Journal of Mathematical Imaging and Vision*, 14(1):39–51, 2001.

16. E.R. Dougherty and Y. Chen. Robust optimal granulometric bandpass filters. *Signal Processing*, 81(7):1357–1372, 2001.

17. E. R. Dougherty. Granulometric size density for segmented random-disk models. *Journal of Mathematical Imaging and Vision*, 17(3):267–276, 2002.

18. E. R. Dougherty and Y. Chen. Logical granulometric filtering in signal-union-clutter model. In J. Goutsias, R. Mahler, and C. Nguyen, editors, *Random Sets: Theory and Applications*, pages 73–95. Springer-Verlag, New York, 1997.

19. E. R. Dougherty and Y. Chen. Granulometric filters. In E. R. Dougherty and J. T. Astola, editors, *Nonlinear Filters for Image Processing*, pages 121–162. SPIE/IEEE Series on Imaging Science & Engineering, 1999.

20. E. R. Dougherty and Y. Chen. Optimal and adaptive design of logical granulometric filters. In P. Hawkes, editor, *Advances in Imaging and Electron Physics*, 117:1–71. Academic Press, New York, 2001.

21. E. R. Dougherty, J. Newell, and J. Pelz. Morphological texture-based maximum-likelihood pixel classification based on local granulometric moments. *Pattern Recognition*, 25(10):1181–1198, 1992.

22. E. R. Dougherty and J. Pelz. Morphological granulometric analysis of electrophotographic images — size distribution statistics for process control. *Optical Engineering*, 30(4):438–445, 1991.

23. E. R. Dougherty, J. Pelz, F. Sand, and A. Lent. Morphological image segmentation by local granulometric size distributions. *Journal of Electronic Imaging*, 1(1):46–60, 1992.

24. E. R. Dougherty and F. Sand. Representation of linear granulometric moments for deterministic and random binary Euclidean images. *Journal of Visual Communication and Image Representation*, 6(1):69–79, 1995.

25. R. Jones and P. Soille. Periodic lines: cascades, and application to granulometries. *Pattern Recognition Letters*, 17:1057–1063, 1996.

26. Y. Balagurunathan, K. Sivakumar and E. R. Dougherty. Asymptotic joint normality of the granulometric moments. *Pattern Recognition Letters*, 22(14): 1537–1543, 2001.

27. B. Li and E. R. Dougherty. Size-distribution estimation in process fluids by ultrasound for particles in the wavelength range. *Optical Engineering*, 32(8):1967–1980, 1993.

28. P. Maragos. Pattern spectrum and multiscale shape representation. *IEEE Transactions on Pattern Analysis and Machine Intelligence*, 11:701–716, 1989.

29. F. Sand and E. R. Dougherty. Asymptotic normality of the morphological pattern-spectrum moments and orthogonal granulometric generators. *Journal of Visual Communication and Image Representation*, 3(2):203–214, 1992.

30. F. Sand and E. R. Dougherty. Asymptotic granulometric mixing theorem: morphological estimation of sizing parameters and mixture proportions. *Pattern Recognition*, 31(1):53–61, 1998.

31. F. Sand and E. R. Dougherty. Robustness of granulometric moments. *Pattern Recognition*, 32(9):1657–1665, 1999.

32. K. Sivakumar and J. Goutsias. Monte Carlo estimation of morphological granulometric discrete size distributions. In J. Serra and P. Soille, editors, *Mathematical Morphology and its Application to Image and Signal Processing*, pages 233–240. Kluwer Academic Publishers, Boston, 1994.

33. K. Sivakumar and J. Goutsias. Discrete morphological size distributions and size densities: estimation techniques and applications. *Journal of Electronic Imaging*, 6:31–53, 1997.

34. K. Sivakumar and J. Goutsias. Morphologically constrained Gibbs random fields: applications to texture synthesis and analysis. *IEEE Transactions on Pattern Analysis and Machine Intelligence*, 21(2):99–113, 1999.

35. R. Sabourin, G. Genest, and F. Preteux. Off-line signature verification by local granulometric size distributions. *IEEE Transactions on Pattern Analysis and Machine Intelligence*, 19(9):976–988, 1997.

36. N. Theera-Umpon, E.R. Dougherty, and P.D. Gader. Non-homothetic granulometric mixing theory with application to blood cell counting. *Pattern Recognition*, 34(12):2547–2560, 2001.

37. N. Theera-Umpon and P.D. Gader. Counting white blood cells using morphological granulometries. *Journal of Electronic Imaging*, 9(2):170–177, 2000.

Chapter 9

Automatic Design of Morphological Operators

The key to successful morphological image processing is the selection of structuring elements. There are a myriad of algorithms for a multitude of imaging applications, but in each and every instance, algorithm performance depends on the structuring elements. The classical approach to morphological processing is to have a human being, or a group of human beings, use intuition and an understanding of the goals to design algorithms based on erosions, openings, hit-or-miss transforms and other basic morphological operators. This approach can work well if the task can be described in elementary geometric terms and the images under consideration are not too complex. It breaks down in situations where satisfactory filtering might require hundreds, or even thousands, of structuring elements. The present chapter introduces automatic algorithm design, where morphological operators are designed based on sample data, structural decomposition, and criteria set by the imaging scientist.

9.1 Boolean Functions

In this chapter, we will exploit the relationship between binary mathematical morphology and Boolean functions. This section is devoted to that relationship.

A binary-valued function $\psi(x_1, x_2, \ldots, x_n)$ of binary variables $x_1, x_2, \ldots, x_n$ is called a **Boolean function**. As a logical function, ψ possesses a logical sum-of-products **disjunctive-normal-form** representation in terms of the n variables $x_1, x_2, \ldots, x_n$:

$$\psi(x_1, x_2, \ldots, x_n) = \sum_i x_1^{p(i,1)} x_2^{p(i,2)} \ldots x_n^{p(i,n)}, \qquad (9.1)$$

where the "sum" denotes OR, the "product" denotes AND, and $p(i, k)$ is either $'$ (prime) or null, depending on whether the variable is complemented or not complemented, respectively. There are at most 2^n products in the expansion, and each product is called a **minterm**. The representation can be (nonuniquely) reduced to a sum of products containing a minimal number of logic gates; that is,

$$\psi(x_1, x_2, \ldots, x_n) = \sum_i x_{i,1}^{p(i,1)} x_{i,2}^{p(i,2)} \ldots x_{i,n(i)}^{p(i,n(i))}. \qquad (9.2)$$

The truth table formulation of ψ corresponds directly to the disjunctive normal form of Eq. (9.1). ψ is defined by a 2^n-row truth table of n variables in which each string $t_1 t_2 \ldots t_n$ of 0s and 1s is assigned a binary value $\psi(t_1 t_2 \ldots t_n)$. The

227

Table 9.1 Boolean function.

x_1	x_2	x_3	$\psi(x_1, x_2, x_3)$
0	0	0	0
0	0	1	1
0	1	0	0
0	1	1	0
1	0	0	0
1	0	1	1
1	1	0	1
1	1	1	0

correspondence between Eq. (9.1) and the truth table is given by the following rule:
the minterm $x_1^{p(i,1)} x_2^{p(i,2)} \ldots x_n^{p(i,n)}$ appears in the expansion of Eq. (9.1) if and only
if there is a 1-valued string $t_1 t_2 \ldots t_n$ in the truth table with $p(i, j)$ null if $t_j = 1$
and $p(i, j) = {}'$ if $t_j = 0$.

As an example, the Boolean function

$$\psi(x_1, x_2, x_3) = x_1 x_2' x_3 + x_1' x_2' x_3 + x_1 x_2 x_3' \tag{9.3}$$

is equivalently defined by Table 9.1. The 1s in rows 1, 5, and 6 correspond to the
second, first, and third terms of Eq. (9.3), respectively. Logic reduction gives the
equivalent representation

$$\psi(x_1, x_2, x_3) = x_2' x_3 + x_1 x_2 x_3'. \tag{9.4}$$

The product set $\{0, 1\}^n$ is composed of binary n-vectors. For instance,

$$\{0, 1\}^3 = \{000, 001, 010, 011, 100, 101, 110, 111\}.$$

A partial ordering of $\{0, 1\}^n$ is defined by $(x_1, x_2, \ldots, x_n) \leq (y_1, y_2, \ldots, y_n)$
if and only if $x_j \leq y_j$ for $j = 1, 2, \ldots, n$. We denote vectors of binary variables by
boldface lowercase letters, such as $\mathbf{x} = (x_1, x_2, \ldots, x_n)$ and $\mathbf{y} = (y_1, y_2, \ldots, y_n)$.
Thus, we equivalently state that $\mathbf{x} \leq \mathbf{y}$ if and only if $x_j \leq y_j$ for $j = 1, 2, \ldots, n$.

A Boolean function ψ is **increasing** if $\mathbf{x} \leq \mathbf{y}$ implies $\psi(\mathbf{x}) \leq \psi(\mathbf{y})$. If ψ
is increasing, then it is called a **positive** Boolean function. ψ is positive if and
only if it can be represented as a logical sum of products having no complemented
variables,

$$\psi(x_1, x_2, \ldots, x_n) = \sum_i x_{i,1} x_{i,2} \ldots x_{i,n(i)}. \tag{9.5}$$

A complementation-free expansion is called a positive expansion. If the variable set in any product of the expansion contains as a subset the set of variables in a distinct product, then whenever the former product has value 1, so too does the latter. Thus, inclusion of the former product in the expansion is redundant and it can be deleted from the expansion without changing ψ. No product whose variable set does not contain the variable set of a distinct product can be deleted without changing ψ. Performing the permitted deletions produces a unique **minimal representation** of ψ. For instance, consider the Boolean function

$$\beta(x_1, x_2, x_3) = x_1'x_2'x_3 + x_1'x_2x_3' + x_1'x_2x_3 + x_1x_2'x_3' + x_1x_2'x_3 + x_1x_2x_3' + x_1x_2x_3.$$

$$(9.6)$$

The function is increasing and possesses the minimal positive expansion

$$\beta(x_1, x_2, x_3) = x_1 + x_2 + x_3. \tag{9.7}$$

It gives the maximum among the input variables. The minimum function has disjunctive normal form

$$\alpha(x_1, x_2, x_3) = x_1x_2x_3, \tag{9.8}$$

which also happens to be the minimal representation. Unless otherwise stated, it is convention that positive Boolean functions are represented by positive expansions.

The set of all input vectors for which a Boolean function has value 1 is called its **kernel**. If the Boolean function is increasing, then the minimal elements of the kernel compose the **basis** of the function. We denote the kernel and basis by $\mathbf{K}[\psi]$ and $\mathbf{B}[\psi]$, respectively, or just $\mathbf{K}$ and $\mathbf{B}$ when not specifying the function. A four-observation increasing function is shown in Fig. 9.1(a). The enclosed vectors compose the kernel, and minimal elements are shown in solid black boxes. Figure 9.1(b) corresponds to a nonincreasing function; again the enclosed vectors compose the kernel. It becomes increasing if 0111 is switched into the kernel.

9.2 Morphological Representation

Boolean functions are used to define translation-invariant windowed operators on binary digital images. To define a windowed operator, let $W = \{w_1, w_2, \ldots, w_n\}$ be an n-pixel window and ψ be an n-variable Boolean function. The corresponding set operator Ψ is defined by

$$\Psi(S)(z) = \psi(S \cap W_z) = \psi(S(w_1 + z), S(w_2 + z), \ldots, S(w_n + z)) \tag{9.9}$$

(Fig. 9.2). Note that we are simultaneously treating S and $\Psi(S)$ as subsets of the digital plane and as binary-valued functions: in the first instance, $S \cap W_z$ is the

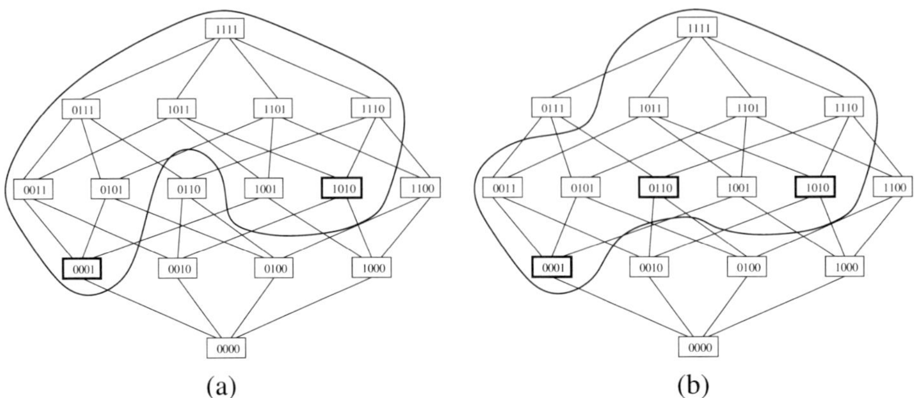

(a)   (b)

Figure 9.1 (a) Increasing, and (b) nonincreasing Boolean functions.

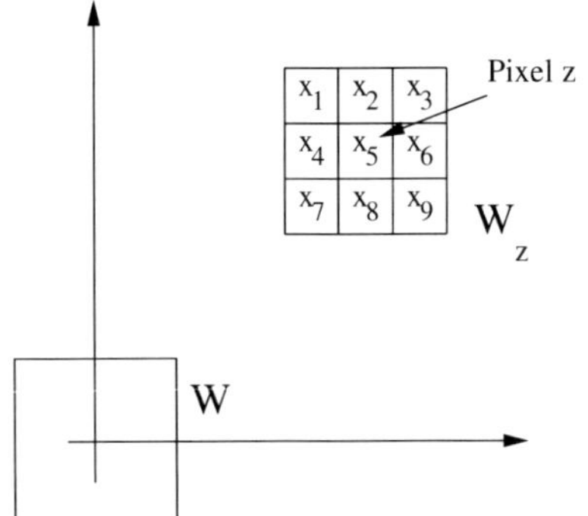

Figure 9.2 3×3 window at pixel z.

intersection between the sets S and W_z; in the second, $S \cap W_z$ is the $\{0, 1\}$-valued function S restricted to W_z. Ψ is translation-invariant because the same Boolean function is applied at every pixel. We call Ψ a **W-operator** and ψ its **characteristic function**. The representation of Ψ corresponds directly to the logical representation of ψ.

A W-operator Ψ_i defined by a single-product positive Boolean function,

$$\psi_i(x_1, x_2, \ldots, x_n) = x_{i,1} x_{i,2} \ldots x_{i,n(i)} \tag{9.10}$$

is an erosion. This is seen by recognizing that $z \in \Psi(S)$ if and only if $\Psi(S)(z) = 1$, which in turn is true if and only if the pixels in the translated window W_z corresponding to the pixels $w_{i,1}, w_{i,2}, \ldots, w_{i,n(i)}$ in W all have value 1 so that the product is 1. The pixels in W_z corresponding to $w_{i,1}, w_{i,2}, \ldots, w_{i,n(i)}$ are $w_{i,1} + z, w_{i,2} + z, \ldots, w_{i,n(i)} + z$, and the product is 1 if and only if all of these pixel translations lie in S. Letting $B = \{w_{i,1}, w_{i,2}, \ldots, w_{i,n(i)}\}$, $z \in \Psi(S)$ if and only if $B_z \subset S$, which is precisely the definition of erosion.

A W-operator Ψ with characteristic function ψ is increasing if and only if ψ

is a positive Boolean function. It follows at once from the logical representation of Eq. (9.5) that a W-operator is increasing if and only if it possesses an erosion representation of the form

$$\Psi(S) = \bigcup_i S \ominus B^i, \tag{9.11}$$

where the correspondence between Eqs. (9.5) and (9.11) is given by

$$x_{i,1} x_{i,2} \ldots x_{i,n(i)} \leftrightarrow B^i. \tag{9.12}$$

The erosion representation is minimal if no structuring element is a subset of another. This corresponds directly to a minimal logical representation and the corresponding structuring elements comprise the basis, $\mathbf{B}[\Psi]$, of Ψ.

Consider the binary moving median M defined over a window W containing an odd number of pixels. The characteristic function μ is defined by $\mu(x_1, x_2, \ldots, x_n) = 1$ if and only if more than $\frac{n}{2}$ pixels in W are 1-valued. Hence a row of the truth table defining μ is 1-valued if and only if at least $\frac{(n+1)}{2}$ of the variables are 1-valued, which means that an n-variable product is a minterm in the disjunctive normal form for μ if and only if at least $\frac{(n+1)}{2}$ variables are noncomplemented. The minimal positive expansion for μ consists of all products of exactly $\frac{(n+1)}{2}$ variables. For instance, suppose W consists of the origin together with the pixels immediately below, above, to the right, and to the left of it. Let x_1, x_2, x_3, x_4, and x_5 denote the five variables corresponding to the five pixels. The minimal representation of μ is given by

$$\begin{aligned}
\mu(x_1, x_2, x_3, x_4, x_5) \;=\; & x_1 x_2 x_3 + x_1 x_2 x_4 + x_1 x_3 x_4 + x_1 x_2 x_5 + x_1 x_3 x_5 \\
+ \; & x_1 x_4 x_5 + x_2 x_3 x_4 + x_2 x_3 x_5 + x_2 x_4 x_5 + x_3 x_4 x_5.
\end{aligned} \tag{9.13}$$

The basis of M consists of 10 structuring elements:

$$\begin{bmatrix} 0 & 1 & 0 \\ 0 & \mathbf{0} & 1 \\ 0 & 1 & 0 \end{bmatrix}, \begin{bmatrix} 0 & 0 & 0 \\ 1 & \mathbf{1} & 1 \\ 0 & 0 & 0 \end{bmatrix}, \begin{bmatrix} 0 & 0 & 0 \\ 1 & \mathbf{1} & 0 \\ 0 & 1 & 0 \end{bmatrix}, \begin{bmatrix} 0 & 0 & 0 \\ 1 & \mathbf{0} & 1 \\ 0 & 1 & 0 \end{bmatrix}, \begin{bmatrix} 0 & 0 & 0 \\ 0 & 1 & 1 \\ 0 & 1 & 0 \end{bmatrix},$$

$$\begin{bmatrix} 0 & 1 & 0 \\ 1 & \mathbf{1} & 0 \\ 0 & 0 & 0 \end{bmatrix}, \begin{bmatrix} 0 & 1 & 0 \\ 1 & \mathbf{0} & 1 \\ 0 & 0 & 0 \end{bmatrix}, \begin{bmatrix} 0 & 1 & 0 \\ 1 & \mathbf{1} & 1 \\ 0 & 0 & 0 \end{bmatrix}, \begin{bmatrix} 0 & 1 & 0 \\ 1 & \mathbf{0} & 0 \\ 0 & 1 & 0 \end{bmatrix} \begin{bmatrix} 0 & 1 & 0 \\ 0 & 1 & 0 \\ 0 & 1 & 0 \end{bmatrix}.$$

Note that including extra 0s in the matrix representation of a structuring element does not change the structuring element.

Morphological representation of arbitrary (nonincreasing) W-operators is achieved via the hit-or-miss transform. If we write the hit-or-miss transform in operator notation, $\Lambda_{E,F}(S)$, where the hit and miss structuring elements are E and F, respectively, then, pixelwise, as a binary-valued function,

$$\Lambda_{E,F}(S)(z) = \begin{cases} 1, & \text{if } E_z \subset S \text{ and } F_z \subset S^c \\ 0, & \text{otherwise} \end{cases}. \tag{9.14}$$

Corresponding to the logical representation of Eq. (9.2) is the **standard morphological representation** for a W-operator,

$$\Psi(S) = \bigcup_i \Lambda_{E^i,F^i}(S). \tag{9.15}$$

The correspondence between Eqs. (9.2) and (9.15) is given by

$$(E^i, F^i) \leftrightarrow x_1^{p(i,1)} x_2^{p(i,2)} \ldots x_n^{p(i,n)}, \tag{9.16}$$

where x_k corresponds to the pixel $w_k \in W$, x_k appears noncomplemented if $w_k \in E^i$, x_k appears complemented if $w_k \in F^i$, and x_k does not appear if $x_k \notin E^i \cup F^i$. If Ψ happens to be increasing, then the hit-or-miss expansion of Eq. (9.15) reduces to the erosion expansion of Eq. (9.11).

A structuring pair is said to be **canonical** if $E^i \cup F^i = W$ and is canonical if and only if the corresponding logical product contains all n variables in the window. The representation of Eq. (9.15) is said to be canonical if all structuring pairs are canonical, in which case it corresponds to the disjunctive normal form of Eq. (9.1). Hence, given the disjunctive normal form of a characteristic function, we at once obtain the canonical morphological representation for the corresponding image operator. As an illustration, consider the Boolean function in disjunctive normal form:

$$\psi(x_1, x_2, x_3, x_4, x_5) = x_1 x_2 x_3 x_4' x_5' + x_1 x_2' x_3' x_4' x_5' + x_1' x_2 x_3 x_4' x_5 + x_1' x_2 x_3 x_4 x_5'. \tag{9.17}$$

If we consider the variables as representing the pixel values in the 4-connected neighborhood of the origin, reading lexicographically, the four canonical structuring pairs defining Ψ are given by

$$\begin{bmatrix} & 1 & \\ 1 & 1 & 0 \\ & 0 & \end{bmatrix}, \begin{bmatrix} & 1 & \\ 0 & 0 & 0 \\ & 0 & \end{bmatrix}, \begin{bmatrix} & 0 & \\ 1 & 1 & 0 \\ & 1 & \end{bmatrix}, \begin{bmatrix} & 0 & \\ 1 & 1 & 1 \\ & 0 & \end{bmatrix}.$$

Using these in Eq. (9.15) provides the canonical morphological representation of Ψ. A reduced expression can be obtained by applying logic reduction to Eq. (9.17) to find noncanonical structuring pairs.

A W-operator is translation invariant and possesses a morphological representation via a union of hit-or-miss transforms, which reduces to an erosion representation if the operator is increasing. In fact, all translation-invariant operators possess such representations. For practical digital image processing, this is achieved by simply choosing a sufficiently large window. In theory, no finite window may do; nevertheless, morphological representation is always possible, albeit, with perhaps structuring elements that are also of infinite extent.

9.3 Optimal W-Operators

The basic image-processing paradigm is to find an algorithm to output a value, vector of values, or another image that gives the best result. To formulate the optimization problem requires giving meaning to the notion as to what is best. If a noisy image is observed, the restoration problem is to find an operator that filters the image in such a way as to estimate the original nonnoisy image. If an image is observed and an edge is desired, the problem is to find an edge detector that operates on the image so as to produce an estimate of the true edge. If it is desired to find a pattern in an image, the problem is to find a pattern recognition algorithm to mark the locations where copies of the pattern are located. These various tasks can be characterized in the framework of taking an observed input image, processing it by an operator, and then comparing the output image to a desired image. The problem is inherently probabilistic because the operator must be applied to a random collection of observed images, and these must be processed to estimate a random collection of desired images. For instance, the desired image class may consist of pages of clean text and the observed images may be noisy versions of these. This means that we must define a probabilistic error measure as a criterion of operator performance.

Suppose the window is translated to pixel z and the n values of the observed image in the translated window W_z form the vector $\mathbf{X}$. We use uppercase vector notation to denote that the vector is random, depending on the collection of images under consideration and the location of the window in the image. Suppose that the value at the pixel in the desired image is Y, where the uppercase notation indicates that the value is random (i.e., a random variable). If Ψ is an arbitrary image operator with characteristic function ψ, then $\psi(\mathbf{X})$ serves as an estimator of Y. There are two possibilities: (1) $\psi(\mathbf{X}) = Y$, in which case there is no error; or (2) $\psi(\mathbf{X}) \neq Y$, in which case there is an error. In the first instance, either $\psi(\mathbf{X}) = Y = 0$, or $\psi(\mathbf{X}) = Y = 1$. In the second, either $\psi(\mathbf{X}) = 0$ and $Y = 1$, or $\psi(\mathbf{X}) = 1$ and $Y = 0$. The error of $\psi(\mathbf{X})$ as an estimator of Y is taken to be the probability of error, $P(\psi(\mathbf{X}) \neq Y)$. This is called the error of the operator Ψ (and of its characteristic function). This probability is equal to the expected (mean) absolute error; that is, the expected value of $|\psi(\mathbf{X}) - Y|$. For this reason it is known as the **mean-absolute error** (**MAE**).

Our goal is to find the operator (filter) that has the minimum MAE for a given estimation problem. There is an implicit assumption that this error is independent of the pixel, which corresponds to the assumption that the operator is translation

Table 9.2 Probability table and resulting optimal filter.

$\mathbf{x}$	$P(\mathbf{x})$	$P(Y = 1|\mathbf{x})$	$P(Y = 0|\mathbf{x})$	$\psi_{opt}(\mathbf{x})$
000	0.20	0.2	0.8	0
001	0.15	0.1	0.9	0
010	0.10	0.1	0.9	0
011	0.05	0.8	0.2	1
100	0.15	0.2	0.8	0
101	0.10	0.8	0.2	1
110	0.20	0.9	0.1	1
111	0.05	0.9	0.1	1

invariant. There may be more than one operator possessing minimum error. Any operator having minimum MAE is called an **optimal filter**. It can be shown that an optimal filter is defined in terms of conditional probabilities by the characteristic function

$$\psi_{opt}(\mathbf{X}) = \begin{cases} 1, & \text{if } P(Y = 1|\mathbf{X}) > 0.5 \\ 0, & \text{if } P(Y = 1|\mathbf{X}) \leq 0.5 \end{cases}. \qquad (9.18)$$

This is an intuitive proposition because it says that if the probability of the ideal value being 1 given the observation $\mathbf{X}$ exceeds $\frac{1}{2}$, then define the value of the filtered image to be 1 at the pixel; on the other hand, if the probability of the ideal value being 1 given the observation $\mathbf{X}$ does not exceed $\frac{1}{2}$, then define the value of the filtered image to be 0 at the pixel. An equivalent way of expressing the matter is also intuitive:

$$\psi_{opt}(\mathbf{X}) = \begin{cases} 1, & \text{if } P(Y = 1|\mathbf{X}) > P(Y = 0|\mathbf{X}) \\ 0, & \text{if } P(Y = 1|\mathbf{X}) \leq P(Y = 0|\mathbf{X}) \end{cases}. \qquad (9.19)$$

For an example, consider the probability table and resulting optimal filter shown in Table 9.2, where $\mathbf{x}$ denotes a specific observation vector, $P(\mathbf{x})$ denotes the probability of $\mathbf{x}$ occurring when the window is observed, and $P(\mathbf{x})$ plays no role in determining the optimal filter, which in this case is the median.

An optimal filter is determined solely by the conditional probabilities according to Eqs. (9.18) and (9.19). But what about the error of the filter? If $\mathbf{x}$ is never observed, then it does not matter how we define $\psi_{opt}(\mathbf{x})$ because no error will ever be made on account of $\mathbf{x}$. For any specific observation vector, there are two contributions to the error: (1) the probability that the vector occurs in the observation image; and (2) the probability of making the wrong decision between 0 and 1 when it is observed. The first contribution is $P(\mathbf{x})$, the probability of observing $\mathbf{x}$; the second is $P(Y = 0|\mathbf{x})$ if $\psi_{opt}(\mathbf{x}) = 1$, and $P(Y = 1|\mathbf{x})$ if $\psi_{opt}(\mathbf{x}) = 0$. Since $\psi_{opt}(\mathbf{x})$ is defined according to Eq. (9.18), the second contribution to the error coming from $\mathbf{x}$ is $P(Y = 0|\mathbf{x})$ if $P(Y = 1|\mathbf{x}) > \frac{1}{2}$, and $P(Y = 1|\mathbf{x})$ if $P(Y = 1|\mathbf{x}) \leq \frac{1}{2}$. Denoting the error of an operator Ψ by $\varepsilon[\Psi]$, we have

$$\varepsilon[\Psi_{opt}] = \sum_{\{\mathbf{x}:P(Y=1|\mathbf{x}>0.5)\}} P(\mathbf{x})P(Y = 0|\mathbf{x}) + \sum_{\{\mathbf{x}:P(Y=1|\mathbf{x}\leq 0.5)\}} P(\mathbf{x})P(Y = 1|\mathbf{x}).$$
$$(9.20)$$

If we list the possible observations $\mathbf{x}$ in a table along with the probabilities $P(\mathbf{x})$ and conditional probabilities $P(Y = 0|\mathbf{x})$ and $P(Y = 1|\mathbf{x})$, then the error is obtained by summing the products of the probabilities and conditional probabilities corresponding to the values (0 or 1) not chosen for ψ_{opt}. For instance, in the median example just considered, $\varepsilon[\psi_{opt}] = 0.15$.

Because images are binary, the expected value of Y given $\mathbf{X}$ is equal to the conditional probability of Y given $\mathbf{X}$; namely,

$$E[Y|\mathbf{X}] = P(Y = 1|\mathbf{X}). \qquad (9.21)$$

Therefore, from Eq. (9.18) we see that

$$\psi_{opt}(\mathbf{X}) = \begin{cases} 1, & \text{if } E[Y|\mathbf{X}] > 0.5 \\ 0, & \text{if } E[Y|\mathbf{X}] \leq 0.5 \end{cases}, \qquad (9.22)$$

which is the binary conditional expectation.

9.4 Design of Optimal W-Operators

In practice, an optimal filter is statistically estimated from sample data. This is obtained by taking a collection of image pairs (ideal and observed), taking from these a set of sample pairs $(\mathbf{x}_1, y_1), (\mathbf{x}_2, y_2), \ldots, (\mathbf{x}_m, y_m)$, and from these estimating the conditional probabilities determining the optimal filter. For each possible vector $\mathbf{x}$ in the window, estimate the conditional probability $P(Y = 1|\mathbf{x})$ by

$$\hat{P}(Y = 1|\mathbf{x}) = \frac{\text{number of times } y = 1 \text{ when } \mathbf{x} \text{ is observed}}{\text{number of times } \mathbf{x} \text{ is observed among the sample pairs.}} \qquad (9.23)$$

The designed estimate, Ψ_{des} (with characteristic function ψ_{des}), of the optimal filter is determined by Eq. (9.18) with $\hat{P}(Y = 1|\mathbf{x})$ in place of $P(Y = 1|\mathbf{x})$ for each $\mathbf{x}$.

The designed filter may or may not be optimal. It is optimal if and only if $\hat{P}(Y = 1|\mathbf{x}) > \frac{1}{2}$ precisely when $P(Y = 1|\mathbf{x}) > \frac{1}{2}$. If the number of sample pairs is very large, then $\hat{P}(Y = 1|\mathbf{x})$ is a good approximation to $P(Y = 1|\mathbf{x})$ and the designed filter can be expected to be close to optimal; however, for insufficiently large samples, $\hat{P}(Y = 1|\mathbf{x})$ has a good chance of differing substantially from $P(Y = 1|\mathbf{x})$. Therefore, for many observation vectors the designed filter will differ from the optimal filter. This means the error for the designed filter will exceed the error of the optimal filter. This increase depends on the number m of sample pairs. The increase also depends on the sample data that have been randomly chosen, so that the increase is random. The expected increase is called the **cost** of design. There is much statistical theory concerning design cost. It does not only depend on the sample size, but also on the characteristics of the ideal and noise images. For small windows, say 3×3, it is often possible to get samples that are sufficiently large to have a negligible design cost. But for even modest sized windows, say 5×5, matters are much worse and the design cost is often so large that the designed filter will have poor performance in comparison to the optimal filter, which is unknown.

In the extreme case, an observation vector $\mathbf{x}$ may never appear in the sample pairs. In this case, there is not even a poor estimate of the conditional probability $P(Y = 1|\mathbf{x})$; there is no estimate at all. In this case, some extra criteria must be employed to decide whether $\psi_{des}(\mathbf{x}) = 1$ or $\psi_{des}(\mathbf{x}) = 0$. This is known as the problem of **generalization**.

For a first example, we design a three-variable operator to minimize additive and subtractive point noise in an image composed of vertical stripes. Figure 9.3 shows samples of the observed and ideal images, along with a diagram with all the stages composing the design process, where "decision" refers to the design of the characteristic function, and "minimization" refers to logic reduction from the disjunctive normal form given directly by statistical estimation of the optimal filter. Table 9.3(a) shows the statistics obtained from the data of the figure and Table 9.3(b) shows the resulting designed filter. Observe that Table 9.3(b) defines a class of four operators that are statistically equivalent. This is because 100 and 010 do not occur in the sample data, and therefore $\psi_{des}(100)$ and $\psi_{des}(010)$ must be defined by some convention external to the sample data. For instance, one may choose to define them so that the resulting filter has minimal computational cost. Using this criterion results in $\psi_{des}(100) = \psi_{des}(010) = 1$. Figure 9.4 shows the application of the designed operator to a different sample image.

Detection of defect lines in an image of a transversal section of a eutectic alloy is a classical problem in mathematical morphology. To apply automatic design, first we have performed a transformation on the image and a shrink so we can observe defect lines in a low-resolution image. The window is the 5×5 square. Figures 9.5 (a) and (b) show the input and the application of the learned operator. The original image has been divided into two halves, the upper half for training and the lower half [shown in part (a)] for testing the result. The size of the training sample was 7,260. There were 3,812 distinct observed pairs: 3,519 pairs $(\mathbf{x}, 0)$ and 293 pairs $(\mathbf{x}, 1)$. Figure 9.5(c) shows the defect lines overlaid on the input image.

When using a window that is too large for the amount of sample images available, the data requirement can be mitigated by selecting the best filter in a subclass of operators instead of from among all W-operators. There is a cost to constrained

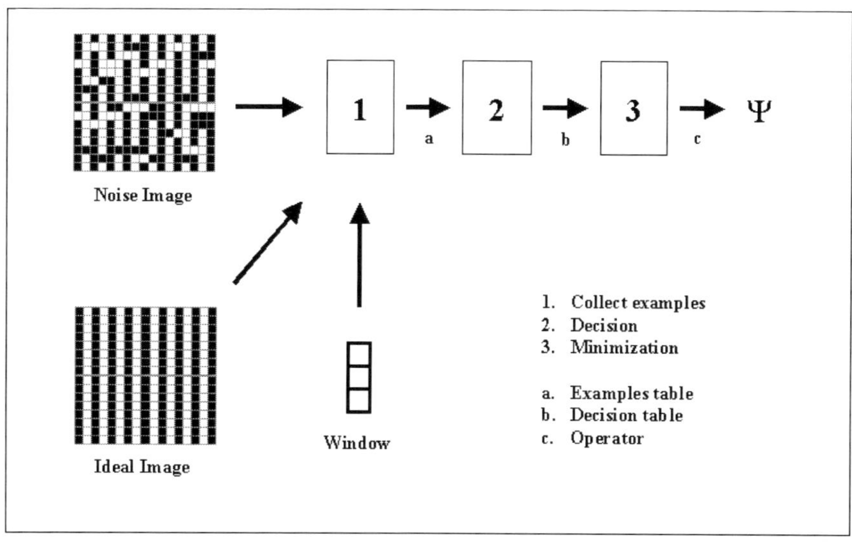

Figure 9.3 Design process.

Table 9.3 (a) Observed statistics, (b) designed filter.

$x_1x_2x_3$	0	1
000	108	0
001	2	0
011	1	18
101	0	19
110	1	18
111	0	71

(a)

$x_1x_2x_3$	$\psi(x_1x_2x_3)$
000	0
001	0
010	X
011	1
100	X
101	1
110	1
111	1

(b)

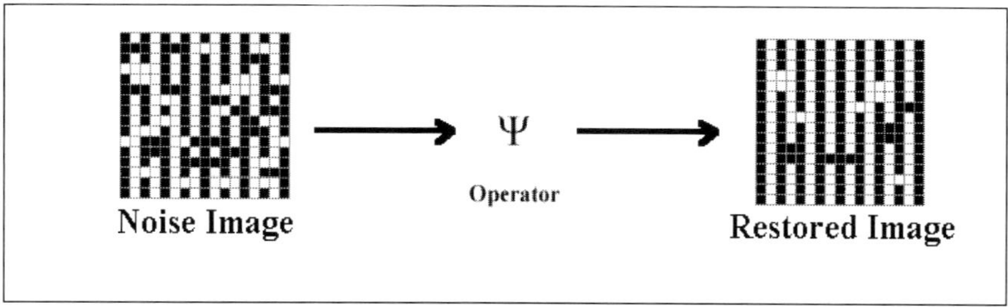

Figure 9.4 Application of designed operator.

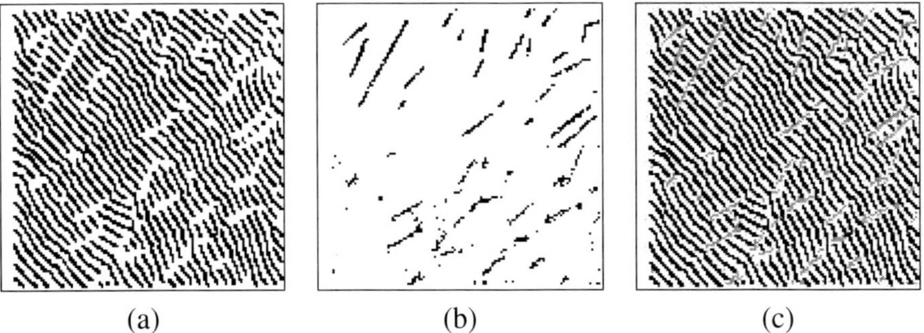

(a)   (b)   (c)

Figure 9.5 Defect lines: (a) input image, (b) upper half for training, lower half for application, (c) application overlaid on input image.

optimization: if Ψ_{con} has the minimum error among all operators in the constrained class, then certainly its error is at least as large as that of Ψ_{opt}, the filter with minimum error among all W-operators. However, the designed estimate of Ψ_{opt} also has a greater expected error than the error of Ψ_{opt}. The difference is the cost of design. If we denote the expected design cost by $\Delta(m)$, m being the sample size, then the relation between the errors is expressed by

$$\bar{\varepsilon}[\Psi_{des}] = \varepsilon[\Psi_{opt}] + \Delta(m), \tag{9.24}$$

where $\bar{\varepsilon}[\Psi_{des}]$ is the expected error of the designed filter (recalling that the designed filter depends on a random sample). A similar relation holds between the optimal constrained filter and its designed estimate, $\Psi_{des-con}$:

$$\bar{\varepsilon}[\Psi_{des-con}] = \varepsilon[\Psi_{con}] + \Delta_{con}(m). \tag{9.25}$$

Because the constrained filter is being selected from among a smaller class of operators, $\Delta_{con}(m) \leq \Delta(m)$. If $\Delta_{con}(m)$ is sufficiently smaller than $\Delta(m)$, then it can be beneficial to optimize over a constrained class of operators.

Comparing Eqs. (9.24) and (9.25), we see that the expected error of the designed constrained filter is less than the expected error of the designed unconstrained filter if and only if the cost of constraint, $\varepsilon[\Psi_{con}] - \varepsilon[\Psi_{opt}]$, is less than $\Delta(m) - \Delta_{con}(m)$, the decrease in design cost from using constraint. For large windows, this is often the case. The difficulty is finding the right constraint. There are many constraints that can be beneficial for filter optimization. The savings in design cost relative to the sample size depends on the class of images under consideration. This means that a constraint may be good in one setting but not in another.

9.5 Optimal Increasing Filters

As with any constraint, there are times when an optimal increasing filter will work quite well in comparison to an unconstrained optimal filter, and there are many times when it will not. This should be evident from the fact that in many situations we employ hit-or-miss operators (nonincreasing) rather than just depend on erosions, dilations, openings, or closings (all increasing).

Two methods have been used successfully to design (estimate) optimal increasing filters. One method is to use the sample data to try to directly find the erosion representation of the optimal increasing filter [which must possess an erosion representation according to Eq. (9.11)]. Another method is to estimate the optimal W-operator, and then switch observation vectors in and out of the kernel until the resulting filter is increasing, with the switching being done so as to minimally increase the error over the optimal W-operator. Here we confine ourselves to the second approach.

If Ψ_{opt} and Ψ_{inc} denote the optimal and optimal increasing W-operators, respectively, then the cost of constraint for requiring an increasing operator is $\varepsilon[\Psi_{inc}] - \varepsilon[\Psi_{opt}]$. Switching refers to a procedure in which we begin with the kernel, $\mathbf{K}[\Psi_{opt}]$, of the optimal filter and iteratively exchange (switch) elements to obtain a sequence of kernels $\mathbf{K}[\Psi_{opt}] = \mathbf{K}_0, \mathbf{K}_1, \mathbf{K}_2, \ldots, \mathbf{K}_m = \mathbf{K}[\Psi_{inc}]$. The cost of switching $\mathbf{x}$ from $\mathbf{K}[\Psi_{opt}]$ to $\mathbf{K}[\Psi_{opt}]^c$ or from $\mathbf{K}[\Psi_{opt}]^c$ to $\mathbf{K}[\Psi_{opt}]$ is the increase in error resulting from the exchange and is given by $|2P(Y = 1|\mathbf{x}) - 1|P(\mathbf{x})$. If the sequence $\mathbf{K}_0, \mathbf{K}_1, \mathbf{K}_2, \ldots, \mathbf{K}_m$ results from switching vectors between $\mathbf{K}[\Psi_{opt}]$ and $\mathbf{K}[\Psi_{opt}]^c$, then vectors with the smallest costs should be switched, since the cost of constraint is the minimum sum of the switching costs.

For any operator, its **inversion set** is composed of all kernel elements having nonkernel elements above them and all nonkernel elements having kernel elements below them. The operator is increasing if and only if its inversion set is null. When determining the optimal switching transformation $\mathbf{K}[\Psi_{opt}] \to \mathbf{K}[\Psi_{inc}]$, only elements in the inversion set of Ψ_{opt} need be considered for switching. If there exists a nonkernel element $\mathbf{y}$ above a kernel element $\mathbf{x}$, then there are two possibilities: either switch $\mathbf{x}$ to 0 (meaning $\psi(\mathbf{x})$ is set to 0) or switch $\mathbf{y}$ to 1 (meaning $\psi(\mathbf{x})$ is set to 1). If $\mathbf{x}$ is switched to 0, then all kernel elements beneath $\mathbf{x}$ (if any) must also be switched to 0; if $\mathbf{x}$ is not switched, then all nonkernel elements above it must switched, and so on. The decision whether to switch the value of an element may depend on the values of elements adjacent to it in the lattice, and also on elements related to those elements, etc.

If S is a set of vectors whose values are switched from their values according to the optimal filter and the result of these switches results in an increasing filter, then S is called a **switching set**. If Ψ is the resulting filter, then the increase in error is given by

$$\varepsilon[\Psi] - \varepsilon[\Psi_{opt}] = \sum_{\mathbf{x}\in S} |2P(Y = 1|\mathbf{x}) - 1|P(\mathbf{x}). \qquad (9.26)$$

Finding an optimal increasing operator is equivalent to finding a switching set that minimizes this increase among all switching sets, and a minimizing switching set is called an **optimal switching set**. Since there always exists an optimal switching set that is a subset of the inversion set, we need only consider subsets of the inversion set when using the switching approach to design an optimal increasing operator.

Figure 9.6 illustrates switching sets: (a) the kernel and nonkernel sets, (b) the inversion set, (c) a switching set, (d) the increasing operator resulting from switching, (e) a nonswitching set, and (f) the nonincreasing operator resulting from switching based on the nonswitching set.

A brute force approach to finding an optimal switching set will only work for small inversion sets. Therefore, efficient algorithms have been developed to find optimal switching sets. Even for these algorithms, for very large windows it may be necessary to take a suboptimal approach.

In practice we have estimates for the conditional probabilities $P(Y = 1|\mathbf{x})$ and these yield a designed filter estimating the optimal filter. Switching is applied to the designed filter to produce a designed increasing filter.

To illustrate increasing filters, we first consider a signal-union-noise model in which we design a nonincreasing filter using five 512×512 training images, and then use the switching approach to find the designed increasing filter. The window is the 21-pixel window centered at the origin with corners absent. Realizations of the relevant random images are shown in Fig. 9.7: (a) an ideal image, (b) the signal-union-noise degraded image, (c) output of the designed nonincreasing filter applied to the degraded image, and (d) output of the designed increasing filter applied to the degraded image. Based on 10 test images, we have estimated the MAEs of the nonincreasing and increasing filters as 0.0006 and 0.0008, respectively, these being in comparison to 0.0734, the MAE of the degraded image. The improvement is excellent for both filters with modest training. While it might seem, therefore, at first glance that the increasing constraint is not useful in the present circumstances, one must look a bit deeper. Even after the maximum amount of logic reduction, the nonincreasing filter still requires more than 2000 hit-or-miss structuring-element pairs, whereas the increasing filter only requires the 4 structuring elements shown in Fig. 9.8.

For another example, we consider restoration of edge-degraded text using the 17-pixel window centered at the origin with the second and fourth pixel in the top row missing, and the corresponding pixels missing on the other three edges of the window. This particular window has proven to be very useful in document processing with a very large savings in complexity, and therefore data requirement, in comparison to the full 25-pixel window. This time design is accomplished with five 1024×1024 training images, with switching again applied to obtain the increasing filter. Realizations of the relevant random images are shown in Fig. 9.9: (a) an ideal image, (b) the edge-degraded image, (c) output of the designed nonincreasing filter applied to the degraded image, and (d) output of the designed increasing filter applied to the degraded image. Based on 10 test images, we have estimated the MAEs of the nonincreasing and increasing filters as 0.0019 and 0.0038, respec-

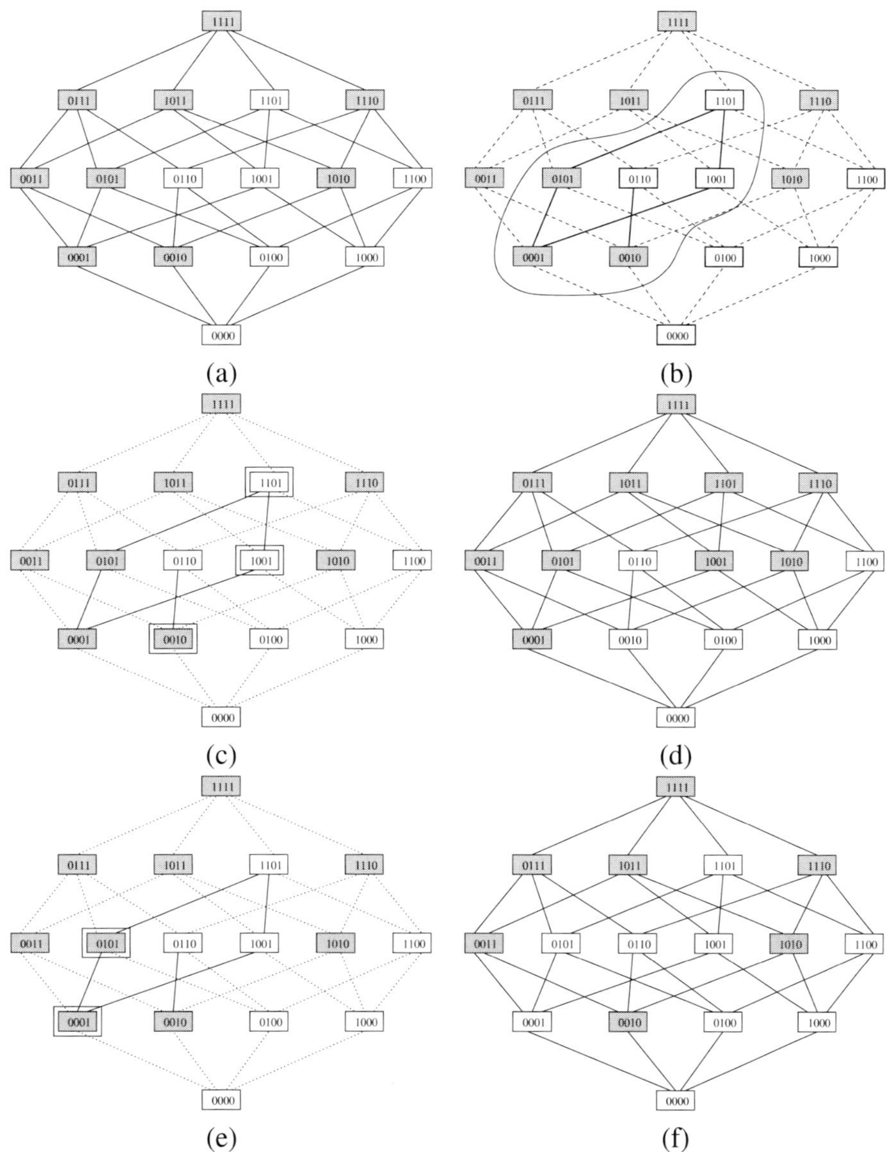

Figure 9.6 Switching sets: (a) the kernel and nonkernel sets, (b) the inversion set, (c) a switching set, (d) the increasing operator resulting from switching, (e) a nonswitching set, and (f) the nonincreasing operator resulting from switching based on the nonswitching set.

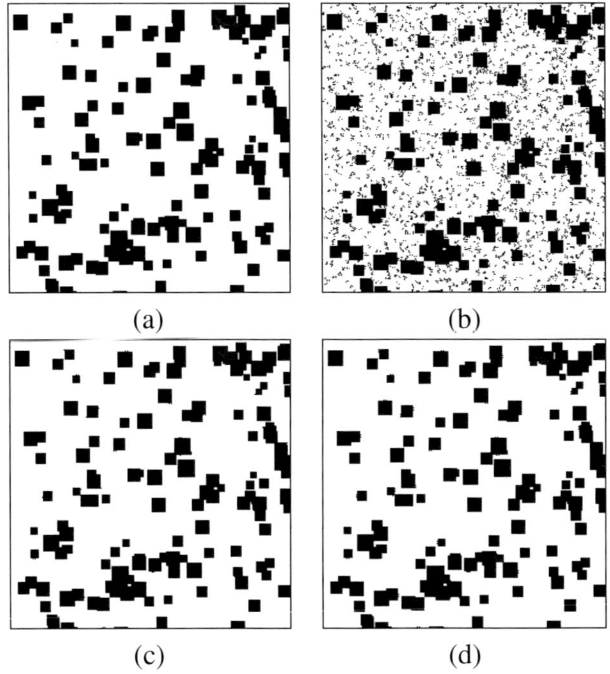

Figure 9.7 Optimal increasing filters: (a) ideal image, (b) signal-union-noise degraded image, (c) nonincreasing filter applied to the degraded image, (d) increasing filter applied to the degraded image.

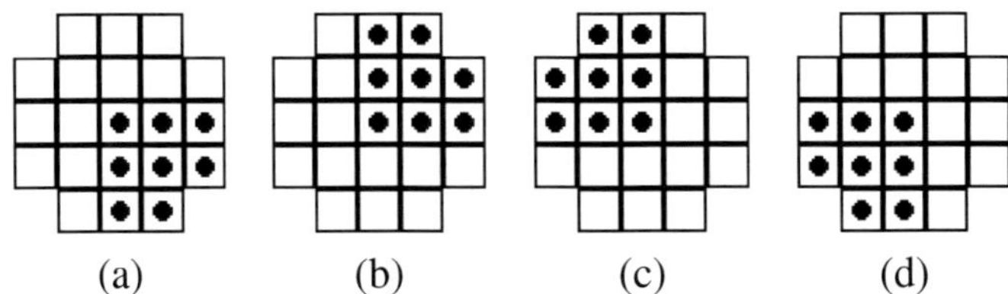

Figure 9.8 Structuring elements of the optimal increasing filter.

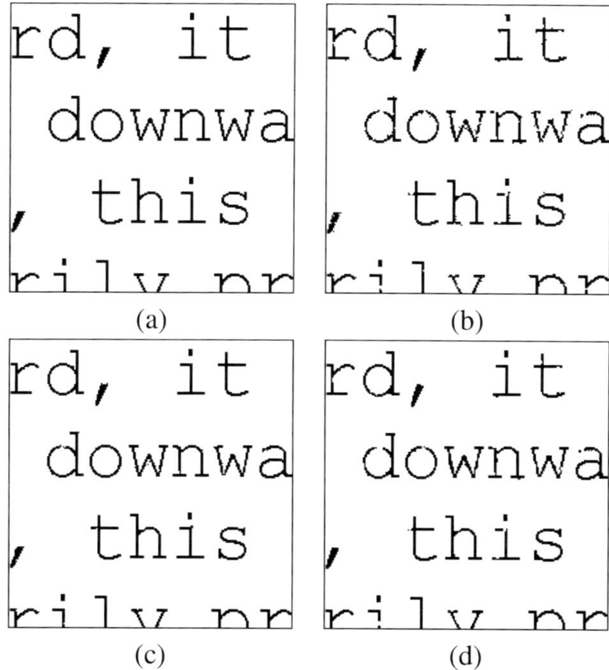

(a)　　　　　　　(b)

(c)　　　　　　　(d)

Figure 9.9 Restoration of edge-degraded text: (a) ideal image, (b) edge-degraded image, (c) nonincreasing filter applied to the degraded image, (d) increasing filter applied to the degraded image.

tively, these being in comparison to 0.0048, the MAE of the degraded image. The nonincreasing filter has far outperformed the increasing filter and has significantly reduced the MAE. Here it must be kept in mind that edge degradation does not produce a high MAE because a low percentage of the pixels in the image are affected; nonetheless, the visual effects are often striking. To help see the better performance of the nonincreasing filter, in Fig. 9.10 we zoom in on the restorations, parts (a) and (b) showing the results of the nonincreasing and increasing filters, respectively. Whereas increasing filters tend to work well for additive and subtractive noise, they typically perform poorly for edge restoration.

9.6　Differencing Filters

If an image is degraded by an antiextensive operation, then the observed image will always be a subimage of the ideal image and the optimally restored image will always result from adjoining pixels to the observed image. Thus, if S is the observed image, the restored image is of the form

$$\Psi(S) = S \cup \Psi_\bullet(S). \qquad (9.27)$$

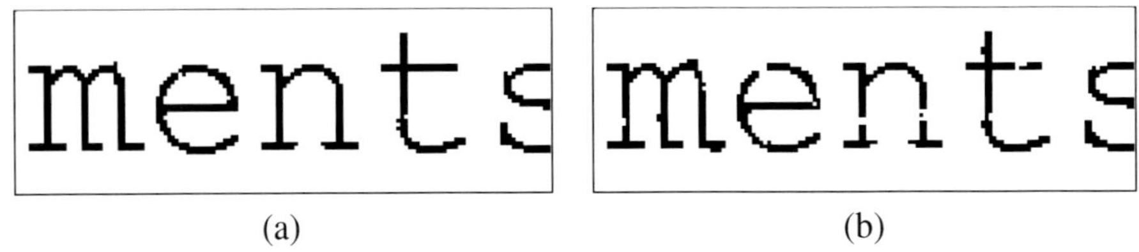

Figure 9.10 Restoration of edge-degraded text: (a) zoom of nonincreasing filter, (b) zoom of increasing filter.

$\Psi_{\bullet}$ possesses a hit-or-miss representation as in Eq. (9.15), but its representation has a special property. Since $\Psi_{\bullet}$ only adjoins pixels, the hit template in a structuring pair (E, F) never contains the origin; if it did, the pixel of interest would already be in S and therefore would not have to be adjoined. Any pair such that E does not contain the origin is called a **thickening template**. Since the hit-or-miss operations in the union of Eq. (9.27) are done in parallel, the operator Ψ is called a **parallel thickening**.

Analogously, if an image is degraded by an extensive operation, then the observed image will always contain the ideal image and the optimally restored image will always result from subtracting pixels from the observed image. Thus, the restored image is of the form

$$\Psi(S) = S - \Psi_{\circ}(S). \tag{9.28}$$

Since $\Psi_{\circ}$ only subtracts pixels, the hit template in a structuring pair (E, F) in its expansion must contain the origin, since if it did not, the pixel of interest would not be in S and therefore could not be removed. Any pair such that E contains the origin is called a **thinning template** and the operator Ψ is called a **parallel thinning**.

Designing parallel thickenings and thinnings only requires considering thickening and thinning templates, respectively. Intuitively, it appears that general restoration could be done via thinning and thickening. Such a filter would take the form

$$\Psi(S) = (S - \Psi_{\circ}(S)) \cup \Psi_{\bullet}(S). \tag{9.29}$$

Under this representation, Ψ is called a **differencing filter**. In fact, the intuition is correct because every translation-invariant operator possesses a differencing representation. Filter design involves finding thinning templates that reduce the error when included among the structuring pairs defining $\Psi_{\circ}$ and thickening templates that reduce the error when included among the pairs defining $\Psi_{\bullet}$.

A key to both efficient design and implementation of differencing filters is to recognize that they can be viewed as a toggling. Let $\mathbf{K}_0$ and $\mathbf{K}_1$ denote the classes of canonical structuring pairs defining the thinning and thickening filters in a differencing representation, respectively, and let Ξ be the filter defined by Eq. (9.15) over the pairs in $\mathbf{K}_0 \cup \mathbf{K}_1$. Let us now compare Ξ and Ψ in Eq. (9.29). Let $S(z)$ be the value, 0 or 1, of the observed image at the pixel z. If $S(z) = 0$, then $\Psi(S)(z) = \Psi_\bullet(S)(z)$. Consequently, $\Psi(S)(z) = 0$ if and only if $\Psi_\bullet(S)(z) = 0$, which is true if and only if $\Xi(S)(z) = 0$, because the template defining $\Psi_\bullet(S)(z)$ is not in $\mathbf{K}_1$. Similar reasoning shows that, if $S(z) = 0$, then $\Psi(S)(z) = 1$ if and only $\Xi(S)(z) = 1$. Now, if $S(z) = 1$, then $\Psi(S)(z) = S(z) - \Psi_\circ(S)(z)$. Consequently, $\Psi(S)(z) = 0$ if and only if $\Psi_\circ(S)(z) = 1$, which is true if and only if $\Xi(S)(z) = 1$ because the template defining $\Psi_\circ(S)(z)$ is in $\mathbf{K}_0$. Similar reasoning shows that, if $S(z) = 1$, then $\Psi(S)(z) = 1$ if and only if $\Xi(S)(z) = 0$. Now let I denote the identity filter, $I(S) = S$. Putting the four cases together shows the following relations:

$$
\begin{array}{llllll}
I = 0 & \text{and} & \Xi = 0 & \Leftrightarrow & \Psi = 0, \\
I = 0 & \text{and} & \Xi = 1 & \Leftrightarrow & \Psi = 1, \\
I = 1 & \text{and} & \Xi = 0 & \Leftrightarrow & \Psi = 1, \\
I = 1 & \text{and} & \Xi = 1 & \Leftrightarrow & \Psi = 0.
\end{array}
\tag{9.30}
$$

This means that

$$
\Psi = I \triangle \Xi,
\tag{9.31}
$$

where $\triangle$ denotes the symmetric difference. $\Psi = 1$ if and only if $I \neq \Xi$. This says that Ξ is a toggle filter: based on the value of the image, Ξ yields a decision to either toggle or not toggle the value.

In cases where there is little degradation, such as in a contemporary document image, there may be very few templates that indicate toggling, so that the toggle filter, and therefore the differencing filter, possess very simple representations. Figure 9.11 provides an example of the savings in logic cost. Ideal and degraded rectangle images are shown in Figs. 9.11(a) and (b), respectively. Differencing representation requires only 4 templates to completely restore the image; whereas direct representation requires 21 templates, as shown in Figs. 9.11(c) and (d), respectively. Differencing typically reduces logic cost for document restoration; however, there are image-noise models in which differencing representation can result in an increase in logic cost.

The issue of generalization is easily handled in the case of differencing filters. If a template is not observed in training, just presume there is no reason to toggle and do not include it in the differencing filter. This is an instance of prior knowledge, where that knowledge tells us that there is very little degradation and, unless there is reason to change a pixel valuc, leave it alone. Indeed, one can go further and require that a template be observed some minimal number of times in training before it can be included among the templates defining the differencing filter.

Figure 9.12 shows the effects of differencing representation using a 5×5 window: (a) an ideal document image at 12-point font and 600 spi (spots per inch), (b) the image degraded by an edge-noise process, (c) restoration via differencing

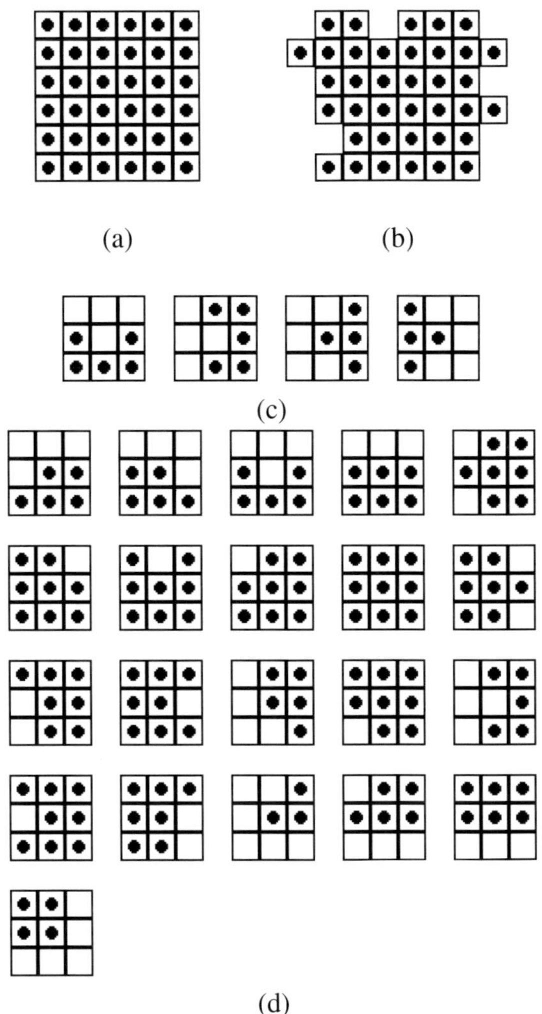

(a)

(b)

(c)

(d)

Figure 9.11 Savings in logic cost of differencing filter: (a) ideal rectangle image, (b) rectangle image with degraded edges, (c) restoration filter using differencing representation, (d) direct representation.

representation, and (d) restoration via direct representation using the same number of training examples and setting the designed filter equal to 0 for canonical templates unobserved in the sample data. In part (d), we see the overabundance of white pixels caused by setting the filter to 0 for unobserved templates. A different generalization might do better, but it too will have bad effects. In this case, the assumption that a pixel should not be toggled without sufficient reason provides beneficial prior knowledge resulting in the superior performance of the differencing filter. A summary of the quantitative results demonstrates this: MAE of the noisy image is 0.02427, MAE after filtering using direct representation is 0.02088, MAE after filtering using differencing is 0.01879, the number of canonical templates in the direct representation is 59,006, and the number of canonical templates in the differencing representation is 15,782.

9.7   Resolution Conversion

Morphological operators are essentially pattern-based decision procedures. For instance, does the structuring element fit within the image when placed at a specific pixel? Based on the answer, yes or no, the pixel is included or not included in the output image. There is no reason that, given an image and a pixel, several questions cannot be simultaneously asked, the result being several output images. In this case we have a vector morphological operator,

$$\boldsymbol{\Psi} = (\Psi_1, \Psi_2, \ldots, \Psi_m). \tag{9.32}$$

For input image A, there are m output images $B_1, B_2, \ldots, B_m$, with $B_j = \Psi_j(S)$ for $j = 1, 2, \ldots, m$. Each component operator, Ψ_j, can be designed using pairs of observations, window in the input image and ideal pixel value in the associated output image. In this section we discuss an important application of vector morphological operators.

Because document images are often bilevel, office printers are generally based on laser/xerographic or ink-jet technologies suited for making binary marks on paper. The sampling resolution for an office printer is usually 240, 300, 400, or 600 spi. Document bit maps created or decomposed for one printer must often be printed on another. This can occur because scanners, decomposers, and printers on a network may have been produced by different manufacturers designing at different resolutions

From a morphological perspective, for m-to-1 resolution conversion, we desire a vector mapping $\boldsymbol{\Psi}$ according to Eq. (9.32) whose input low-resolution image must be operated upon to yield m output images that are interleaved to produce a single high-resolution image. Each pixel z in the low-resolution image occupies the same printing region as m pixels $y_1, y_2, \ldots, y_m$ in the high-resolution image. If $\boldsymbol{\Psi}$ is based on a window W, then W is translated to z, and the operators $\Psi_1, \Psi_2, \ldots, \Psi_m$ determine the values at $y_1, y_2, \ldots, y_m$ in the high-resolution

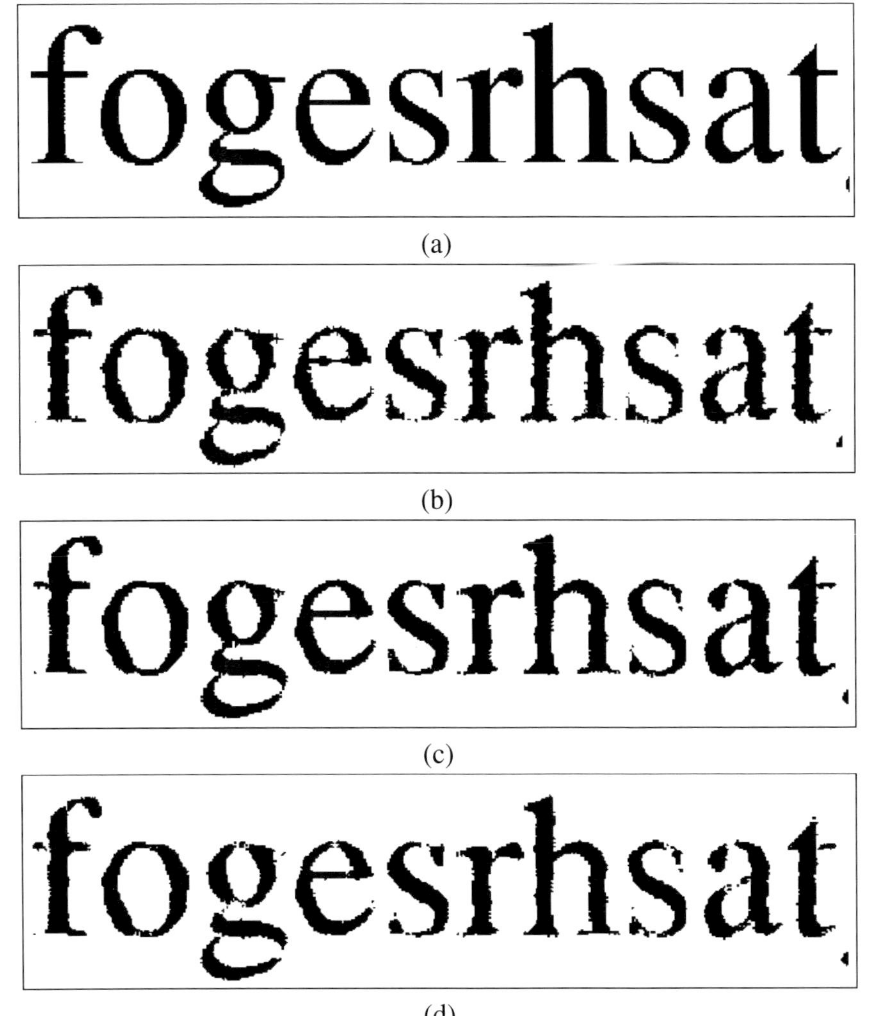

Figure 9.12 Effects of differencing and direct representation: (a) ideal document image, (b) image with edge noise, (c) restoration filter using differencing representation, (d) restoration filter using direct representation.

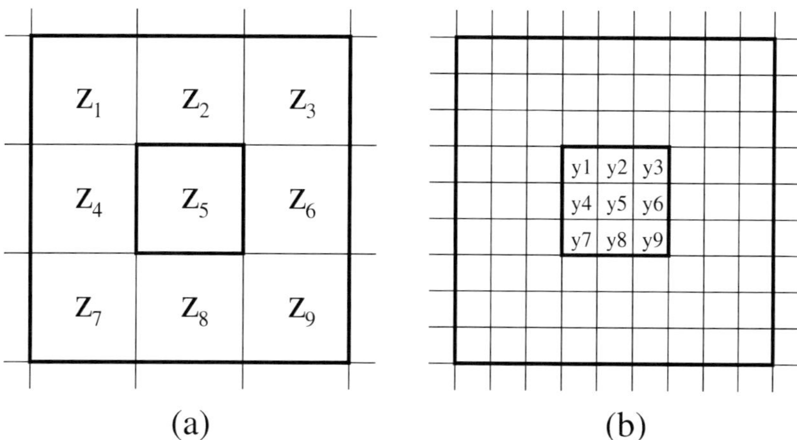

(a)   (b)

Figure 9.13 Grid mapping geometry, 9-to-1 resolution conversion using 3×3 observation window: (a) low-resolution input image, (b) high-resolution output image.

image. Figure 9.13 illustrates the window structure for 9-to-1 resolution conversion using a 3×3 observation window. Each pixel y_i in the high-resolution image depends on the pixels $z_1, z_2, \ldots, z_9$ of the low-resolution image.

A bit map obtained from some form of input scanner or electric camera is known as a reprographic bit map. The problem of reprographic 300 spi to 600 spi resolution conversion is encountered when a document bit map has been digitized and thresholded by a 300 spi scanner operating in a text mode and a 600 spi printer is available for output. We have 4-to-1 resolution conversion. The window is translated to z and the operators $\Psi_1, \Psi_2, \Psi_3, \Psi_4$ determine the values at y_1, y_2, y_3, y_4 in the high-resolution image. Consider as an image class one of the blue pages from a telephone book. Suppose a page has been digitized by a desktop scanner at 300 spi and, for the purpose of printing the list of numbers on a high-resolution printer or for compatibility with a particular character recognition algorithm, the bit map must be converted to 600 spi. We employ a 5×5 window. As a training set of images, a page from a telephone book has been digitized on a desktop scanner capable of both 300 and 600 spi resolution. In our example, a section of text consisting of approximately 200,000 pixels at 300 spi and 800,000 pixels at 600 spi has been employed as the source of the training set. The joint statistics used in filter design are obtained by a multistage acquisition process beginning with the 600 spi image being subsampled in a manner that generates four interleaved bit maps, each at 300 spi. Figure 9.14 shows a 300 spi input bit map and the corresponding 600 spi bit map obtained from an automatically designed 4-phase differencing filter.

Here we have considered integer conversion, meaning that the ratio of the high-to-low resolutions is an integer. Noninteger resolution conversion can be performed in a somewhat similar manner, albeit, with somewhat more complexity.

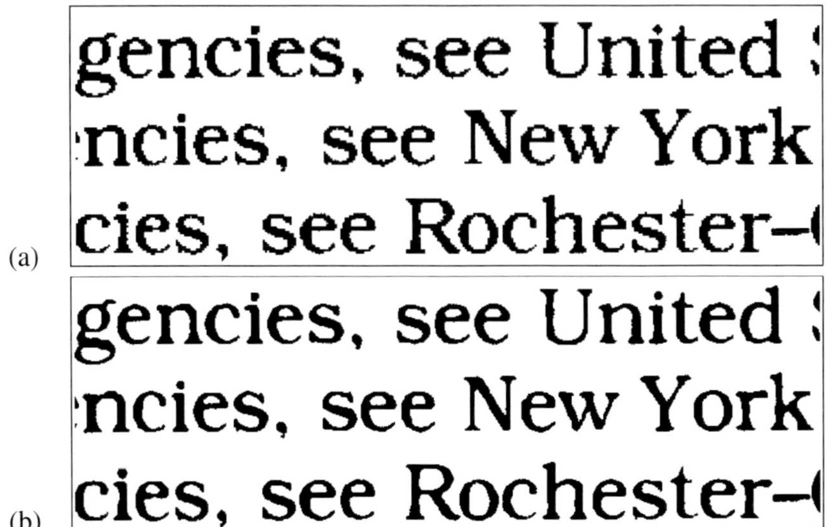

Figure 9.14 Resolution conversion of scanned text: (a) 300 spi input bit map, (b) 600 spi bit map obtained from 4-phase filter.

9.8 Multiresolution Analysis

In principle, optimally filtering an image at high-resolution is better than optimally filtering at low resolution; however, increased resolution brings an increase in the number of variables for a fixed size domain and a concomitant increase in error when estimating the optimal filter. For a given amount of sample data, it is advantageous to design a filter at lower resolution if the decrease in design error more than offsets the error increase owing to resolution constraint. More generally, we can use pyramidal (multistage) resolution reduction to arrive at a multiresolution analysis, and quantify constraint costs and design errors throughout the pyramid to determine a resolution for which the sum of the constraint cost and estimation error is minimal. While the theory is quite general and there are many kinds of resolution constraint, we will focus on one straightforward constraint, down-sampling.

Consider discrete-to-discrete subsampling in which each pixel $u \in W_1$ corresponds to a q-pixel subwindow $W_u \subset W_0$, the n subwindows are disjoint, and their union equals W_0, so that the subwindows form a partition of W_0. A down-sampling is defined by the mapping $W_u \to u$. More precisely, the observation vector in W_0 is compressed via a set of mappings, $\xi_1, \xi_2, \ldots, \xi_n$: if $\mathbf{x} = \mathbf{x}_1 \mathbf{x}_2 \cdots \mathbf{x}_n$ is the vector formed by concatenating the observation vectors in the subwindows and $\mathbf{z} = (z_1, z_2, \ldots, z_n)$ is the observation vector in W_1, then resolution reduction is characterized by a mapping $\mathbf{z} = \rho(\mathbf{x})$, where ρ is defined via mappings $\xi_1, \xi_2, \ldots, \xi_n$ according to $z_i = \xi_i(\mathbf{x}_i)$ for $i = 1, 2, \ldots, n$ (Fig. 9.15). Mathematically, one is free to define the resolution mapping ρ in any way. Relative to filtering, if one wishes to estimate Y based on the region covered by W_0, filters

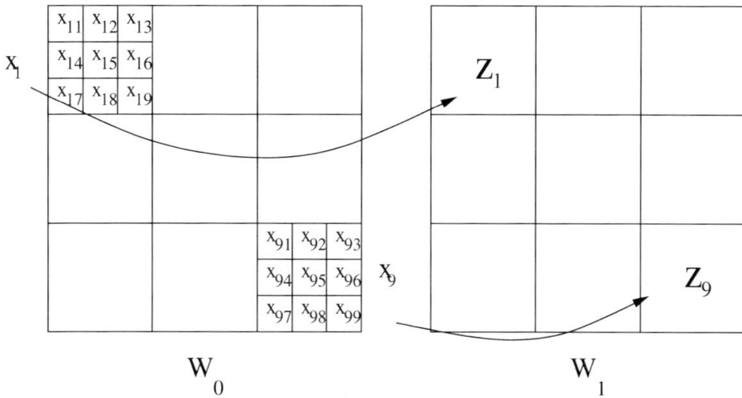

Figure 9.15 Resolution mapping from 9×9 window to 3×3 window.

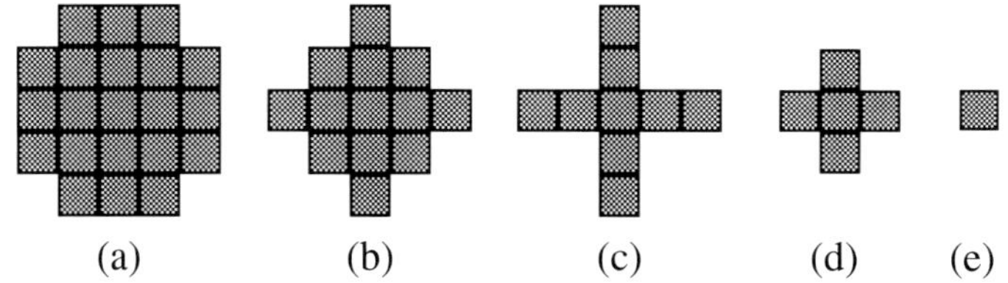

(a)   (b)   (c)   (d)   (e)

Figure 9.16 Window sequence for multiresolution analysis.

can be designed based on the high-resolution sampling of W_0 or the low-resolution sampling of W_1. Relative to W_0, filtering based on W_1 represents resolution constraint. The construction of ρ via $\xi_1, \xi_2, \ldots, \xi_n$ and the partition of W_0 represents a specific approach; however, there are others, for instance, overlapping subwindows and subwindows of different sizes. Down-sampling can be concatenated to yield a multiresolution analysis in which filtering can be considered at various resolutions, with the best resolution depending on the choice of resolution mapping, the amount of sample data, and the images under consideration. Finding suitable resolution mappings is the key difficulty.

Perhaps the easiest-to-use form of resolution constraint is subwindowing. Figure 9.16 shows successive constraint of the 21-pixel window $W : W = W_0 \rightarrow W_1 \rightarrow W_2 \rightarrow W_3 \rightarrow W_4$. If we raster scan W_0 and W_1 and label their values $x_1, x_2, \ldots, x_{21}$, and $z_1, z_2, \ldots, z_{13}$, respectively, then the resolution-constraint mapping ρ_1 is defined by $z_1 = \xi_1(x_1, x_2, x_3) = x_2, z_2 = \xi_2(x_4, x_5) = x_5, z_3 = \xi_3(x_6) = x_6, z_4 = \xi_4(x_7, x_8) = x_7, \ldots, z_{13} = \xi_{13}(x_{19}, x_{20}, x_{21}) = x_{20}$. The mappings ρ_2, ρ_3, and ρ_4 are analogously defined to complete the multiresolution analysis. For image restoration, we consider binary images degraded by edge noise. Figures 9.17 (a) and (b) show a realization of the ideal image and a degraded version of that realization, respectively. By using a large number of realizations, we are able to obtain a very good estimate of the optimal filter for each window. Figure 9.18 shows the MAEs for the various designed filters as functions of increasing

An initial test has been m	An initial test has been m
a result of a microarray s	a result of a microarray s
radiation (REF), 30 gene	radiation (REF), 30 gene
characterize the responsiv	characterize the responsiv
examined by blot assays	examined by blot assays
sulfonate, MMS), or ultra	sulfonate, MMS), or ultra
deficient cells would be a	deficient cells would be a
were made for each gene	were made for each gene
(a)	(b)

Figure 9.17 Realizations of ideal and degraded image: (a) ideal image, (b) degraded image.

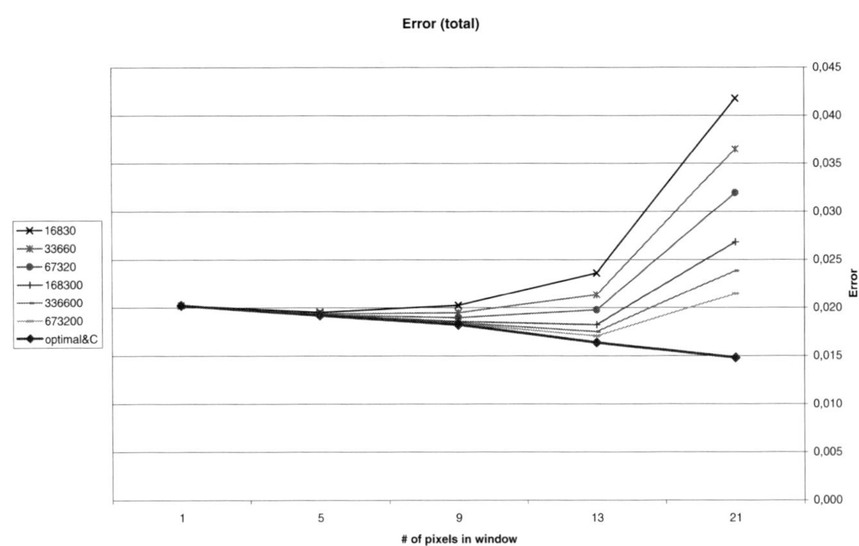

Figure 9.18 MAE for filters designed from different number of examples as a function of the number of pixels in the window.

resolution. The error of the optimal filter decreases steadily for increasing resolution but the errors for the designed filters begin to increase. The increase and the point at which it begins depend on the sample size. For each amount of training (sample size), there is an optimal window (resolution) for restoration, this window being given by the minimum on the curve. Noting that the MAE of the degraded image is the MAE for the single pixel (1 on horizontal axis), we see that all training sizes are sufficient to achieve essentially optimal filtering for the 5-pixel window, and that none is sufficient to achieve improved MAE using the 21-pixel window. This indicates the importance of choosing a resolution appropriate for the sample size.

In the preceding methodology, we have data at all resolutions but only use data at the optimal resolution. By pursuing a hybrid approach, we can use data from all resolutions. Let us first suppose we have two resolutions, high and low. Consider a particular observation vector $\mathbf{x}$ at high resolution and suppose $\mathbf{z} = \rho(\mathbf{x})$ under the resolution mapping. If $\mathbf{x}$ is observed a sufficient number of times in the data, then we have a good estimate $\hat{P}(Y = 1|\mathbf{x})$ of the conditional probability $P(Y = 1|\mathbf{x})$, and it would be prudent to estimate Y based on $\hat{P}(Y = 1|\mathbf{x})$; however, if $\mathbf{x}$ has been rarely (or never) observed, then it can be beneficial to estimate Y based on the estimate of the conditional probability $P(Y = 1|\mathbf{z})$. Letting ψ_0 and ψ_1 denote the optimal filters at the higher and lower resolutions, respectively, we define the **multiresolution filter**

$$\psi_{(0,1)}(\mathbf{x}) = \begin{cases} \psi_0(\mathbf{x}), & \text{if } N(\mathbf{x}) \geq t \\ \psi_1(\mathbf{x}), & \text{if } N(\mathbf{x}) < t \end{cases}, \tag{9.33}$$

where $N(\mathbf{x})$ is the number of times $\mathbf{x}$ appears in the sample data, t is a threshold determining the minimum number of times $\mathbf{x}$ must be observed to apply the high-resolution filter to $\mathbf{x}$, and the subscript $(0, 1)$ indicates that data at both resolutions are being used. One must be careful in choosing the threshold because too high of a threshold will force filter design at too low of a resolution. At minimum, $t = 1$. Multiresolution filtering can be employed in a pyramidal fashion using data at several resolutions.

For illustration, consider the successive subwindow constraint of the 25-pixel square window W, $W = W_0 \rightarrow W_1 \rightarrow \cdots \rightarrow W_{11}$, shown in Fig. 9.19. One 256×256 image is used for training, and errors are estimated from five such images. Figure 9.20 shows some image realizations: (a) realization of the ideal image degraded by salt-and-pepper noise, (b) the realization restored by the filter designed over the full window W, (c) the realization restored by the best single operator (W_7), and (d) the realization restored by the multiresolution filter. Both direct design and the multiresolution filter improve up to W_7, but after that direct design gets worse rapidly while the multiresolution filter continues to improve. In this case, the best MAE for direct design is 0.0078, which is achieved with W_7, and the best MAE for the multiresolution filter is 0.0068, achieved using windows

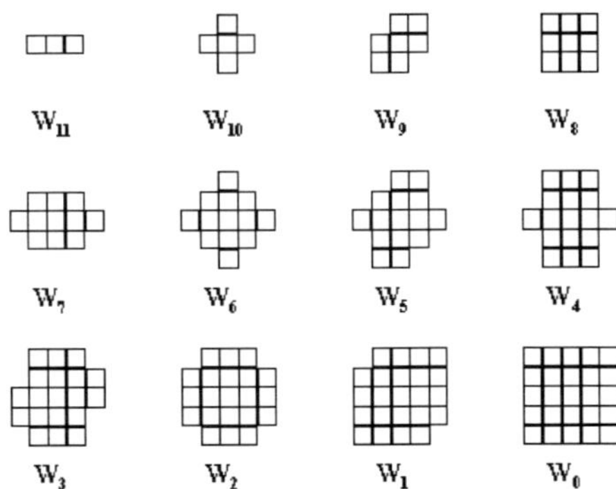

Figure 9.19 Window sequence for hybrid multiresolution filter design.

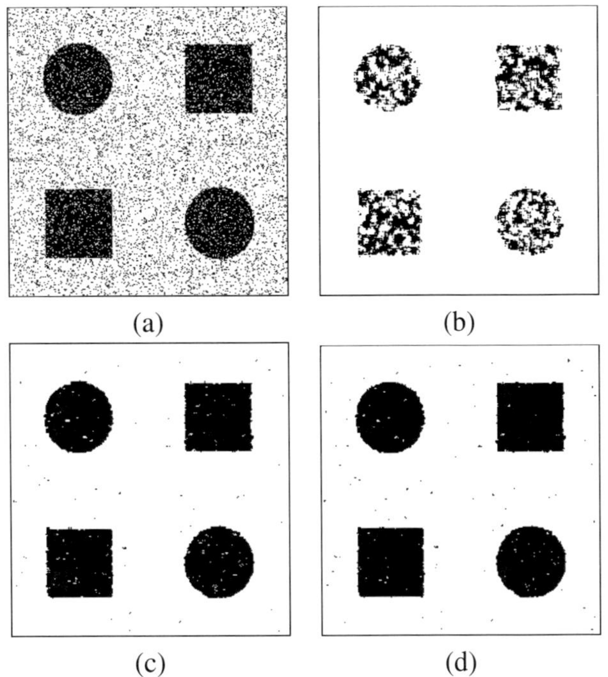

Figure 9.20 Hybrid multiresolution filter design: (a) ideal image degraded by salt-and-pepper noise, (b) restoration by the filter designed over the full window, (c) restoration by the best single operator, (d) restoration by the multiresolution filter.

W_{11} through W_4. There is no gain in the fourth decimal place by using greater resolution than W_4.

9.9 Envelope Filters

The essential reason for constraint in filter design is to reduce the size of the class from which the designed filter is selected. A constraint must be sufficiently tight so that there is a significant reduction in the amount of training required for a specific window size, and it should be such that the optimal constrained filter is close to the optimal filter. With envelope constraint, this is (hopefully) achieved by using two human-designed filters, α and β, such that $\alpha \leq \beta$, and requiring that the designed filter lie in the **envelope** created by α and β, namely,

$$\alpha \leq \psi_{des-env} \leq \beta \tag{9.34}$$

Depending on the closeness of α and β, the constraint is either tight or loose. For one extreme, $\alpha = \beta$; for the other extreme, $\alpha = 0$ and $\beta = 1$, identically. The first case is degenerate in the sense that the imaging scientist is selecting the filter without using the data; the second case is degenerate in the sense that there is no constraint.

To design an envelope-constrained filter, we design the filter ψ_{des} in the ordinary unconstrained manner, and then let

$$\psi_{des-env} = (\psi_{des} \vee \alpha) \wedge \beta \tag{9.35}$$

This design method is motivated by the fact that the optimal envelope-constrained filter, ψ_{env}, is given in terms of the optimal filter by

$$\psi_{env} = (\psi_{opt} \vee \alpha) \wedge \beta \tag{9.36}$$

If the optimal filter lies in the envelope, then envelope constraint is beneficial, meaning the error of $\psi_{des-env}$ is no greater than the error of ψ_{des}. Since one wishes to choose an envelope sufficiently tight to reduce design cost, it may well be that the optimal filter does not lie entirely within the envelope, in which case there is the usual trade-off between lowering the design cost while increasing the constraint cost. All of this can be analyzed quantitatively; however, we leave that to the literature. One point is certain: successful application of the method requires a good choice of envelope.

We illustrate the use of envelope design by designing a filter to detect edges for images degraded by randomly placed additive-subtractive noise. We will compare three approaches: (1) application of an unconstrained statistically designed filter over a 3×3 window, (2) application of an envelope-constrained statistically designed restoration operator followed by a heuristically designed edge operator, and (3) application of an envelope-constrained statistically designed restoration operator followed by an unconstrained statistically designed edge operator. The first approach is purely statistical, whereas the others demonstrate different hybrid human-machine interactions.

The first four parts of Fig. 9.21 show: (a) a ground image, (b) the ground image degraded by noise, (c) the ideal internal-boundary edge image resulting from the ground image, and (d) the edge image detected from the noisy ground image by using an unconstrained filter, ψ, designed over a 3×3 window using four training images.

Next, using four training images, we design an unconstrained noise-restoration operator ξ over a window created by self-dilating a 3×3 diamond ($2E_4$). The MAE of ξ is 0.0028. Figure 9.22(a) shows the result of applying ξ to the image in Fig. 9.21(b). Rather than restoring the degraded noisy image by an unconstrained designed filter, we can apply an envelope filter. We let α be the ASF formed by opening, closing, and then opening by the 5×5 square, E_{24}, and β be the ASF formed by closing, opening, and then closing with the same structuring element. The resulting envelope-constrained filter, φ, over a 3×3 square (using the same amount of training) has MAE 0.0013, and application of φ to the image of Fig. 9.21(b) is shown in Fig. 9.22(b).

Following application of the envelope filter φ, we apply the standard 8-connected internal gradient operator ρ to obtain an edge image. Application of ρ to the image of Fig. 9.22(b) is shown in Fig. 9.22(c). Next we design an unconstrained edge operator ζ that operates on the output of φ to detect the ideal edge. ζ has a 3×3 window and is designed using four training images. The result of applying ζ to the image of Fig. 9.22(b) is shown in Fig. 9.22(d). Comparing the errors of the three designed edge operators, we have $\varepsilon[\psi] = 0.0065$, $\varepsilon[\rho\varphi] = 0.0027$, and $\varepsilon[\zeta\varphi] = 0.0024$.

We close this section by noting that envelope constraint can be used for gray-scale images.

9.10 Aperture Filters

Thus far we have focused on the automatic design of window-based operators for binary images. One can also consider automatic window-based operators for gray-scale images; however, owing to a far greater number of possible gray-scale configurations over a window of fixed size as compared to the number of binary configurations, the data requirement is enormously increased. Hence, constraint becomes even more critical. In this section we consider the design of gray-scale filters using windowing in both the domain and the range.

Not only are the observations constrained to the domain window W, but the values of the observations are constrained to a range window $K = \{-k, -k + 1, \ldots, k\}$. If $\mathbf{x} = (x_1, x_2, \ldots, x_n)$ is the gray-scale vector observed in the window W, for each component x_j, define the truncated value

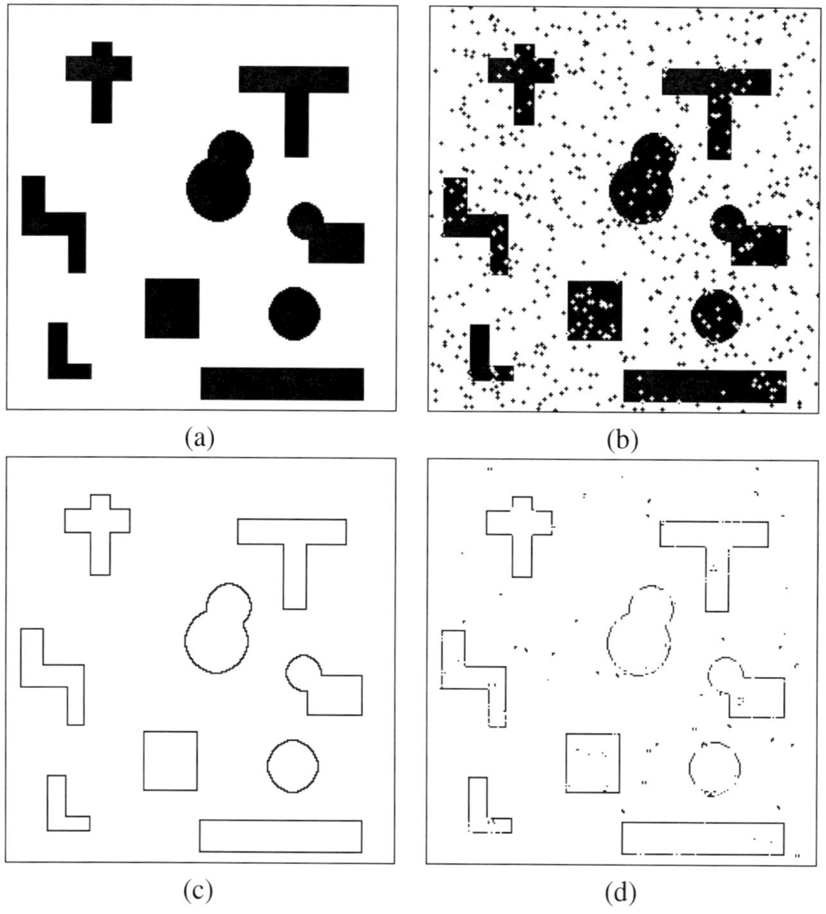

Figure 9.21 (a) Ground image, (b) noise on ground image, (c) ideal edges image, (d) edges detected by a machine-designed operator.

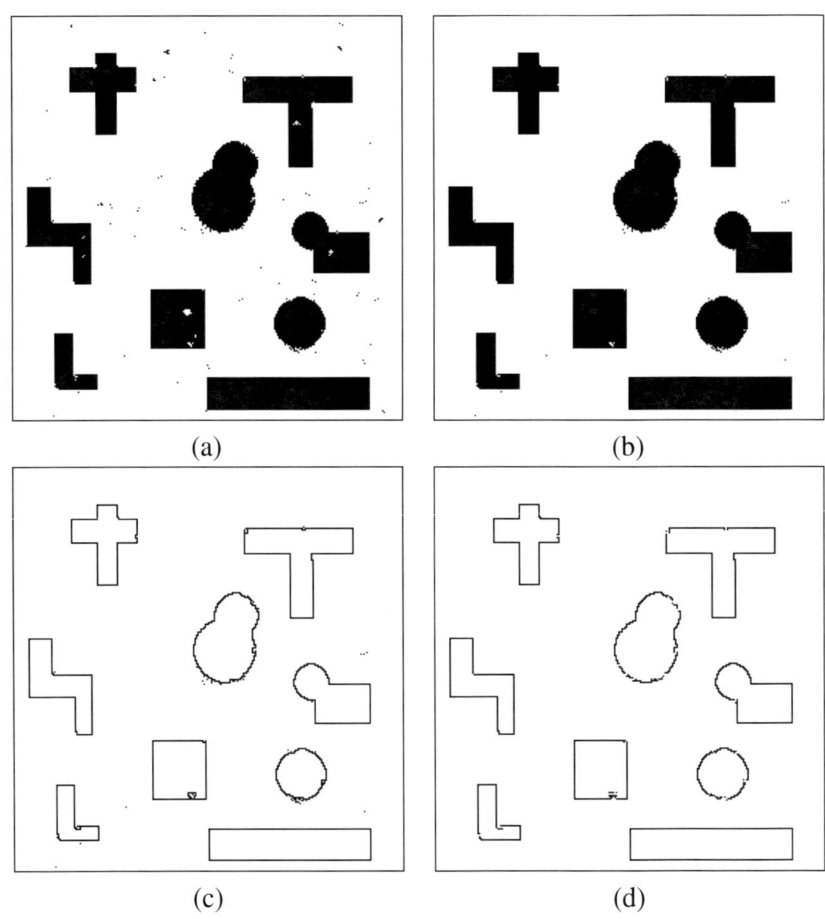

Figure 9.22 (a) Restoration by a machine-designed operator, (b) restoration by a human-machine-designed operator, (c) internal edge detection of part (b), (d) edges detected by a machine-designed operator trained from part (b).

$$\overline{x_j} = \begin{cases} x_j, & \text{if} & -k \leq x_j \leq k \\ k, & \text{if} & x_j > k \\ -k, & \text{if} & x_j < -k \end{cases}, \qquad (9.37)$$

and let $\overline{\mathbf{x}} = (\overline{x_1}, \overline{x_2}, \ldots, \overline{x_n})$. A filter of the form $\psi(\overline{\mathbf{x}})$ is called an **aperture filter** with aperture A being the product set $W \times K$. Geometrically, observations within the aperture are unchanged, whereas those outside the aperture are projected vertically into the boundary of the aperture (from above or below).

Equation (9.9) shows how a W-operator Ψ is defined via its characteristic function ψ. The situation is different for aperture filters because the aperture must be translated vertically as well as in the plane. We explain matters using signal operators. Consider input signal h and point t at which the aperture filter $\Psi_A(h)(t)$ is to be defined. A range point z must be chosen so that the signal is observed through the aperture $W_t \times K_z$, where W is horizontally translated to t and K is vertically translated to z. There is a significant difference here between aperture filters and ordinary windowed filters because the vertical translation depends on the values of the signal within the window, and is therefore random. It would be unreasonable to fix the translation of K, since this could cause extreme constraint error if the signal does not pass through the aperture.

Aperture placement is illustrated in Figs. 9.23 and 9.24, which show additive noise and blurring, respectively. In each figure, parts (a) and (c) give the ideal signal and parts (b) and (d) give the observed (corrupted) signal. Signals are shown as solid dots, $\times$ marks the center of the aperture, and shadowed dots show vertical projections of the signal points into the aperture. In parts (a) and (b) of Fig. 9.23, the aperture is placed vertically at the observed value [based on part (b)], $\overline{\mathbf{x}}$ differs from $\mathbf{x}$ at three points [in part (b)], and the value of the ideal signal at t does not lie within the aperture [as seen in part (a)]. In parts (c) and (d) of Fig. 9.23, the aperture is placed vertically at the median of $\mathbf{x}$ [based on part (d)], $\overline{\mathbf{x}}$ differs from $\mathbf{x}$ at one point [in part (d)], and the value of the ideal signal at t lies within the aperture [as seen in part (c)]. Analogous considerations apply to Fig. 9.24, where aperture placement is at the observed value in parts (a) and (b), and at the median of $\mathbf{x}$ in parts (c) and (d).

Based on mean-square error, the optimal gray-scale filter for predicting Y based on the observation $\mathbf{x}$ is the conditional expectation of Y given $\mathbf{x}$. Aperture constraint results in ψ_A being defined as the conditional expectation of Y given $\overline{\mathbf{x}}$. Constraint error depends on the amount by which these two conditional expectations differ across the set of all possible observation vectors. As a last point, we note that aperture filter design is inherently computationally expensive, and therefore efficient algorithms must be used in their design.

For an illustration of aperture filters, consider the image of random pyramidal grains in Fig. 9.25(a), where maximum gray values are taken whenever two grains overlap. The corresponding blurred image is shown in Fig.9.25(b), blurring being accomplished with a 3×3 convolution kernel. Filters have been trained on 10 images and tested on 10 independent images. The MAE of the original blurred images is close to 0.40, and the optimal linear filter [Fig. 9.25(c)] over a 7×7 window can only bring this down to about 0.35. Much better performance is achieved with an aperture filter over the 17-pixel window [Fig. 9.25(d)] used previously for document restoration, and having a gray range of 5. This filter brings the error down to approximately 0.20. The linear filter needs less than 50,000 sample pairs in training to achieve its best performance, whereas the aperture filter requires 600,000 pairs to achieve an MAE of 0.20; however, even at 50,000 sample pairs, the aperture filter error is only 0.26.

Aperture-filter performance is especially striking in relation to linear filtering when visual effects are considered owing to better edge restoration. Figure 9.26 shows a zoomed portion of the image and its restorations: (a) original image, (b) output of optimal linear filter applied to the blurred image, and (c) output of the aperture filter applied to the blurred image. These images demonstrate the superior edge restoration of the aperture filter and its superior visual quality.

9.11 Relation to Pattern Recognition

The dimension of the output of a W-operator need not be the same as that of the input. For instance, the image to be operated upon may have 256 gray levels, whereas the output is binary, examples of this phenomenon being edge detection and target location. In the previous section we used aperture filters to reduce the range dimensionality for the purpose of facilitating design precision; however, for low-bit images one might not use aperture constraint and simply allow the window values to assume the full range. So as not to complicate the discussion with the specifics of aperture reduction, we take this view in the current section. In fact, this approach has been used for low-bit document images. In this unconstrained setting, we consider operators of the form $\psi(x_1, x_2, \ldots, x_n)$, where $x_1, x_2, \ldots, x_n$ take values in the gray range of the input image and $\psi(x_1, x_2, \ldots, x_n)$ is equal to

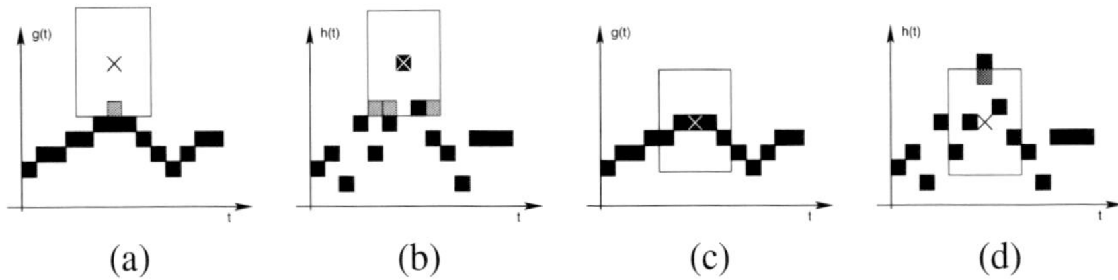

(a)   (b)   (c)   (d)

Figure 9.23 Aperture placement for additive noise: (a) aperture on ideal signal with placement at observed value, (b) aperture on observed signal with placement at observed value, (c) aperture on ideal signal with placement at median, (d) aperture on observed signal with placement at median.

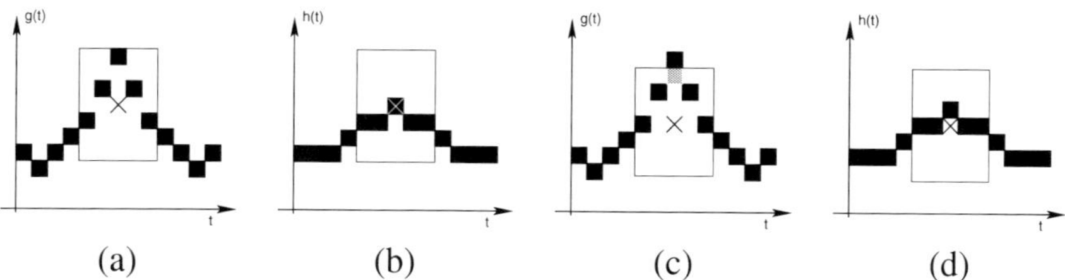

Figure 9.24 Aperture placement for blurring: (a) aperture on ideal signal with placement at observed value, (b) aperture on observed signal with placement at observed value, (c) aperture on ideal signal with placement at median, (d) aperture on observed signal with placement at median.

0 or 1. Letting $\mathbf{X} = (X_1, X_2, \ldots, X_n)$, optimization is achieved by minimizing the error $\varepsilon[\psi] = P(\psi(\mathbf{X}) \neq Y)$. If we consider 0 and 1 as class labels, then this optimization problem is identical to the basic two-class problem in pattern recognition, with $\varepsilon[\psi]$ being the Bayes error and the optimal classifier being given by Eq. (9.18). The design method we have employed by using the estimation of Eq. (9.23) is called the "plug-in rule" in pattern recognition and the problem of constraint, as quantified by Eqs. (9.24) and (9.25), appears there in exactly the same form. Various constraints are used in image processing owing to the nature of images. Other methods such as neural networks are commonplace in pattern recognition. Concepts such as the VC dimension apply to the automatic design of W-operators.

An important issue for operator design is the behavior of the design for large samples (as the sample size $m \to \infty$). Recalling Eq. (9.24), a design rule is said to be **consistent** for a distribution of $(\mathbf{X}, Y)$ if $\Delta(m) \to 0$ as $m \to \infty$. For a consistent rule, the expected design cost can be made arbitrarily small for a sufficiently large amount of data. A design rule is **universally consistent** if $\Delta(m) \to 0$ as $m \to \infty$ for any distribution of $(\mathbf{X}, Y)$. The plug-in rule is universally consistent.

A host of design rules have been developed for pattern classification. Their performance depends on the pattern classes and sample sizes under consideration. For instance, a classical rule is the *k-nearest-neighbor (kNN) rule*. For k odd, the k observed vectors closest to $\mathbf{x}$ are selected and $\psi(\mathbf{x})$ is defined to be 0 or 1 according to which is the majority among the labels of these points. If $\mathbf{x}$ is observed at least k times in the sample data, then $\psi(\mathbf{x})$ is defined in exactly the same manner as we have done so far, but if it is observed less than k times, then a sufficient number of the points closest to $\mathbf{x}$ in n-dimensional space are taken so that there are at least k observations in the data, and $\psi(\mathbf{x})$ is 0 or 1, depending on which is the majority value for y among those observed vectors. The kNN rule is universally consistent in the following sense: under the conditions that $k \to \infty$ and $k/m \to 0$ as $m \to \infty$, $\Delta(m) \to 0$.

While the preceding binary-output (two-class) considerations can be extended

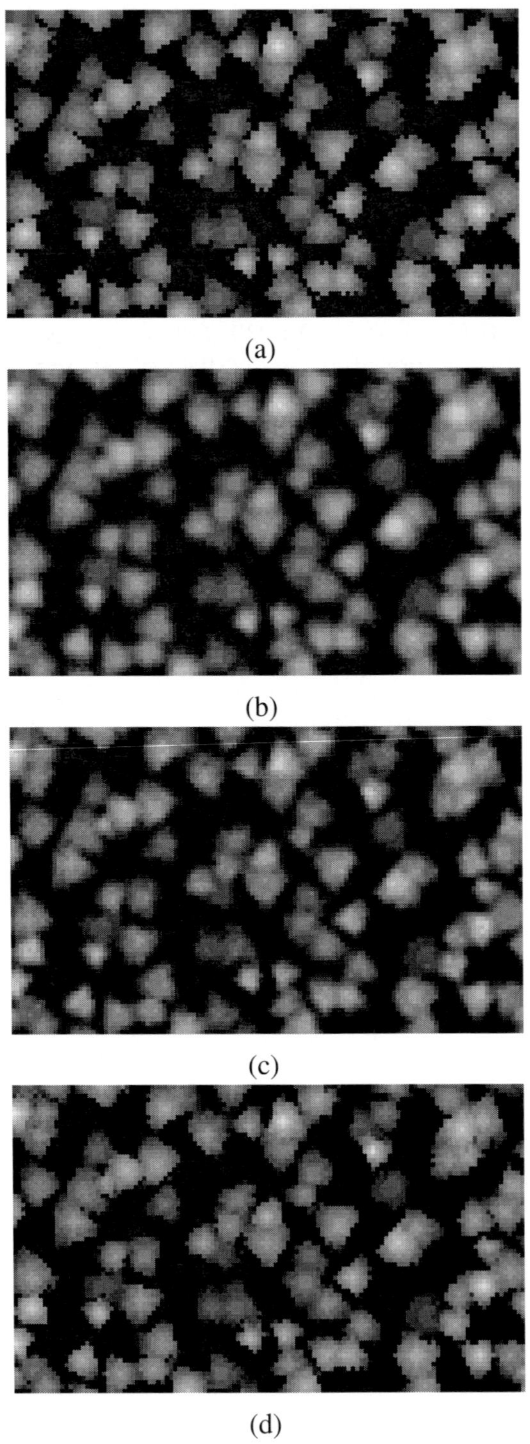

Figure 9.25 Blur restoration: (a) synthetic model, (b) degraded image using 3×3 kernel for blurring, (c) restored with an optimal linear 7×7 windows, (d) restored with a 17-pixel window aperture filter.

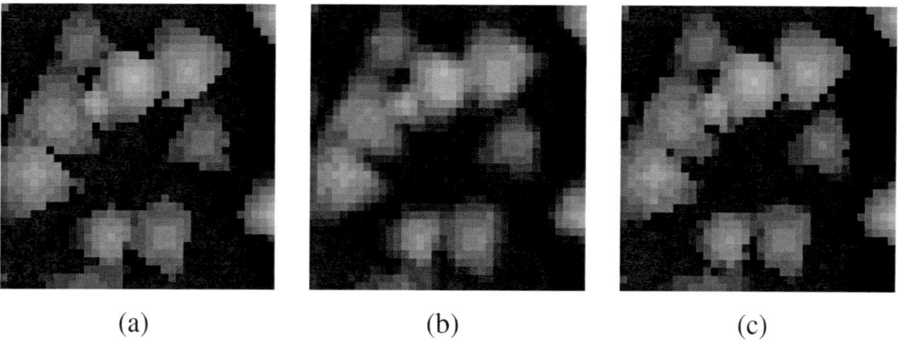

(a) (b) (c)

Figure 9.26 Zoomed portions of Fig. 9.25: (a) input image, (b) optimal linear restoration, (c) aperture filter.

to gray-scale (multi-class) problems, they can be applied directly in the case of stack filters. We have seen in Eq. (5.53) how threshold decomposition can be used to compute the dilation of a gray-scale signal by a flat structuring element. This approach can be applied much more generally. Corresponding to Eq. (5.51), for any numerical value x, define the threshold (binary) variable $x^t = 1$ if $x \geq t$ and $x^t = 0$ if $x < t$. Then, operator ψ satisfies the **threshold decomposition property** if there exists an increasing binary operator ζ_ψ such that

$$\psi(\mathbf{x}) = \max\{t : \zeta_\psi(\mathbf{x}^t) \neq 0\}, \qquad (9.38)$$

where $\mathbf{x}^t = (x_1^t, x_2^t, \ldots, x_n^t)$. An operator satisfying this property is called a stack filter (see Sec. 5.7), and it must be increasing. Since ψ is defined via a single binary operator, it is in effect a binary operator on gray-scale signals. For instance, gray-scale flat erosion is a stack filter and, according to Eq. (5.53), it satisfies Eq. (9.38) with ζ_ψ being binary erosion by the same structuring element. A complete characterization of stack filters is given via morphological representation: an increasing operator always possesses a representation as a maximum of erosions; in the case of a stack filter the representation can be expressed using flat erosions. The design cost for stack filters is reduced in comparison to general increasing filters, but the constraint error is increased. In filter design it is assumed that all threshold random vectors $\mathbf{X}^t$ are identically distributed and all the data from the different thresholds are pooled, thereby greatly enhancing the amount of data available for design.

We close by noting that there has been little research into the automatic design of image operators using design rules from pattern recognition (an exception being the use of decision trees for aperture filters). It may well be that filter design can be improved in different situations by abandoning the plug-in rule in favor of some other design rule, just as in the case of general pattern recognition.

9.12 Exercises

1. Consider the ideal image

$$\begin{pmatrix}
1 & 0 & 1 & 0 & 1 & 0 & 1 & 0 & 1 & 0 & 1 \\
0 & 1 & 0 & 1 & 0 & 1 & 0 & 1 & 0 & 1 & 0 \\
1 & 0 & 1 & 0 & 1 & 0 & 1 & 0 & 1 & 0 & 1 \\
0 & 1 & 0 & 1 & 0 & 1 & 0 & 1 & 0 & 1 & 0 \\
1 & 0 & 1 & 0 & 1 & 0 & 1 & 0 & 1 & 0 & 1 \\
0 & 1 & 0 & 1 & 0 & 1 & 0 & 1 & 0 & 1 & 0 \\
1 & 0 & 1 & 0 & 1 & 0 & 1 & 0 & 1 & 0 & 1 \\
0 & 1 & 0 & 1 & 0 & 1 & 0 & 1 & 0 & 1 & 0 \\
1 & 0 & 1 & 0 & 1 & 0 & 1 & 0 & 1 & 0 & 1 \\
0 & 1 & 0 & 1 & 0 & 1 & 0 & 1 & 0 & 1 & 0 \\
1 & 0 & 1 & 0 & 1 & 0 & 1 & 0 & 1 & 0 & 1
\end{pmatrix}$$

and noisy image

$$\begin{pmatrix}
1 & 0 & 1 & 0 & 1 & 0 & 1 & 0 & 1 & 0 & 1 \\
0 & 1 & 1 & 1 & 0 & 1 & 0 & 1 & 1 & 1 & 0 \\
1 & 0 & 1 & 0 & 1 & 0 & 1 & 0 & 1 & 1 & 1 \\
0 & 1 & 1 & 1 & 1 & 1 & 0 & 1 & 0 & 0 & 0 \\
1 & 0 & 1 & 0 & 1 & 0 & 1 & 0 & 1 & 0 & 1 \\
0 & 0 & 0 & 1 & 0 & 1 & 0 & 1 & 1 & 1 & 0 \\
1 & 0 & 1 & 0 & 1 & 0 & 1 & 0 & 1 & 0 & 1 \\
0 & 1 & 0 & 1 & 0 & 1 & 1 & 1 & 0 & 1 & 0 \\
0 & 0 & 1 & 1 & 1 & 0 & 1 & 0 & 1 & 0 & 1 \\
0 & 1 & 1 & 1 & 0 & 1 & 1 & 1 & 0 & 0 & 1 \\
1 & 0 & 1 & 0 & 1 & 0 & 1 & 0 & 1 & 0 & 1
\end{pmatrix}.$$

From these, design an optimal filter over the 5-pixel window consisting of the origin together with its 4-connected neighbors. Find the error of the filter based on the estimated conditional probabilities based on the two sample images. This error is only an estimate of the actual error of the designed filter. How can a better estimate of filter error be obtained?

2. Show that the gray-scale median is a stack filter by showing it satisfies Eq. (9.38), with ζ_ψ being the binary median over the same window.

3. Show that dilation by a set is a stack filter by showing it satisfies Eq. (9.38), with ζ_ψ being the binary dilation by the same structuring element.

References

1. J. T. Astola and P. Kuosmanen. Representation and optimization of stack filters. In E. R. Dougherty and J. T. Astola, editors, *Nonlinear Filters for Image Processing*. SPIE/IEEE Presses, Bellingham, WA, 1999.

2. G. J. F. Banon and J. Barrera. Minimal representation for translation-invariant set mappings by mathematical morphology. *SIAM Journal on Applied Mathematics*, 51:1782–1798, 1991.

3. G. J. F. Banon and J. Barrera. Decomposition of mappings between complete lattices by mathematical morphology: Part I. General lattices. *Signal Processing*, 30:299–327, 1993.

4. J. Barrera, E.R. Dougherty, and M. Brun. Hybrid human-machine binary morphological operator design: an independent constraint approach. *Signal Processing*, 80(8):1469–1487, 2000.

5. J. Barrera, E. R. Dougherty, and N. S. T. Hirata. Design of optimal morphological operators using prior filters. *Acta Stereologica*, 16(3):183–200, 1997.

6. J. Barrera, E. R. Dougherty, and N. S. Tomita. Automatic programming of binary morphological machines by design of statistically optimal operators in the context of computational learning theory. *Journal of Electronic Imaging*, 6(1):54–67, 1997.

7. E. R. Dougherty. Optimal mean-square n-observation digital morphological filters — part I: Optimal binary filters. *Computer Vision, Graphics and Image Processing: Image Understanding*, 55(1):36–54, 1992.

8. E. R. Dougherty. Optimal mean-square n-observation digital morphological filters — part II: Optimal gray-scale filters. *Computer Vision, Graphics and Image Processing: Image Understanding*, 55(1):55–72, 1992.

9. E. R. Dougherty and J. Barrera. Pattern recognition theory in nonlinear signal processing. *Journal of Mathematical Imaging and Vision*, 16(3):181–197, 2002.

10. E. R. Dougherty, J. Barrera, G. Mozelle, S. Kim, and M. Brun. Multiresolution analysis for optimal binary filters. *Journal of Mathematical Imaging and Vision*, 14(1):53–72, 2001.

11. E. R. Dougherty and R. P. Loce. Efficient design strategies for the optimal binary digital morphological filter: probabilities, constraints, and structuring-element libraries. In E. R. Dougherty, editor, *Mathematical Morphology in Image Processing*, chapter 2, pages 43–92. Marcel Dekker, New York, 1993.

12. E. R. Dougherty and R. P. Loce. Optimal mean-absolute-error hit-or-miss filters: morphological representation and estimation of the binary conditional expectation. *Optical Engineering*, 32(4):815–823, 1993.

13. E. R. Dougherty, Y. Zhang, and Y. Chen. Optimal iterative increasing binary morphological filters. *Optical Engineering*, 35(12):3495–3507, 1996.

14. E. R. Dougherty. Minimal representation of τ-openings via pattern bases. *Pattern Recognition Letters*, 15:1029–1033, 1994.

15. E. R. Dougherty and J. Handley. Recursive maximum-likelihood estimation in the one-dimensional discrete Boolean random set model. *Signal Processing*, 43(1):1–15, 1995.

16. E. R. Dougherty, S. Kim, and Y. Chen. Coefficient of determination in nonlinear signal processing. *Signal Processing*, 80(10):2219–2235, 2000.

17. E. R. Dougherty and R. P. Loce. Precision of morphological-representation estimators for translation-invariant binary filters: increasing and nonincreasing. *Signal Processing*, 40:129–154, 1994.

18. E. R. Dougherty and R. P. Loce. Optimal binary differencing filters: design, logic complexity, precision analysis, and application to digital document processing. *Journal of Electronic Imaging*, 5(1):66–86, 1996.

19. M. Gabbouj and E. J. Coyle. Minimum mean absolute error stack filtering with structuring constraints and goals. *IEEE Transactions on Acoustics, Speech and Signal Processing*, 38:955–968, 1990.

20. A. M. Grigoryan and E. R. Dougherty. Robustness of optimal binary filters. *Journal of Electronic Imaging*, 7(1):117–126, 1998.

21. A. M. Grigoryan and E. R. Dougherty. Design and analysis of robust optimal binary filters in the context of a prior distribution for the states of nature. *Journal of Mathematical Imaging and Vision*, 11(3):239–254, 1999.

22. C.-C. Han and K.-C. Fan. A greedy and branch and bound searching algorithm for finding the optimal morphological erosion filter on binary images. *IEEE Transactions on Signal Processing Letters*, 1:41–44, 1994.

23. N. R. Harvey and S. Marshall. The use of genetic algorithms in morphological filter design. *Signal Processing: Image Communication*, 8(1):55–71, 1996.

24. N. S. T. Hirata, E. R. Dougherty, and J. Barrera. Iterative design of morphological binary image operators. *Optical Engineering*, 39(12):3106–3123, 2000.

25. N. S. T. Hirata, E. R. Dougherty, and J. Barrera. A switching algorithm for design of optimal increasing binary filters over large windows. *Pattern Recognition*, 33(6):1059–1081, 2000.

26. R. Hirata Jr., M. Brun, J. Barrera, and E. R. Dougherty. Multiresolution design of aperture filters. *Mathematical Imaging and Vision*, 16(3):199–222, 2002.

27. R. Hirata Jr., E. R. Dougherty and J. Barrera. Aperture filters. *Signal Processing*, 80(4):697–721, 2000.

28. V. G. Kamat, E. R. Dougherty, and J. Barrera. Multiresolution Bayesian design of binary filters. *Journal of Electronic Imaging*, 9(3):283–295, 2000.

29. P. Kraft, N. R. Harvey, and S. Marshall. Parallel genetic algorithms in the optimization of morphological filters: A general design tool. *Journal of Electronic Imaging*, 6(4):504–516, 1997.

30. P. Kuosmanen and J. Astola. Optimal stack filters under rank selection and structural constraints. *Signal Processing*, 41(3):309–338, 1995.

31. P. Kuosmanen and J. Astola. Breakdown points, breakdown probabilities, midpoint sensitivity curves, and optimization of stack filters. *Circuits, Systems, and Signal Processing*, 15(2):165–211, 1996.

32. P. Kuosmanen, P. Koivisto, H. Huttunen, and J. Astola. Shape preservation criteria and optimal soft morphological filtering. *Journal of Mathematical Imaging and Vision*, 5(4):319–335, 1995.

33. J. H. Lin, T. M. Sellke, and E. J. Coyle. Adaptive stack filtering under the mean absolute error criterion. *IEEE Transactions on Acoustics, Speech, Signal Processing*, 38:938–954, 1990.

34. R. P. Loce and E. R. Dougherty. Facilitation of optimal binary morphological filter design via structuring-element libraries and observation constraints. *Optical Engineering*, 31(5):1008–1025, 1992.

35. R. P. Loce and E. R. Dougherty. Optimal morphological restoration: The morphological filter mean-absolute-error theorem. *Journal of Visual Communication and Image Representation*, 3(4):412–432, 1992.

36. R. P. Loce and E. R. Dougherty. *Enhancement and Restoration of Digital Documents: Statistical Design of Nonlinear Algorithms*. SPIE Press, Bellingham, WA, 1997.

37. A. V. Mathew, E. R. Dougherty, and V. Swarnakar. Efficient derivation of the optimal mean-square binary morphological filter from the conditional expectation via a switching algorithm for the discrete power-set lattice. *Circuits, Systems and Signal Processing*, 12(3):409–430, 1993.

38. P. Salembier. Adaptative rank order based filters. *Signal Processing*, 27(1):1–25, 1992.

39. P. Salembier. Structuring element adaptation for morphological filters. *Journal of Visual Communication and Image Representation*, 3(2):115–136, 1992.

40. O. Sarca, E. R. Dougherty, and J. Astola. Secondarily constrained Boolean filters. *Signal Processing*, 71(3):247–263, 1998.

41. O. Sarca, E. R. Dougherty, and J. Astola. Two stage binary filters. *Journal of Electronic Imaging*, 8(3):219–232, 1999.

42. D. Schonfeld and J. Goutsias. Optimal morphological pattern restoration from noisy binary images. *IEEE Transactions on Pattern Analysis and Machine Intelligence*, 13:14–29, 1991.

43. I. Tabus, D. Petrescu, and M. Gabbouj. A training framework for stack and Boolean filtering — fast optimal design procedures and robustness case study. *IEEE Transactions on Image Processing*, 5(6):809–826, 1996.

44. S. S. Wilson. Training structuring elements in morphological networks. In E. R. Dougherty, editor, *Mathematical Morphology in Image Processing*, chapter 1, pages 1–41. Marcel Dekker, New York, 1993.

45. L. Yin. Optimal stack filter design: A structural approach. *IEEE Transactions on Signal Processing*, 43(4):831–840, 1995.

Index

Edward Dougherty is a professor in the Department of Electrical Engineering at Texas A&M University in College Station. He holds a Ph.D. in Mathematics from Rutgers University and an M.S. in Computer Science from Stevens Institute of Technology. He is author of eleven books and editor of four others. He has published more than 100 journal papers, is an SPIE fellow, and has served as editor of the *Journal of Electronic Imaging* for six years. Prof. Dougherty has contributed extensively to the statistical design of nonlinear operators for image processing and the consequent application of pattern recognition theory to nonlinear image processing. His current research is focused in genomic signal processing, with the central goal being to model genomic regulatory mechanisms. He is head of the Genomic Signal Processing Laboratory at Texas A&M University.

Roberto A. Lotufo obtained the Electronic Engineering Diploma in 1978 from Instituto Tecnologico de Aeronautica, Brazil; the M.Sc. degree in 1981 from the University of Campinas (UNICAMP), Brazil; and the Ph.D. degree in Electrical Engineering in 1990 from the University of Bristol, U.K. He is a professor in the Department of Computer Engineering and Industrial Automation at the University of Campinas, where he has worked since 1981. His principal interests are in the areas of image processing and analysis, mathematical morphology, image segmentation, and medical imaging. He is one of the main architects of two morphological toolboxes: MMach for Khoros, and SDC Morphology Toolbox for MATLAB. Prof. Lotufo has published more than 50 refereed conference and journal papers.